Kulturstiftung Sibirien

Studies in Social and Cultural Anthropology

Alexandra Lavrillier and Semen Gabyshev

An Arctic Indigenous Knowledge System of Landscape, Climate, and Human Interactions

Evenki Reindeer Herders and Hunters

Verlag der Kulturstiftung Sibirien
SEC Publications

This publication has been made possible with the assistance
of the French National Research Agency (ANR).

Bibliografische Informationen der Deutschen Nationalbibliothek:
Die Deutsche Nationalbibliothek verzeichnet diese Publikation in der Deutschen Nationalbibliografie: detaillierte bibliografische Daten sind im Internet über <http://dnb.d-nb.de> abrufbar.

Cover graphic:
Alexandra Lavrillier and Semen Gabyshev

Typesetting:
Kulturstiftung Sibirien gGmbH, Fürstenberg/Havel

Printer:
Books on Demand GmbH, Norderstedt

Electronic edition: www.siberian-studies.org/publications/arcticiks.html

ISBN: 978-3-942883-31-3

Printed in Germany

CONTENTS

Observing and Predicting Norms, Anomalies, and Transformation

LIST OF ABBREVIATIONS

ANR – French National Research Agency.

BRISK – Bridging Indigenous and Scientific Knowledge about Global Changes in the Arctic: Adaptation and Vulnerability of Societies and the Environment, a project funded by the ANR and coordinated by the first author (CEARC, UNESCO, LMD, CNRS-MNHN).

C-B TO – Community-based Transdisciplinary Observatory.

CEARC – Research Centre: Cultures, Environment, Arctic, Representations, Cultures, OVSQ, UVSQ, France.

EK – Ecological Knowledge.

EU – European Union.

ESCAPNES – Environmental and Social Climate Change Adaptation Plans for Northern Eurasian Societies / project proposal submitted to the EU.

Evenki C-B Observatory – Evenki Community-Based Observatory, or Evenki Community-Based Transdisciplinary Observatory.

GDR – Groupe de recherche – Research Group (inter-institutional group of French researchers).

IEK – Indigenous Ecological Knowledge.

IK – Indigenous Knowledge.

IP – Indigenous Peoples.

IPBES – Intergovernmental Platform on Biodiversity and Ecosystem Services.

IPEV – Polar Institute of Paul Emile Victor.

IPK – Indigenous Peoples' Knowledge.

LINKS – Local and Indigenous Knowledge Systems (UNESCO).

OVSQ – Observatory of Versailles Saint Quentin en Yvelines, France.

POLARIS project – Cultural and Natural Heritage in Arctic and Sub-Antarctic Regions for a Cross-Cultural and Sustainable Valorisation Process and Tourism Development: Siberia, Lapland and Patagonia (FP7-PEOPLE-2012-IRSES).

TEK – Traditional Ecological Knowledge.

UNEP – United Nations Environmental Program.

UNESCO – United Nations Educational, Scientific and Cultural Organization.

UVSQ – University of Versailles Saint-Quentin en Yvelines (University Paris-Saclay), France.

LIST OF TABLES AND DIAGRAMS

PREFACE

Co-written by an anthropologist and a reindeer herder (BRISK project[1] co-researcher) on the basis of their field materials, this book offers documentation and analysis of complex traditional environmental knowledge. It presents the emic typologies and concepts the Evenki use for understanding norms and anomalies, observing and predicting changes, and adaptating. This book is one of the results of the Evenki community-based transdisciplinary observatory established by the authors for monitoring climate and environmental changes with herders (2012–2016) and funded by the ANR and IPEV (cf. Introduction).

Lavrillier conceived of the initial structure of the book and its content on the basis of the obtained co-production material and analysis. She then formalised the final content via regular writing sessions and consultations with S. Gabyshev and the herders concerned. The analytical sections were written either collectively or individually: we indicate the authorship at the top of each section or with names in parentheses within the text. We believe that it is important for epistemological reasons to make these annotations in such a cooperative piece. According to the same logic, while the diagrams were co-authored by S. Gabyshev and A. Lavrillier, we have placed the name of the person who conceived of each diagram first: the year in which each diagram was initially completed is also provided.

All photographs are copyrighted by Alexandra Lavrillier, except pp. 12 (bottom), 116–118, 138 (top), 146–151, 216 (left), 334-345, 438 by Semen Gabyshev, pp. 12 (left), 66, 94, 136, 172, 205, 216 (right), 369 (bottom) by Vasilii Gabyshev, and pp. 84, 115 by Oleg Iakovlev. For permission to use these photographs, please contact A. Lavrillier (Alexandra.lavrillier@uvsq.fr).

We wanted the book to be accessible not only to scientists, but also to the Evenki, regardless of whether they speak the Evenki language or not: thus, most of the book is in Evenki, Russian, and English. Of course, the translations required a great deal of work, especially since IPK is very difficult to translate into Western languages. Because of the specificities of the TEK, the content of this book is understandable only if one reads it from the beginning to end. For instance, the reader will only be able to understand Part III if he or she reads the introduction and Part II first, since these sections explain the typologies and concepts needed for comprehending the discussion in Part III.

The texts in Evenki are by S. Gabyshev: some of them were transcribed by L. Egorova and A. Lavrillier. All the translations into Russian were made by Lavrillier and checked by Gabyshev: the same is the case for the translations from Evenki and Russian into English. Egorova corrected the Russian text, while James White corrected the English translations.

All Evenki texts are in the local dialects. The observers and authors chose a means of writing Evenki that would allow them to represent the real pronunciation of local

1 For more details on the BRISK project, cf. Introduction.

dialects. It was decided to use Cyrillic as a transcription system, the same system used for the Yakut alphabet. In order to express vocalic and consonantal harmonies, we avoid the Russian letters я (ia), ю (iu), ё (io), and е (ie) throughout, replacing them with йа (ia), йу (iu), йо (io), йэ (ie) (as we would do if we were transliterating the Latin alphabet). Secondly, we wanted to use Cyrillic so the text will be accessible to Evenki readers. We are conscious that this system will not satisfy specialists in the Evenki standard language (which has its own orthography). We understand their position and apologise: we are ready to rewrite our text and publish another book if they so desire, but for this book it was the choice of the nomads themselves to write down their dialects as they are actually pronounced. Indeed, the herders with whom we worked, like many Evenki speakers in Eastern Siberia, neither read nor understand the standard Evenki language. Finally, the Evenki dialect speakers with whom we worked have an oral understanding of their language, i.e. for them, it is not obvious how to represent their language in a written form.

Acknowledgements

We thank the following institutions for their generous funding or support:
- ANR – French National Research Agency (project BRISK – ANR-12-SENV-0005).
- IPEV – Polar Institute of Paul Emile Victor (project BRISK's OBS 1127).
- CEARC – Research Centre: Cultures, Environment, Arctic, Representations, Cultures, OVSQ, UVSQ, France.
- OVSQ – Observatory of Versailles Saint Quentin en Yvelines, France.
- UVSQ – University of Versailles Saint-Quentin en Yvelines (University Paris-Saclay), France.
- The Department for Scientific Cooperation of the French Embassy in Moscow, especially Mr Alexis Michel and Michel Balazard.
- POLARIS project – Cultural and Natural Heritage in Arctic and Sub-Antarctic Regions for a Cross-Cultural and Sustainable Valorisation Process and Tourism Development: Siberia. Lapland and Patagonia (FP7-PEOPLE-2012-IRSES).
- UNESCO.
- Kulturstiftung Sibirien, Fürstenberg/Havel.

To the following herder-hunter families from the Amur region and Yakutia, we wish to express especial thanks for their trust, interest, and invaluable participation in the project: the Gabyshevs, the Vasilevs, the Savins, the Kolesovs, the Pavlovs, the Egorovs, the Safronovs, the Iakovlevs, the Nikolaevs, the Andreevs, the Abramovs, the Kurbaltunovs, the Isakovs, the Rostolovs, the Trenkins, the Kirillovs, the Maksimovs, and the Neustroevs.

We also thank Olga Kukharenko, Nikolai Kukharenko, and Olga Morozova from the Pedagogical University of Blagoveshchensk and the University of the Amur Region for their support and for organising our contacts with the Blagoveshchensk and Tynda meteorological centres and stations. We acknowledge the following organisations for their support in the process of providing long-term invitations so that the herders (including S. Gabyshev) could enter France: Science-Accueil and Association Franco-Evenke Sekalan. In addition, we thank our families for their support and great patience.

We give many thanks to N. V. Esina and O. B. Byl'eva from the Tynda weather station (Amur region), our colleagues in the BRISK project, the International Centre for Reindeer Husbandry ICR in Kautokeino for their invitation, the English corrector James White, and the Russian corrector Liudmila Egorova. We also thank our colleagues of the CEARC, in particular Jan Borm, UVSQ, and Erich Kasten, Kulturstiftung Sibirien.

The Evenki Transdisciplinary Community-based Observatory: Conditions of the Co-production

Semen Gabyshev (left)

Alexandra Lavrillier (right)

1 INTRODUCTION
by A. Lavrillier

A transdisciplinary community-based observatory: Methods and co-productions

The topic of 'climate change' is quite widely studied by anthropologists and sociologists in the western areas of the Arctic (Canada, Alaska, Greenland) (Krupnik and Jolly 2002; Krupnik and Ray 2007; Krupnik et al. 2010; Ford et al. 2006, 2007; Huntington et al. 2004; Nuttall et al. 2005; ACIA 2005; Berkes and Jolly 2001; among many others). However, this does not apply, as far as I know, to Siberia in general. There are a few exceptions with leading studies such as those by Forbes, Stammler, Stammler-Gossmann, and Vlasova carried out among the Yamal Nenets (Forbes et al. 2006; Forbes 2008; Stammler-Gossmann 2010; among others); among the Nenets and Sami (Nuttall et al. 2005); among the Nenets and Ienisseysk Evenki (Vlasova 2006); among the Even and Itelmen of southern Kamchatka (Sharakhmatova 2011); a little information about the Yukaghir (Shadrin 2009); and a short report on the Chukchi (Kavry and Boltunov 2005–06). Regarding Far Eastern Siberia, besides works by S. Crate in Central Yakutia among the Yakut horse and cows herders and hunters (2008, among others), by Bogoslovskaia among the Chukchi and local Inuit (Bogoslovskaia et al. 2008), and by M. Rojo (some analyses of Tuva herders, villagers, and townspeople in his 2016 PhD thesis), there is almost no participatory (as far as we know) anthropological research on climate change. Regarding the documentation of Evenki EK, there are the impressive tomes about flora and fauna entirely in standard Evenki (Pikunova and Pikunova 2004, 2008); however, we were unable to understand the text and their research goals were also incommensurate with our own (see glossary).

This book represents a specific scientific development within the transdisciplinary BRISK project, particularly the Evenki part that attempts to merge the social sciences (anthropology) with indigenous knowledge (IK) and the natural sciences (climatology, geography, biology). It is important to publish this step forward in transdisciplinary co-production between indigenous and anthropological knowledge in order to maintain its identity and authorship before it fuses with other types of knowledge. As noted by D. Nakashima (UNESCO specialist in co-production of knowledge) during his long transdisciplinary experience (Nakashima et al. 2012; UNESCO 2009), it is crucial for each knowledge or science to keep its identity.

The knowledge and analysis in this book should be further developed in a transdisciplinary way in other publications, where it can be merged with insights from remote-sensing, climatology, and biology, as well as with the IPK of other peoples involved in the BRISK project.

The BRISK project (Bridging Indigenous and Scientific Knowledge) focuses on global changes in the Arctic, examining social and environmental adaptation and vulnerability. It was conceived of by A. Lavrillier, D. Nakashima, M. Roué, and C. Claud. It has the objective of enabling innovative assessments of environmental, economic, political, and social effects, vulnerabilities, and adaptive strategies. It documents the cutting-edge literature with respect to scientific and indigenous methods of observing global change. BRISK allows us to make comparisons at several levels: it examines, in different socio-political contexts, human – natural environment relationships by comparing different types of reindeer herding in Eurasia, and it considers the notion of 'extreme meteorological events' from the differing perspectives of climate scientists and indigenous peoples. In order to bring together indigenous and scientific knowledge and observe global change (climatic, environmental, industrial, and social), Community-Based Transdisciplinary Observatories (C-B TO) have been jointly planned and established by scientists (natural and social) and indigenous peoples. D. Nakashima added the term 'transdisciplinary' to indicate that the C-B observatory works with both science and IEK while also being 'community-based'.[1]

This book is devoted exclusively to the results of the Siberian Evenki Community-Based Transdisciplinary Observatory (henceforth, Evenki C-B observatory). The idea for this Evenki observatory was born in 2011 at the end of an anthropological study on the perception of climate and environmental changes, which Lavrillier led from 2006 until 2012. The Evenki C-B observatory at the centre of this book was mutually conceived and established in January 2013 by Evenki reindeer herders and scientists from both the social and environmental sciences. The methodologies and definitions of co-production, as well as the main issues for investigation, were adapted and defined *in situ* by local non-native (Tynda and Blagoveshchensk stations) and indigenous weather forecast specialists (L. Egorova), the reindeer herders and hunter-observers Semen Gabyshev, Vasilii Gabyshev, and Albert Kolesov, other herders who took part more occasionally in the project, and of course the anthropologist herself. The observations (still ongoing within a new project) offer four years of records that monitor global changes according to criteria derived from both climatology and indigenous ecological knowledge (further IEK). In addition to this, biannual anthropological field trips (in the winter and summer for a total of 16 months) jointly led by a social anthropologist (Lavrillier) and herders studied the impact of global changes and the way in which herders adapt to them; equally, IEK was documented and partially published in this

1 Cf. www.arcticbrisk.org

book. Finally, we mapped nomadic roads, camps, and sacred places while also taking measurements.

Thus, our aim in this book is not to provide a state-of-the-art on IEK, Arctic climate change, cognitive anthropology, or Evenki ethnography, subjects on which there is already a vast and rich body of scientific literature, but to publish part of our results and to share some of our current analyses. In addition, we wanted to give a considerable amount of attention to the development of the Evenki TEK we collected, which is almost impossible to do in short papers. Due to this commitment, we have been unable to address the ethnographic literature on the Evenki in detail (see, among many others, Shirokogoroff 1929, 1935; Vasilevich 1969; Mazin 1984; Sirina 2006, 2012, etc.).

According to Berkes and Nadasdy (1999, 1999), interest in TK experienced an impressive surge during the 1980s, with numerous conferences, seminars, and workshops dedicated to it. Around 15 years later, it reached the western part of the Arctic (Canada, Alaska, and Greenland). According to Gearheard and Shirley (2007: 63), by the end of the 1990s 'an emphasis on local and traditional knowledge and participatory research has helped create a paradigm shift: scientific studies can no longer take place in the Canadian Arctic without some communication and consultation with a local community'. On these topics, see (among many others) Nelson 1969, Robbe 1994, Hungtington et al. 2004, 2005, Nadasdy 1999, ACIA 2005, Helander-Renvall 2005, Macaulay et al. 1998, D'Andrade 1995, Quinn 2005, Nakashima et al. 2012, Roturier and Roue 2009, UNESCO 2009. However, this is not the case for Siberia (see below).

1.1 Indigenous knowledge and science

by A. Lavrillier

1.1.1 A specific TEK approach

The hypothesis at the basis of our study derives from my 20 years of experience: it demonstrates that indigenous knowledge is not only 'practical knowledge', 'knowledge in practice' (these expressions are often used as synonyms of TEK (Helander-Renwall 2005: 4, et passim), or a 'way of life' (as many publications stress like Nadasdy, 1999: 4; Berkes 1999: 8; or Descola 1986, see also below), but is also conceptual (that is to say, it contains many theoretical concepts). It does not only correspond to the practices of indigenous peoples; more precisely, it corresponds both to the elements of the environment that the nomads use/exploit and to the surroundings (insects, plants, vegetal cover, etc.) that they observe during their movements. In 2014, Oleg Iakovlev, a reindeer herder and hunter, provided the following definition: 'Everything is written in the natural environment, you just need to observe all around you and to remember everything you have observed'. Tamara Andreeva, another herder, stated in 2007 that the 'natural environment knows everything, when you learn to listen to it, it tells you everything'.

Equally, in contrast to what has been written previously about this subject (Berkes 1999; Nadasdy 1999; among many others), indigenous knowledge is transmitted not only through practices, but also through what I call the 'nomadic seminar', an equivalent to scientific seminars (cf. below). What I mean here is that the transmission of knowledge occurs through discourse and the collective analysis of case studies. Among the Evenki, such seminars often take place in the evening, when nomads share their experiences after a day spent hunting or herding. Such knowledge is shared both within the same encampment and between different encampments. It is not only adults who participate in such 'nomadic seminars': teenagers and pre-teens also join as students. They learn theoretical knowledge and listen to stories about how to escape difficult situations. Valentina Enokhova, the daughter of an Evenki shaman, mentioned in the 1990s (prior to the substantial development of the notion of ecological indigenous knowledge in Arctic studies of climate change) that 'in the taiga, all our nomads are professors, like university professors: they know everything about the natural environment and they can survive in it'.

One of the first contacts I had with this phenomenon was in 1994 during my first nomadic field trip. I spent each day with the herders under the command of a reindeer herder with the nickname 'Captain' (Egor Maksimov). I was surprised that, after explaining the behaviour of domestic reindeer, bears, and other animals in detail, the 'Captain' elaborated on the annual life cycle of the two main ant species of the local area, despite the fact that the Evenki have no use for or interaction with ants. Among Evenki groups of neighbouring regions (Northern and Southern Transbaikalia and Manchuria), Shirokogoroff has already made similar very interesting insights (Shirokogoroff 1929: 310–311).

The knowledge of this indigenous people is so vast that mastery of it allows them to survive in the forest, herd reindeer, organise hunting, and establish an annual nomadic route; thus, it takes a long time to acquire. Nomads able to set up a new annual cycle of nomadisation must ensure that it passes through all the necessary kinds of landscape and contains a large enough animal population to allow for reindeer herding and food and fur hunting: such individuals need to be at least 30 years old, since younger people will not have acquired sufficient knowledge (Lavrillier 2005–2006). Thus, nomads believe that a person who has not spent his/her childhood in the forest is incapable of leading herding or hunting in such areas. The fact that this knowledge is transmitted over such a long period and during 'passive' stages like childhood (when one does not always take an active part in hunting and herding) stresses the fact this knowledge is not only 'practical' in nature. Many Evenki nomads are worried about the current state of knowledge transmission. Although such concerns were rarely mentioned in interviews in 2013, it was frequently stated in the interviews held during this project that the young are not properly receiving indigenous knowledge. Another piece of proof regarding the sheer scale of this knowledge is that, despite six years of fieldwork studying Evenki society and its management of the environment, I discovered in 2013 that I knew almost nothing about it. It was as if I had spent many years passing before a closed door: it was only when observing the environment together with herders that the door was opened, revealing treasure behind it.

The theoretical features of IEK are revealed in the work we did with herders in the BRISK Evenki C-B observatory. It analyses existing situations and produces models and hypotheses for the evolution of, for example, vegetal covers and the landscape. As we will see within the present book, the nomads have a great capacity for developing predictions into theories and translating their knowledge and passion for analysing the interactions between elements of the environment into diagrams. However, we should state now that the nomads themselves have no need for diagrams to explain their knowledge to each other, since their language facilitates the circulation of the concepts within which all the necessary knowledge is contained. Nevertheless, when explaining this knowledge to Western scientists,[2] nomads are also skilled at translating these concepts into diagrams. (cf. below and the conclusion)

Before going further into the detailed presentation of the methodology we developed (see below), let us now focus on some aspects of co-production.

1.1.2 Specificities and co-productions

There are numerous definitions of TEK, beginning with the one provided by Berkes: 'TEK is defined as a cumulative body of knowledge, practice and beliefs about the dynamic relationship of living beings with one another, and with their environment, which has evolved by adaptive processes, and has been handed down from generation to generation' (Berkes 1999).

Huntington defines it as 'a system of experiential knowledge gained by continual observation and transmitted among members of a community. It is set in a framework that encompasses both ecology and the interactions of humans and their environment on physical and spiritual planes' (Huntington 1998).

As pointed out above, TEK is often seen as being exclusively 'practical', supremely concrete collectively inherited, and very different in terms of the production of knowledge from Western sciences. For instance, Nadasdy (1999: 2) wrote, 'in contrast to TK, which is assumed to be qualitative, intuitive, holistic, and oral, science is seen as quantitative, analytical, reductionist, and literate'.

2 In this book, the expressions 'Western sciences', 'Western scientists', etc., refer to the academic scientific world in all continents and countries. Even if this is far from an ideal expression, it is widely used with this meaning in most of the publications and discussions about TEK, where it is counterpoised to 'indigenous sciences', also known as 'indigenous knowledge systems'.

According to our research, indigenous ecological knowledge is not just a block of information and practices transmitted from generation to generation. It is made up of elements of knowledge experienced by different individuals in different ways during each generation: it is not intuitive, but analytical (cf. Part III). Thus, the transmission of indigenous knowledge and science can be characterised both as 'collective' (inherited from previous generations) and 'individual' (obtained by various individuals and developed through experimentations).

One of the key issues of indigenous knowledge is its **ownership**. According to our observations, ownership appears to be both collective and individual simultaneously. This knowledge is '**collective**' in the sense that it is inherited from previous generations and learned and shared from early childhood. At the same time, it is '**individual**' in that each individual tests, questions, experiments, innovates, and adds some new elements, thereby bringing an individual touch (Berkes 1999; Lavrillier 2008, fieldwork observation, and interviews from 1994, see also the table here below). This question has become very sensitive among some Arctic peoples like the Sami (cf. below Gunn-Britt Retter).

The following co-productions were completed during four years of intensive work by S. Gabyshev, A. Lavrillier, but also by the heavily involved herder-hunters Vasilii Gabyshev, Albert Kolesov, Oleg Iakovlev, and the herders who gave interviews and more occasional information (around 20 families) (cf. Preface). First of all, it was indispensable to perform documentation of the Evenki TEK in the Evenki language in order to identify their environmental typologies. Secondly, together with the herders, we produced a set of maps. This had several aims. Firstly, we wanted to document scientifically the ancestral uses of these landscapes by nomadic families and clans (ethno-expertise). Secondly, we wanted to obtain a better understanding of the logic deployed to manage environmental resources (animal, vegetal, and topographic) and the ways in which the micro-climates are used. Thirdly, the Evenki, situated in different nomadic groups and in different administrative regions of the Amur region and south Yakutia, provided 48 months of daily observations: these were placed into tables, the categories of which were jointly determined by anthropologists, climatologists, and herders. Fourthly, both the herders and the anthropologist produced a large number of pictures and films (all geo-positioned and classified according to the typologies or the process whereby changes were monitored in the environment). In addition, the pictures were compared with satellite images (remote sensing) with the collaboration of S. Gadal: however, we have not yet completed this work (Dal Molin 2013). Fifthly, Lavrillier and Gabyshev made many diagrams (some of which are included in this book) that explain some indigenous ecological concepts, observed processes of changes, and the hypotheses and predictions made by nomads about the future climate and environmental transformation. All of this helped us to come to an understanding about the Evenki's complex cognitive system regarding the landscape, climate, and human interactions. (cf. below in 1.2.3 and 1.2.4)

In the process of creating this co-production of knowledge and undertaking collective scientific work, questions of **intellectual property (authorship rights) appeared with regards to** this traditional knowledge. We found it necessary to confirm this knowledge as intellectual property and thus oblige other scientists to quote it in an appropriate manner. As Gunn-Britt Retter recognises in her 13 principles of IP TEK (2015), IP TEK is often excluded in terms of the recognition of intellectual rights. The question of the recognition of IP TEK by scientists as a valuable and science-like system of knowledge that needs to be properly quoted is now omnipresent in climate and environmental studies as well as in the international political scene (UNESCO, UNEP, the Arctic Council, IPBES and so on). Often, scientists seem to believe that IP TEK does not need to be attributed to authors or quoted according to scientific rules. For instance, many papers or presentations on research which include IP TEK never mention the names of native collaborators and very rarely cite them as co-authors. (see also here below in the introduction 1.2.4)

A recent definition of TEK, published by G.-B. Retter of the Saami council and agreed upon by the peoples involved in the Indigenous Secretariat of the Arctic Council, answers several of the points mentioned above: 'TK is a systematic way of thinking and knowing that is elaborated and applied to phenomena across biological, physical, cultural and linguistic systems. TK is owned by the holders of that knowledge, often collectively, and is uniquely expressed and transmitted through indigenous languages. It is a body of knowledge generated through cultural practices, lived experiences including extensive and multigenerational observations, lessons and skills. It has been developed and verified over millennia and is still developing in a living process, including knowledge acquired today and in the future, and it is passed on from generation to generation'. Then she identifies 13 principles that must be attached to TEK, including the notion of authorship and rules of collaboration with sciences (Retter 2015).

1.1.2 Evenki conditions for knowledge production versus science

Below, the reader can find a table attempting to compare the epistemological specificities of science on the one hand and IP TEK on the other in terms of the conditions for knowledge production. This comparative table was formulated and presented by Lavrillier in 2012 at a UNESCO conference in Paris before the start of BRISK and updated in 2016 at the end of the project. Three criteria have been changed. This table will be further developed in future studies.

Criteria	'Western sciences' 'Scientific knowledge'	'Indigenous sciences' 'Indigenous knowledge'
Time scale & methods of observation	Mostly occasional and global	Continuous and local
Transmission of knowledge (K) & Documentation	Written documents (paper and electronic)	Oral (discourses and direct practices), constantly changing
K production conditions	Carrying out experiments, observations	Carrying out experiments (testing), observations
Practice of K	Put into practice by others (mostly)	Put into practice by the observers and experimenters themselves and their relatives
Production of K depends on ...	? Funding	Everyday activities (economic or otherwise) and observation
Type of epistemological cognition & K type	Deductive reasoning Inductive reasoning 'Knowing that', 'knowing how'	Deductive reasoning Inductive reasoning 'Knowing that', 'knowing how'
Non-practical K	YES	YES (Nomads know many things that they do not use ≠ 'practical knowledge')
Measurements	YES	2012: ? (in case of anomaly anatomic study) 2016: YES (cf. Part III of this book)
K questioning	YES	YES (= the basic condition for adaptation)
Theoretical construction	YES	YES (= economic activities, strategies, worldview; * cf. Part III of this book)
(2016) Modelling	YES	YES (cf. Part III)
Use of specific language	YES (maths for instance)	2012: ? topography, snow, ice, reindeer terminology 2016: YES (= the typologies in this books and reindeer herding terminology)
Special instruments for K production	YES	NO
? Age	Several centuries	Several millennia
Notion of K authorship	YES	YES

Table 1: Table comparing the specificities of knowledge management and production between science and indigenous knowledge (© A. Lavrillier, 2012, conferences, the * mark the adaptation of the table made in 2016 according to the authors' results).

1.2 A transdisciplinary project: origin and developments

1.2.1 From the point of view of the social anthropologist
by A. Lavrillier

The idea for this project first appeared in 2006 when I went back to the fieldwork that lay at the basis of my PhD monograph: this took place five years after my last experience of field research. My PhD was focused on nomadic economies and ritual practices and their adaptation to urban life. However, from 2006, the discussions of the nomads (spontaneous ones which were not raised by me) began to focus on changes in the climate and the environment (the theme of 'climate change' almost did not exist in the Russian media). These new changes, together with an expansion in industrial development, became a new and important source of anxiety.

Indeed, the Evenki have been noticing climate and environmental changes for several decades, such as a rise in both winter and summer temperatures, but these changes have been increasing more rapidly over the last 7–12 years. A single word in Evenki sums up the main trend of weather change in narratives: *okollen* ('it's getting hotter'). This has a strong connotation of danger for people. Being 'people of cold weather' is part of Evenki identity. They say emphatically: 'in contrast with other peoples in the world, the Evenki cannot physically stand warmth and can survive only in cold weather, like reindeer' (Mazin 1984; Nuttall et al. 2005). They have noticed that the coldest part of winter is now two months shorter than it was 30 years ago and is also warmer than it was then: therefore, the snow period is much shorter. There were regional differences in snow cover: it is thin in southeastern Siberia but deep in Yakutia (Crate 2008) and in Kamchatka. In addition, the Evenki link the warming with an increase in forest fires, a fact that Yamal Nenets also mention (Nuttall et al. 2005: 678).

The Evenki also link the changes in climate with various observed changes in flora and wild fauna, like the extinction of some plant and animal species and the appearance of new species of birds (originally native to warmer regions) and insects. Some Evenki have also noticed that sable fur (their main source of income) is not as thick as it used to be. They also link the major and unexplained changes affecting their domestic reindeer to climate change: reindeer are suddenly dying during very hot weather, parasitic illnesses are developing, and new species of flies are laying larvae in antlers that cause infection and sometimes death. In other words, the Evenki perceive climate change as a great anomaly, just as Orlove has suggested for other parts of the world (Roncoli et al. 2009; Lavrillier 2013: 263).

This first documentation of the effects of climate change was not deep enough, however: it became evident that we needed another approach, one which would involve more IEK and IEK holders.

In addition, during the 2006–2012 fieldwork, some nomads demanded the creation of a collective project to help them understand the unexpected environmental and climatic changes and anomalies about which they were worried.

The notion of the 'co-researcher'

This is the context in which the idea to involve Evenki herder-hunters directly in the research process and monitoring appeared. With the herders (including S. Gabyshev), we started to design a project (ESCAPNES) that we submitted to the EU (without success).

Later during the Polar Wolds conference (organised by the GDR CNRS 'Mutation polaires'), I met M. Roué, who proposed that we join our efforts with those of D. Nakashima to create a common project centred on community-based observatories among Evenki and Sami reindeer herding communities. Then, jointly with other scientists and students (C. Claud, M. Rojo, V. Masson-Delmotte, S. Gadal, M. Dal Molin, etc), we submitted the BRISK project to the French National Research Agency. In September 2012, the project proposal was forwarded to the French National Research Agency (ANR).

The entire BRISK team decided to consider the indigenous herders and hunters involved throughout the research process as researchers. Since they were heavily involved in the scientific process and in the co-production of knowledge, we decided to designate them with the term 'co-researcher'. As with all projects involving participatory research (the so-called methodology of 'citizen science' 'or public participation in scientific research', where non-scientists take part to scientific observation), this caused an important change in the research process, since indigenous peoples are no longer considered to be the 'observed' (I refer here to the great divide in anthropology between 'us' and 'we', 'observers' and 'observed', well demonstrated by Lenclud 1996 among others); instead, they are 'co-observers' of the natural environment and its inhabitants (humans and animals).

This status of 'co-researchers' was recognised by the ANR (which allowed for the allocation of salaries), the University of Versailles, the CEARC (which in its official invitations recognised the herders as researchers and Gabyshev as a laboratory fellow), UNESCO, and MNHN (which invited the herders to various conferences and expert workshops). This book follows this principle, as all oral and textual productions from this Evenki C-B observatory officially mention the IP co-researchers as co-authors.

I have noted that this method, in comparison with those of classical anthropology, gives much richer knowledge and allows for better relationships with the herders.

I have observed that, thanks to this new method, I gained access to 80 per cent of the IEK that I had previously ignored due to the influence of classical anthropology on my earlier field research. Equally, it allows for the real co-production of knowledge, as shown in this book. Nevertheless, this method can only be used once one has obtained a solid knowledge of the indigenous society and its language. It is also crucial to know the individuals in this society well and with whom it is possible to work. All of this is possible only through the methods of classical anthropology or real fieldwork experience.

1.2.2 From the point of view of Evenki reindeer herders
by S. Gabyshev

As indicated above, the idea for the project emerged from several reindeer herders, who asked for support from scholars working in the social and environmental sciences. It is important to note that, as we said before, in contrast with Alaska, Greenland, or Canada, where projects on climate change with similar approaches are common, this theme was relatively novel in Far Eastern Siberia (except for Crate 2008 and later publications); equally, collaborative research with indigenous peoples is still not the scientific approach generally chosen in Russia. Thus, a transdisciplinary project with the participation of herders was a real local innovation.

Some years brought very significant anomalies that triggered real panic among the herders. For instance, 2007 saw a pan-Siberian phenomenon of rain falling on top of snow: this killed thousands of reindeer. Another bad year was 2011, when the nomads suffered from a lack (if not a complete absence) of snow: this produced a much harsher cold. The anxiety reached such a high level that, during the following summer, Lavrillier was able to gather many eschatological narratives (Lavrillier 2013).

In September 2012, the project proposal was forwarded to the French National Research Agency (ANR); however, we had already begun to discuss the implementation of the project with A. Lavrillier in the summer of 2012.

I proposed the idea to create a base of observation in Iranik forest (250 km from the village): this location is at a high altitude, is integrated into a system of different micro-climates and snow levels, and is surrounded by various landscapes. From an economic point of view, the locale is at the heart of the hunting-herding-fishing system: it is here where the herders keep reindeer, fish, and hunt. It is on the path of the migration routes of many animals (migratory birds, wild reindeer, roe deer, elks, sable). In addition, a wide range of plants and trees can be found here.

As a reindeer herder, I have never been involved in any scientific project, but the first global project I thought about was to provide an account of the climate and environmental changes I witnessed from my first experiences of hunting in the 1980s to the present day. In addition, I also wanted the industrial world to hear about the changes in the natural environment we face.

I wondered if the project would be successful and feared that our traditional knowledge would not be taken seriously. First of all, it is important to note that before taking part in this project, I had never been in a big town (and very rarely in small ones): I spent all my life in the forest with reindeer or in small villages.

Growing knowledge from immersion in nature

I was born in 1974 in a small Siberian village and spent all my childhood with my parents and siblings in the forest, nomadising with the reindeer herd. I went to school from the age of eight. It was a shock for us to be separated from our parents and taken out of the taiga forest: we missed the singing of the birds, the reindeer, the natural landscape, fishing, and so on. We lived in the boarding school while our parents, nomadic reindeer herders and hunters, were in the forest with the reindeer. Despite school and continuous advice from teachers to 'learn and study', our link to the natural environment was very strong. In school, we were always willing to run off to the herd. On many occasions, we ran by foot away from the school to the forest and the closer nomadic camps (around between 8–20km from the village): people do not always live in these camps, but one can always find tents and food. Usually, after three days, the personnel from the boarding school came to take us back by car.

When I was ten years old, times became difficult: my father went blind and so my parents took me with them into the forest to hunt. From that time, I was hunting in the forest for my family from the autumn to the New Year; from the New Year to the summer, I studied at school. I spent all my summers in the forest. I was 'the eyes of my father' (his own expression), since he took me hunting and guided me with his detailed memories: tree by tree, stump by stump, mountain by mountain, river after river, road after road. He taught me how to recognise tracks (specie, gender, age), where to pass mountains, how to find roads, and where to cross rivers: I also learnt from him how to perform rituals. He instructed me on what species to hunt, how much and when, for what purpose, and how to read and use the landscape. I learned from him how to survive in the forest, to drive sledges, and so on. At 14 years old, I gave up school and started to work as a hunter-herder on the state farm: I owned a firearm and hunted wild reindeer. The local Soviet newspaper wrote about the 'hunting child' in a propaganda piece.

After that, I spent all my life in the forest, very rarely going into local villages or towns. I worked in various state reindeer herding brigades in the Amur region and in the southern region (Aldan, Aldakai). Thus, I had the opportunity to visit many places in the forest and to work with different Evenki kin groups, all of which possess different kinds of knowledge. Throughout my life (and especially during my time as a nomad), the Evenki language was my main language. I would even say that I am more fluent in Evenki than in Russian. When I went to school at eight years old, I did not know a word of Russian. So we think about our knowledge in Evenki even today: that is why it was essential to work with a social anthropologist who knows the Evenki language well in this project.

During these years, I gained this knowledge day by day, year by year: it was a lifelong process of acquiring knowledge about the behaviour of animals, micro-climates, and landscapes as an entire system that can be adapted to most of Siberia.[3] For instance, thanks to the Evenki knowledge system of the landscape, I can guess the locations of rocks, rivers, passes, and so on, even if I do not know the specific place (cf. part II).

Our project undertook many expeditions into the nomadic forest, led by Lavrillier as the social anthropologist: we also travelled to Russian cities and to the West. My first big town was Blagoveshchensk (224 000 inhabitants in 2016): my village has about 580 inhabitants. Initially, I went into panic mode: I had the impression that I had gone too far away from home. The air was difficult to breathe. In the streets, my head spun because there were too many people: I asked myself how I would be able to work in such a crowded town. I was asked to present the project at the State University of Amur. I was worried and feared that I might be rejected: as a simple reindeer herder, I had no idea how to conduct a talk at a university. My first experience was a bit simpler, since it was at the local school 'Arktika' for indigenous youth. Another exotic experience was the trip to Moscow (around 8000 km from my home), a megalopolis where I found an overly-crowed atmosphere, like that of an anthill. I was wondering whether I would be rejected by the French authorities as a simple reindeer herder trespassing in the realm of science; however, I was made very welcome by the team in charge of scientific cooperation at the French embassy in Moscow and received support from them. I especially thank Mr Alexis Michel and Michel Balazard. In addition to their interest in my knowledge and support for our project, they allowed me to obtain my first publication (Gabyshev interview 2014).

After this, we had many meetings and conferences in France with biologists, climatologists, and geographers: there, I initially felt like a small child at whom everybody would laugh and dismiss my knowledge as uninteresting and inapplicable. However, everybody seemed interested in my experience and wanted to develop collaborative projects.

In the course of different meetings and successful presentations of the project, I was invited to take part to various scientific events and expeditions, such as an expedition to the Chinese Evenki organised by a French university, a conference at Vilnius University, a project with linguists in the Amur region (lead by O. Morozova), and workshops and a conference at UNESCO in the framework of IPBES ILK (Intergovernmental Platform on Biodiversity and Ecosystem Services – Indigenous and Local Services or the COP21). With Lavrillier and the climatologists M. Rojo and

3 This perhaps explains why the Evenki are the only Siberian people spread over such a vast area (the entirety of geographical Siberia) and why they were the preferred guides for most of the first Russian explorers (Lavrillier 2005–2006).

C. Claud, I have co-written several oral presentations (cf. list of references) and two papers, one of which was published in 2016 (Lavrillier, Rojo, Gabyshev 2016).

To me, it was also very pleasant to discover and observe the Western world and the scientific realm.

Now, after several years of monitoring and working with other scientists, I can better estimate the conditions for hunting during the winter, in particular with regards to the snow cover.

Besides my intellectual interest in participating in such research, I think these projects and methodologies can improve the transmission of TEK (one of the aims of this book), compel our youth to consider their TEK positively, and encourage our state to better recognise our way of life and our presence in ancestral lands. In addition, there is a real need to study environmental changes further. (see also Introduction 1.2.4)

1.2.3 From the point of view of Evenki villagers
by L. Egorova

When I became involved in the BRISK project and took part in this book, I discovered a new realm, one of reindeer herders and hunters. It would seem that I did not know what a complex and interesting life it is!

As a villager and a fluent speaker of the Evenki language, it was a real discovery for me. For instance, in addition to the fact that I previously did not know the majority of the terms of the typologies, I opened myself to a completely new interpretation (for me) about the different states of the snow. I felt the thin frontier between the snow type *chegha* (чэҕа) and the snow type *buldo* (булдо).

I also understand that nomadisation is a real science! One must calculate exactly when one must nomadise and to which specific place for reindeer herding and hunting. Nomadising is not simply travelling wherever your feet take you, but travelling at a specific time and to a specific place at each moment of the year. This forced me to think about our place in the world. How is it that my generation, or more precisely that part of my generation which lives in villages, is so distant from the life and knowledge of our parents? Our parents nomadised in the taiga for generations. The distance appeared because our parents were obliged (of course, with the best of motives) to give their children to the well-known boarding schools for many long months. On the one hand, we received a good education, but, on the other, we were detached from the nomadic world and the traditional economies of our parents, e.g. from reindeer herding and hunting knowledge. Nowadays, the youth is slowly returning to our indigenous way of life, and this is right.

I am convinced that it is a highly correct approach to directly involve indigenous persons in a project on the study of climate and changes within it. Indeed, this is interesting in many aspects: one can learn about and discover a lot of issues, and such projects attract the youth (this is very important). It can trigger a desire to recover their roots. I would like to thank everybody who allowed us to rediscover this forgotten knowledge, to become conscious of its richness, to regain what is so important to us, and to give our children and grandchildren the opportunity to receive this knowledge.

1.2.4 Development of a transdisciplinary method
by A. Lavrillier and S. Gabyshev

General outlines

Throughout the entirety of the BRISK project, we made the following choices for the various C-B observatories. In Paris, we first made some rough outlines with D. Nakashima, S. Roturier, the climatologists C. Claud and M. Rojo, and the geographer S. Gadal (but without any IP representative). This helped to define some general issues and methodological points, such as the importance of ensuring that each type of knowledge maintains its own identity during the process of knowledge co-production between IEK and Western sciences. Second, we had long discussions with climatologists about the instruments and measures they wanted installed and used *in situ*. For a better understanding of global changes in the Arctic and in order to meet the various requirements from the IP, global changes were studied according to a common main methodology adapted to each C-B TO, thus giving voice to different local contexts and local IP interests.

Evenki C-B observatory methods

Within the BRISK team, the first C-B TO was established among the Evenki in January 2013. Our methodology and analysis were developed step by step, with a great deal of feedback between Paris and Siberia and between IP knowledge, social anthropology, climatology, and ecology. Our choices for the Evenki C-B Observatory were the following.

Adapting science and monitoring to the nomadic world

During the first year, we tried to satisfy the climatologists' wishes by requiring daily weather measurements to be taken at one fixed point from the nomads; however, it quickly transpired that this was impossible for two reasons. Firstly, the nomads were not able to change their way of life and stop moving around for the sake of the project alone. Secondly, fixed measurements were of little use for recording data about a nomadic people who use local micro-climates for reindeer herding and hunting.

Nevertheless, the climatologists needed fixed temperatures over several years for their analysis. We found two ways of doing this. Firstly, access was granted to a land-based weather station (Rojo 2016). Secondly, inspired by the hydrologist E. Gautier (Costard et al. 2014), a colleague of Lavrillier, we used simple thermo-buttons.

A set of thermo-buttons was installed (after discussions between A. Lavrillier and indigenous co-researchers) along the main nomadic road according to the micro-climates used in nomadisation and snow measurements. Three key locations were selected by the herders and S. Gabyshev for the thermo-buttons: the Iranik, Imakta Ahike, and Tungurcha Solokit rivers inside the nomadic territory of the 'Yakutia-Amur' region. The thermo-buttons provided four daily observations (one every six hours: 3 am, 9 am, 3 pm and 9 pm local time) for the period from October 2013 to February 2016.

Regarding the fixed measurements, a frequent question from colleagues and funding institutions was why we did not use automatic and self-alimented weather stations like *Campbell Scientific GRWS100* (4000 € cost). There were two reasons: we could not be sure about their reliability without external power in the context of frosty weather (inferior/colder from-50°C) and little sun and they were too fragile for the wild forest environment. Indeed, the professional thermometers offered by the Blagoveshchensk Meteo Centre in winter 2013 and installed among the Evenki in spring 2013 were destroyed by bears (along with one of the thermo-buttons in 2014). Nevertheless, as we see in our publications, the climatologists M. Rojo and C. Claud managed to complete nomadic measurements with professional data from land stations,[4] despite the fact that they were very remote from each other and the nomadic area: they provided more complete data over a longer period of time (between 30 and 50 years, according to the sites). Moreover, we must not forget that the richness of the project depends to a large degree on indigenous observation and knowledge.[5]

While the climatologists in Paris guaranteed us that simple domestic thermometers and humidity measurers would be sufficient, the weather forecast specialists in Russia were gobsmacked that we were prepared to use such basic and imprecise instruments. They gave us more professional equipment and provided training to both the social anthropologists and the herders on how and where to install them and how to collect measurements.

We had to adapt to other constrains imposed by the nomadic way of life (short periods of daylight, many exhausting activities, no phone or internet connections,

4 The first data were kindly provided by N. V. Esina and O. B. Byl'eva of the Tynda weather station (Amur region).

5 With Rojo, Claud, we wrote a transdisciplinary paper that uses indigenous knowledge about micro-climates, observations, and climatologic analysis of thermo-buttons and land-based station data (Evenki and Tuva). Although the scientific realm may wish to develop transdisciplinarity and interdisciplinarity, scientific publication formats rarely allow for the publication of real transdisciplinary papers where each science can equally demonstrate their methods and arguments. Thus, this paper is still unpublished.

remoteness from villages and towns, differing levels of interest for being permanently involved in the project) and find an easy way to make observations and maintain the motivation of the nomad observers.

In order to adapt our methodology, we worked with Liudmila Danilovna Egorova, who represents in her person combined types of knowledge: some IPK from her reindeer-herding family and scientific knowledge from her background as a weather forecaster. With her and S. Gabyshev, we established a daily observation table. It was crucial not to confuse the two types of knowledge and not to predetermine the information by making the categories of the observation tables too rigid (the recommendation of D. Nakashima). As such, a significant degree of freedom was given to the nomads when they made their observations. The observation table begins with basic meteorological measurements (temperatures, humidity, precipitation, winds, etc.) and continues with Evenki criteria about the behaviour of wild and domestic animals, changes in the vegetal cover, etc.; additionally, a cell for free comments often provided information about bodily sensations (like 'warm', 'cold', 'normal', 'good', 'sleepy', 'bones pain'), which function as 'barometers' and tools for weather prediction (cf. Evenki climatology). The indigenous co-researchers (S. Gabyshev, V. Gabyshev, A. Kolesov, and O. Iakovlev) filled in the table of observations (in Evenki and/or Russian) twice a day at the same times that the thermo-buttons took their measurements. This table was adapted several times during the project according to the requirements of various herders. Notably, they added entries regarding events related to industrial development.

Lavrillier, Egorova, and Gabyshev developed and used these methodologies and tables of observation in a new project, PARCS (funded by Chantier Arctique Français 2015–2017). This is being conducted with atmospheric physicians (K. Law) and focuses on the perception of the environment in Siberia in terms of a forest-town comparison.

From the side of the herders, we saw from the very beginning that they were unaccustomed to measuring the temperature in C°; instead, they did so according to the state of the surrounding natural elements (water, ice, snow, fog, noise, and so on) (cf. Evenki climatology). Nevertheless, S. Gabyshev and V. Gabyshev, who had never worked on computers before, quickly became accustomed to taking measurements and using Word, Excel, PowerPoint, Elan, and Google Earth.

Midway through the project, it became clear that Evenki knowledge understands and reads the environment according to a grid made up of different typologies: this constitutes a complex cognitive system.

The results and types of data

All in all, thanks to the methods used for making daily observations from 2013 to 2016 and the collective expeditions led by the nomads and the social anthropologist, we produced the following data and analyses:

- Meteorological monitoring data (according to Western and nomadic systems): 1) temperatures (mobile and fixed), humidity, precipitation (including snow cover measures), wind variation; 2) information about the evolution and behaviour of wild fauna and flora and domestic reindeer.
- Abnormal winter and summer analysis as well as analysis of events considered 'extreme' by the nomads (conducted with reference to both TEK and social anthropology). This demonstrated frequent anomalies in the evolution of the snow and ice covers and significant variations in different topographies. In addition, these anomalies seem to be causing significant changes in the vegetal cover. (cf. Evenki climatology, snow and ice typology, Part III)
- Evenki TEK documentation. Gabyshev and Lavrillier (with other herders) developed co-productions about TEK, environmental changes, and land use by mapping the many nomadic roads in this huge area (partly published in this book). Some of these maps were used and merged with remote-sensing analysis by S. Gadal, M. Dal Molin, S. Gabyshev, and A. Lavrillier.
- The socioeconomic impact of these changes were studied in detail (sometimes collectively with herders). Nomads adapt by modifying some herding/hunting practices and utilising their considerable mobility. These adaptive practices are today threatened by the local development of extractive industries.

The anthropological studies, partially published (Lavrillier et al. 2016; Lavrillier and Gabyshev 2017), focused on analysing changes in economic activities and in relationships between nomads and villagers, the worries of the nomads, legal changes and regional policies, adaptive practices (including modern techniques), and beliefs and ritual practices. Emic notions of climate and environmental change, climate, and adaptation are also studied. Together with the rest of the data quoted in this book, this represents an enormous amount of materials, but one should keep in mind that this is only the beginning: the authors will continue following the routes discovered here in future projects. In this research, the classical methods of anthropology (participatory observations, analysis of spontaneous discourse and semi-structured interviews) were used in addition to the co-production approach. This is because a good background knowledge of Evenki society and language is indispensable.

While the nomads volunteered to participate in the Evenki C-B observatory and actively contributed to the scientific process (project co-planning, data gathering, production and co-analysis, and measurements), it was impossible to ask them to perform actions that would disturb their usual nomadic lives. So, in order to be successful, the project had adapt to the society it sought to study and not vice versa.

If we compare these methodologies and the characteristics of the data they produce, we see that the results of classical anthropological methods are collective in nature, in that many individuals are observed and interviewed. In contrast, the data from co-production emerges from a few individuals belonging to the indigenous society and are produced especially for the scientific project. In this case, the indigenous society is both 'observing' and the 'observed', while the data itself bears a more individual character.

Thus, the new methodology we developed with the Evenki allows for the co-production of knowledge, combining social and cultural anthropology, indigenous cognition, and some elements of meteorology. Climatological elements were added during the collective and transdisciplinary analysis in Paris (cf. table here below).

SIBERIAN EVENKI C-B OBSERVATORIES TYPES OF DATA (JAN 2013 – DEC 2016) Produced data and analysis (some still on going)		
Co-production Climato + Anthrop. - IPK	Climatological Knowledge	Co-production Anthrop + IP knowledge
Thermo-button measurements along nomadic roads (topography-microclimate-temperatures)		**Typology of landscape** - Topographic* - Vegetal cover* - Flora and fauna
		Typology of snow and ice*
Interpretation of climatic variability	- Models: *Climatological and weather simulations*	**IP climatology** - Compass rose system* - IP forecasting system (short and long term)*
Daily observation tables (IP & scientific climatology criteria)	- Land-based stations: *temperatures, precipitation* - Satellite imagery	**Analysis of norms** - Snow/ice installation and melting process* - Seasonal cycles* - Landscape evolution*
		Analysis of anomalies* - Climate in recent decades (approximately) - Detailed info for the last 2–5 years - Landscape changes - Snow/ice cover (year-by-year) installation/melting
Maps of nomadic territories		**Landscape management/uses (mapping and diagrams):** - Diachronic and chronic grazing, hunting, and camps - Industrial development

Table 2: Table presenting the various forms of co-production of knowledge at the Evenki C-B Observatory (© A. Lavrillier, 2014, conferences at UNESCO). The * indicate that the data are partly published in this book.

We thus worked in parallel on **two paths of study: one centred on a systematic study of the IP knowledge system and the other on observations of changes**. We simultaneously examined normal processes and anomalies.

Entering the realm of the Evenki knowledge system

Upon first arriving in the field, we asked the following question: 'We are studying changes in the climate and environment, but what the climate and the environment mean for the Evenki?'

We started to gather the elements of the vast Evenki knowledge system: this included typologies of topography, snow, ice, vegetal cover, and so on (cf. Part II). Year by year, these typologies became bigger and bigger, and were further enriched by pictures, interviews, and audio recordings for the purposes of linguistic documentation.

In this process, as well as in the process of studying changes, we developed a specific method of imposing colour drawings and diagrams onto pictures (see here below). One of the best examples of this method can be seen in its application to explain the Evenki landscape concept of *arbun ediyu/arbun solgu, арбун эдиӻу/арбун солгу* (cf. Topographic typology). After three years of work, we have gathered and analysed various parts of the knowledge system, some of which we present in this book (see table above).

Despite the fact that we now understand Evenki indigenous knowledge as a complex and organised reading grid containing different typologies and conceiving of elements of the landscape as an interactive system, no herder or hunter will ever tell an academic the entirety of the knowledge system or all the typologies. It is very difficult to access this knowledge.

First, the knowledge is encapsulated in the Evenki language and is often very difficult or impossible to translate. Second, this knowledge is so vast that not every nomad possesses it in its entirety: villagers, townspeople, and even linguistic specialists often remain ignorant of it. Third, herders and hunters are not very talkative. Fourth, the memory of this knowledge has some temporal aspects; for instance, the Evenki can easily remember the terms for the state of the natural elements in a given season, but sometimes temporarily forget the terms used during other seasons.

Fifth, indigenous knowledge is often sporadically delivered by the nomads, like small pieces of a giant invisible puzzle: initially, they often provide a very laconic explanation from which the academic understands little or nothing. It takes hours (or even days) of further explanation, misunderstanding, clarification, and diagrams to create a co-production accessible to Westerners. To obtain knowledge, it is often necessary to wait for nomads to meet some specific kinds of snow, ice, or species and then catch a new word or phrase in their discussions. In order to increase the speed with which the knowledge was being delivered, Lavrillier discovered a new method. When alone in the forest, she took pictures of everything; when back in the tent during the evening, she showed the nomads her pictures and asked for explanations. We then developed the aforementioned method of imposing colour drawings and diagrams onto the pictures in order to explain and express indigenous knowledge. With time, the herder-observers themselves started to propose pictures, drawings, diagrams, and other pieces of knowledge. Among the nomads, some individuals like S. Gabyshev got caught up in the game of scientific investigation and proposed their own research ideas, concepts, and hypotheses: they also became engaged in co-writing (cf. Conclusion).

In order not to be biased when studying or presenting Evenki TEK, we did not consider existing scientific materials (even simple ones) about snow, ice, vegetal cover, and climatology. It was only upon writing the book with complete data that we looked for translations of some Evenki terms (*pingo*, an icing blister, for instance). Thus, everything here is from Evenki EK. We also did not study in detail the publications with the results from other participatory research projects on similar topics, like those for the EALAT or SIKU[6] (Krupnik et al. 2010); Lavrillier only accessed some oral presentations and general papers (Oskal et al. 2009 for instance). She did not reveal the papers to Gabyshev in order to avoid external influences. So, while we were slightly influenced by other projects, we do not know if our book resembles publications on similar topics and approaches. This was an epistemological choice based on the following question: do similar approaches to similar societies give similar results or forms of results? What can various environmental sciences take from such a book and such IP knowledge? These questions are still open.

Another interesting fact is that, at the beginning of the Evenki C-B observatory, we did not plan to work on snow and ice in particular; we had much more global intentions. However, we realised later that the largest and more profound part of the material gathered is about snow and ice (second place is occupied by climatology).

Thus, the work we present here is the result of eight expeditions lasting around 16 months and three years of indigenous observation: it took years of patience and work from both the anthropological and indigenous sides. It is an attempt to put together at least a small part of the giant brainteaser of the indigenous knowledge system.

Before going into detail about the Evenki knowledge system, allow me (Lavrillier) to present some inputs from our transdisciplinary experiences.

1.2.5 Difficulties of mutual understanding between the sciences (social and environmental) and IEK

by A. Lavrillier

The interactions and collective work with the natural sciences were difficult and full of misunderstandings; nevertheless, they were enriching. It was much easier working with the biologists, since they (Roturier and Nakashima) have been participating in transdisciplinary research with anthropologists for several dozen years.

6 Note from the editor: About SIKU, see also Krupnik and Bogoslovskaia 2017.

When we sought to establish transdisciplinarity within the BRISK project between anthropology, IPK, geography, biology, and climatology, we noticed that some problems arose from differences in vocabulary. This in turn raises questions about the status and value of the data at the basis of this research (see below). For instance, anthropologists talk about 'indigenous **observation data**' (considering it valuable observation data). In contrast, climatologists use the terms 'indigenous perception' or 'indigenous **feeling**', thereby suggesting that it is very rough data and perhaps not good enough for environmental science. Thus, between the sciences, there are also different terms for designating the same things: it takes time to establish mutual understanding.

Status and ownership of data in different sciences

Depending on the science concerned and the relationship between sciences and IPK, the status and ownership of information can be considered very differently. This is because of the specific features of sciences themselves (and of the societies where the sciences were developed); again, one must work at establishing common ground.

For example, the research results of the anthropologists were initially considered by the geographers to be 'simple data': like automatic temperature measurements that have yet to be analysed, such data, belongs to everyone and must be shared without limit. These 'data' are treated as raw and anonymous data to which geographers add their scientific input. They also did not pay attention to the potential ethical problems raised by widely sharing anthropological data. In the same way, an anthropologist will tend to consider a satellite image as pure data (i.e. without scientific and personal input or meta-data), forgetting that calculations for revealing the features of the ground and vegetal cover have been contributed. Equally, some geographers tend to consider information about the landscape given by the herder-hunters as raw data (i.e. without personal input or ownership) (from discussions during collective working sessions between Lavrillier, Gabyshev, Claud, Rojo, Roué, and Gadal).

These differences in the recognition of the value and status of the data among the sciences themselves and between the sciences and IPK are likely to raise tensions, maybe even conflicts, between members of a project or between researchers and IP; it represents a real dilemma.

For instance, anthropologists also tend to consider indigenous knowledge as 'data': they regard it as something that is collective and transmitted from generation to generation: it therefore belongs to international cultural heritage. However, this neglects the individual input of the informants. Therefore, 'data' from indigenous knowledge have very often been used by anthropologists (including Lavrillier) extremely widely and without mention of the informant's name(s) in a copyright label.

There are several reasons why anthropologists often do not directly indicate the names and locations of the informants: many people give the same information or do the same things, making it difficult to quote everybody; sometimes, information is given with a demand for anonymity; it is occasionally necessary to hide places and peoples from view; the publisher may force authors to erase names in order to save space; and since anthropology is the science of society, not mentioning the names of people involved is a way of distancing the study from an individual level. In such cases, informants are partially dispossessed of their authorship (although they are usually mentioned in the acknowledgments). Between ethical reason and scientific reason, a middle way must be found. This book is a humble attempt to do so.

Regarding the value attributed by different sciences to indigenous information, I used my experiences with linguists at the Max Planck Institute (Leipzig, Germany) and climatologists. Linguists almost always mention informants' names and individual details with each record and in their main publication, thereby proving the source of the information. Thus, individual indigenous authorship gives scientific value to the data. A slightly different case is that of the PhD student in the BRISK project (M. Rojo): possessing an economic and climatological background, he (among others) is studying the perceptions and effects of climate change in the Tuva Republic (Russia). As a young scientist trained to work on the basis of large-scale numerical data sets, he initially doubted whether anthropological field data (interviews and observations) could really be strong enough for building scientific analysis. He was especially concerned when he needed to work with a limited source base of three-to-five persons. For a scientist accustomed to quantitative data, information from indigenous peoples (i.e. qualitative data) was questionable, especially given that it came from only a few individuals.

The co-researchers with whom we worked thought that, in general, scientists tend to disseminate their data too widely. On the basis of this concern, we have concealed some of the information provided by the nomads (like that related to hunting game, for instance).

Discussing with our colleagues (anthropologists, climatologists, and geographers), we understood that, depending on the science, the term 'data' does not refer to the same type of information or knowledge, to the same level of integrated analysis, or to the same level of scientific investment. Consequently, the relationships and attachments that researchers in different sciences have with their 'data' differ.

For instance, for anthropologists, 'fieldwork data' and post-fieldwork analysis (like a study with maps or the description of an emic concept) are not 'raw data' devoid of scientific or individual input or authorship. All anthropologists know this, but it is still important to stress these aspects in a transdisciplinary context. As already stated above, anthropological field data are extremely difficult to obtain. The investment required for fieldwork is tremendous: one needs to learn languages and know the country concerned, as well as make other. Even this is not enough to approach local people and convince them to work with you. As the data are gathered piece by piece, it takes months, if not years, to deduce IPK from observed practices. In addition, fieldwork is expensive, physically and psychologically demanding, dangerous, and lonely. Many anthropologists are subjected to life-long physical trauma and struggles with funding: in additional, they might find it difficult to obtain a position because of their time in the field. Finally, field data are obviously the result of pre-selecting the focus of the research, the people to ask, and the methods: thus, 'raw data from the field' already bear scientific and personal input.

All of this explains the value and heavy 'cost' of anthropological data and the attachment of anthropologists to it. Most of the time, others sciences are not conscious of these investments and do not understand this attachment. When this is not recognised by other scientists and data are considered to be 'raw', misunderstandings between members of the team are likely.

In the same way, the anthropologist often disappears behind the indigenous co-researcher in the framework of IPK studies. This makes it possible to forget that the delivered information/material (provided by an IP and an anthropologist or by the anthropologist alone) is the result of considerable work, as explained in this introduction. These results are often perceived to be the work of the indigenous co-researcher alone or as if they were delivered in a publishable state in the field.

As shown in Gabyshev's text above, Siberian indigenous peoples know the value of their knowledge for the nomadic world and way of life, but wonder whether it is worthy of serious consideration and inclusion within scientific research. They also think that their knowledge is too often used without proper mention of their names and that their data are not regarded with sufficient respect. They may also feel that they vanish behind the figure of the researcher, becoming little more than an 'exploited tool'. Their 'data' are also the result of considerable effort and personal involvement: they learn from the age of 11 to hunt alone in frozen and dangerous forests, to ask elders about technical knowledge, to continually observe the natural environment, and to make deductions and hypotheses about it. Their 'data' are, as proven in this book, both inherited and developed individually. They want their authorship and the value of 'their data' to be recognised (Gabyshev). This is one of the reasons why Gabyshev and the other nomads are co-authors of all the Evenki C-B observatory presentations and papers.

The different paths of transdisciplinary co-production

The following diagram, made and presented by Lavrillier and Gabyshev in March 2014 at a conference in Kautokeino (Norway), illustrates the different types of knowledge co-production from the viewpoint of the present book. It also considers the uses of such co-production for IP (cf. From the point of view of Evenki reindeer herders).

So, this book is an attempt to prove the quality and show the results of co-production between IPK and anthropology, thereby demonstrating its potential use for the environmental sciences. The interest of social anthropologists in such research is not in how IEK corresponds with environmental scientific results, but in understanding and revealing the hidden cognitive structure of this IEK system. There is always a general scientific curiosity about whether environmental sciences and IEK agree or not, but this question is not important for social anthropology. It is in this way that social anthropologists can be good co-production partners between IP and environmental sciences (see the table below).

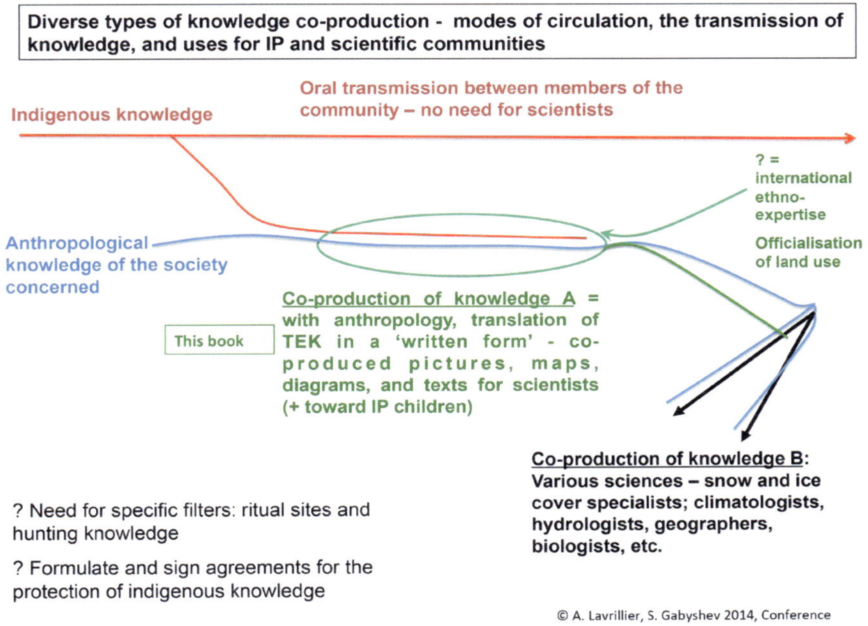

Indigenous knowledge system for weather and climate forecasting

According to the analysis produced by Lavrillier and Gabyshev, the Evenki have their own system for weather/climate observation and prediction. It is marked by two characteristics.

First, it works on a specific understanding of annual variation, which balances between norms and anomalies. We noticed that within phenomena considered 'norms', yearly variations are included to a certain extent: if these variations are too significant or too regular, they are considered 'anomalies'. For instance, if the cold winter is delayed by two weeks on one occasion, it is still considered 'within the normal variation'. In contrast, if the cold winter is delayed by two or three weeks repeatedly over the course of several years (as we see in our paper on sable, Lavrillier et al. 2016), it is considered an 'anomaly'. This is also the case with considerable jumps in the temperature during the winter and perturbations in the snow cover, which the nomads consider 'big anomalies'. In summary, the climate has lost its logic for the nomadic Evenki and has become very difficult to predict.

Second, the Evenki system of weather/climate observation and prediction is based on knowledge about interactions between elements of the environment. Lavrillier's analysis of the Evenki observation tables shows that the Evenki predict the weather by observing birds, domestic reindeer, the vegetal cover, and human bodies and moods. They also note changes in the wind and the noises made by wood fires. For example, some remarks often note that when an elder present in the camp complains of pain in his back or in his bones, snow falls the next day. Other examples include a sleepy mood among all the members of a camp or tiredness among the domestic reindeer: these events predict snowfall. Bird song announces rises in temperature, while seals or ravens flying high in groups mean the weather will get worse (storms). When domestic reindeer fight each other quietly, a period of cold weather is coming. If the local sables run around a lot (leaving many tracks in the snow) to make food reserves, it means that heavy snowfall will occur. In contrast, if there are few sable tracks on the snow, it means that it will get colder. If the moon or sun has a halo (*diularen/ дйуларэн,* meaning literally 'made a house'), it will be very cold. If the bushes and trees are small and their branches curve downwards (*smorshilis'/сморшились*), it will get much colder. The observation tables prove the efficiency of this prediction system. (cf. Indigenous science of climate)[7]

Misunderstandings and mutual understanding between the social and environmental sciences and with indigenous knowledge

In the course of our work in the BRISK Evenki C-B observatory, we had to face several misunderstandings between the environmental sciences, indigenous knowledge, and anthropology. One of the more illustrative examples concerns the notion of 'first snow'. At the beginning of the project, the Evenki herders wanted the environmental scientists (climatologists and geographers) to analyse the changes in the 'first snow', since they had noticed (as explained in this article) that the snow had been arriving late. After several months, the analysis of data from climatologists (Rojo and Claud) arrived. To our surprise, it showed a result that completely contradicted the observation of the nomads: the 'first snow' arrived much earlier than before. The key to this contradiction was two different understandings of the expression 'first snow'. For the climatologists, 'first snow' designates the first flakes of snow (even if it melts immediately), while 'first snow' for the Evenki means the first installation of the snow cover (they do not even notice the fall of the first temporary flakes and are exclusively interested in the installation of the snow cover, since the latter is the basis for hunting activities in the late autumn and winter). In contrast, the climatologists were not so

7 In the Evenki C-B observatory, A. Lavrillier and S. Gabyshev systematically documented this predictive system: this will be published in another book.

interested in this installation, but rather in the first snowflakes: they struggled to find annual data for the installation of the snow cover. *In fine*, there was no contradiction: the first snowflakes are appearing earlier, while the installation of the snow cover is occurring later and later. This misunderstanding allowed us to discover an interesting fact: one phenomenon (e.g. 'first snow') can be understood in several ways according to different knowledge systems: this leads to different approaches and analyses. This can create apparent contradictions where in reality there are none (Lavrillier and Gabyshev 2017).

Complementarity between sciences (SHS and EnvSci) and traditional knowledge – two systems of observation

Sometimes, indigenous knowledge can teach Western science. For instance, thanks to the Evenki knowledge about micro-climates and their choice of locations for thermobuttons, the climatologists had to pay attention to the considerable differences in temperatures over short distances (Lavrillier; Rojo; Gabyshev, unpublished paper).

We also noticed that the nomads considered some scientific documents or tools as incomplete or incapable of representing the environment as they know it. This was the case for the 'cloud atlas' given by the National Centre for Weather Forecasting of Blagoveshchensk, the official book of all weather forecasting centres in Russia. The nomads were supposed to use it to fill in the tables of observation, but they unanimously declared that 'most of the clouds here are not ours!' or 'we do not have such clouds'. With Egorova, we made a digest with a selection of existing clouds for the nomad-observers on an A4 document with Russian names and pictures. One of the Russian names designating the cumulonimbus can be literally translated as 'cumulative rainy cloud' (kuchevo-dozhdevaia oblaka, кучево-дождевая облака). However, in the tables of observation for the winter, the nomads wrote the name 'cumulative snowy cloud' (kuchevo-snegovaia oblaka, кучево-снеговая облака). To them, it was unthinkable to have 'rainy' clouds in winter, so they adapted the name.

Throughout the BRISK project, we see that, in comparison with nomads, the climatologic information about the periods concerned takes into account far fewer parameters as drivers of changes (i.e. mean minimum and maximum air temperatures, precipitations), but provides detailed analysis of global and regional trends. It provides a more detailed account of changes in temperatures and precipitations than the Evenki observations, but nevertheless confirms the observations of the nomads. In contrast, as shown in the examples of winter sable hunting (Lavrillier et al. 2016) or in this book, traditional knowledge takes into account a larger number of parameters (i.e. jumps in the air and ground temperatures, precipitation, the Evenki typology of the evolution of the snow cover, the weather and environmental conditions of the previous seasons, the life cycle of the vegetal cover, and so on), but does not measure them mathematically. In order to approach so many parameters through Western scientific knowledge, we would need contributions from a vast number of scientific disciplines. In addition, social anthropology and economics demonstrate the interplay between climate change and the economic, political, social, and cultural drivers, which produces various (direct) effects (Lavrillier and Gabyshev 2017).

Nonetheless, different types of knowledge and science are complementary to each other and allow for mutual learning. The nomads learned about similar observations of the climate and environmental changes from other Arctic peoples. The Evenki herders who actively took part in the project stress that they are now making deeper and more detailed observations for themselves. Indeed, having enjoyed the intellectual exercise of transdisciplinary observation and analysis, they have started to build more hypotheses about future changes in the climate and environment (S. Gabyshev and V. Gabyshev). In addition, this demonstrates to climatologists that climate change is a human and economic concern (and not just a matter of numbers and chemical formulae) with effects on flora and fauna. It highlights to social anthropologists the importance of including the environmental sciences in their research, since there are some aspects of indigenous knowledge with which they cannot work unless they enrol the help of other sciences. As indicated above, the social anthropologist also learns that, by placing indigenous knowledge holders on the levels of 'observer', 'co-observer', and 'co-analyser' (instead of 'observed' or 'analysed', as in classical anthropology), the research results are much richer, deeper, and more systemic, especially with regards to traditional knowledge about the environment, the interplay of various drivers, and the interaction of humans with their environments (Lavrillier and Gabyshev 2017).

Outline description

At this stage in the research, we conceived of the following structure, which combines analytical texts (traditional for anthropology) and other formats, less traditional for Western sciences. These less traditional formats include abstract diagrams with explanations in Evenki, Russian, and English, diagrams on pictures, and encyclopaedic entries with pictures and tri-lingual explanations provided in such a way that it is almost like the herders are themselves explaining the forest to a visitor (cf. Preface).

In Part II, after a presentation of the Evenki regional group and their calendar, this book will present elements of the Evenki ecological knowledge system in the form of typologies: we will start with the typologies of the natural landscape. It will focus on topography and vegetal cover. Second, it will explain indigenous climate science and provide a description of weather forecasting methods. Then it documents the cloud, precipitation, and wind and air typologies. We continue with the complex snow and ice typology. The third part is dedicated to the system of observing and predicting norms, anomalies, and transformations, with a discussion of the phenomena considered as 'norms'/'anomalies' by the nomads. We also consider the way the Evenki produce hypotheses and models. The anomalies analysed were observed between between 2012 and 2016: they are categorised according to indigenous ecological knowledge.

2 THE SYSTEM OF EVENKI ECOLOGICAL KNOWLEDGE AND ITS TYPOLOGIES

2.1 The Evenki group concerned and their calendar

by A. Lavrillier

The Evenki are a minority indigenous people of Russia (with 38 396 individuals in 2010). 18 232 Evenki live in Yakutia and 1501 in the Amur region (the two regions with which this book is concerned).[1] Particularly mobile, these people live in small groups spread across a vast area defined by the Yenissei River in the west and Sakhalin Island in the east: it stretches from the coast of the Arctic Ocean to northern China.

In the Russian classification system, the type of reindeer herding practised by the Evenki of southern Yakutia and the Amur region is known as the taiga, Orochen, Evenki, or Tungus type (Vasilevich and Levin 1951: 5).[2] The Evenki have a dual economy and a dual logic of subsistence between hunting and reindeer herding, with seasonal interplay between the two. They keep small herds of reindeer for transportation purposes, but also in order to have a 'stock of meat' in case of a shortage of food game. Each species of fur or food game is hunted in accordance with a rigid seasonal calendar and diverse strategies. Thus, hunting is carefully planned so that every species can breed successfully (Lavrillier 2005, 2011a).

The collectivisation of herding and hunting was implemented during the 1960s in this region. They became employees of state farms (sovkhoz), while their reindeer were turned into the property of these institutions. Fur and food hunting were intensified, as well as reindeer herding, to allow for greater levels of meat and fur production. The Soviet authorities sold meat, antlers, and fur products on national and international markets. During the Soviet period, the nomads were partly settled in purpose-built villages.

After the collapse of the Soviet Union and the concomitant economic crisis, the closure of state farms meant that the hunter-herders and most villagers stopped receiving salaries. From the beginning of this crisis in the early 1990s, many Evenki in

---

1 Census of the Russian Federation 2010: Nationalities. www.gks.ru/free_doc/new_site/perepis2010/croc/perepis_itogi1612.htm [21 December 2015]

2 The supposedly different reindeer herding types 'Evenki' and 'Saian' (the latter of which is practised by the Tozhu, Tofa, and Dukha) (Vasilevich and Levin 1951) are more similar than they are different. Both use reindeer for transportation purposes (sledging, pack carrying, and riding) and for milk: they very rarely slaughter the reindeer (Ermolova 2003).

the region concerned returned to their former nomadic lifestyles and recovered traditional forms of reindeer herding and sustainable hunting, thus restoring the subsistence economy: they also reorganised their pastures and nomadic roads. Today, only approximately 30 per cent of the population still lead a truly nomadic lifestyle. The others now live in villages and towns. However, even for villagers and some townspeople, fur and food game hunting represents important economic input. Indeed, the nomads provide crucial support to their relatives in the villages by sharing their hunting products or incomes (Lavrillier 2005).

The southeastern Evenki inhabit a territory characterised by small mountains and natural forests (larch, pine, fir, birch, cedar), often with a rich under-layer of vegetation consisting of lichens, mosses, and berry bushes. The continental climate is quite variable (< −50°C / + 30°C). Parts of the terrain contain fast-flowing rivers and streams, while some of the wider valleys include extensive bogs and meadows, which provide ideal summer pastures for the reindeer. In order to meet the requirements of both herding and hunting, subsistence activities in the landscape can only be achieved through a high level of mobility (1500 – 2000 km yearly). This ability to move and sustainably manage the environment (despite its variability) is allowed by deep indigenous knowledge and cognition of the environment and its micro-climates.

The southeastern Evenki have been in contact with extractive industries (gold and coal mines) since the end of the 19th century, but recently there has been a significant growth in industrial projects (pipelines, dams, roads and railways) either directly in their nomadic areas or close to them. It is important to note that Siberian peoples have no property rights to their ancestral lands.

Nowadays, the reindeer are owned by three types of economic units: enterprises that appeared after the collapse of Soviet authority and the transformation of former state farms (*sovkhoz, совхоз*) which offer salaries for pastoral work; family cooperatives (also called clan communities), which are indigenous mini-companies recognised by the Russian government that receive a fee for each living reindeer; and private herders, who lack an administrative form, official recognition, and a regular income. For Evenki people from both family cooperatives and the private sector, fur and hunting (mostly sable) represent a vital source of income.

The Evenki C-B TO is based within a nomadic community which lives across the regional frontier between the Amur region and southern Yakutia. This is an ancestral nomadic space where interrelated kin groups nomadise. Before 1933, it was known as the Evenki Vitimo-Olekminsk district (including the rivers: mid Olëkma, Niukzha, Aldan, Timpton, Gonam, Sutam, Gynym, and Algoma) of the Eastern Siberia Region. In 1933, a decree redrew the regional boundaries across the territory of this nomadic kin group, dividing these Evenki into two administrative jurisdictions: Yakutia and the Amur region (Ob izmenenii… 1933). This brought no specific changes to the practices of the nomads, who (ignoring the frontier) used the same ancestral roads and continued their exchange and kinship relationships. Nevertheless, at the administrative level it caused continuous negotiations between the institutions of the two regions. This is one of the biggest nomadic Evenki areas of Russia, with a surface area of 7000 km² and around 15 000 reindeer led by 250 reindeer units. This is a considerable concentration of reindeer and human population for this kind of reindeer herding: small herds (40–100 reindeer per unit) and hunting form the basis of the subsistence economy.

Map of the research area (2011-2016)

This map is intentionally non-specific and lacking in detail in order to respect the justified concerns of the indigenous herder-hunters.

© Map from Max Planck Insititute, adapted by Lavrillier.

The Evenki New Year – summer (diuya, дйуҕа)[1]

The year starts with the best season, the summer. Several nuclear families (three to ten) join together just before the beginning of the season in order to nomadise together throughout the summer.

The most important activity is reindeer herding. At this time, the calves are still fragile and dependent on their mothers: the camp rings out with their husky calls to their mothers. Without snow and the tracks left in it, it is difficult to keep the herd gathered in one place: hungry from the long summer, they spread out to search for food. The herders spend their days walking all around the forest in an attempt to assemble the herd; they eagerly anticipate the appearance of the horseflies that push

1 For the detailed terminology of the Evenki calendar, cf. Seasonal chain with expected changes (pp. 164–172).

the reindeer towards the smoke fires in the camps (which protect them from this kind of insect). During the period when there are many horseflies, midges, blackflies, and mosquitoes, men, women, and children must keep the smoke fires (*samŋin, самңин*) burning.[1] The tents are installed in a circle or semi-circle around the reindeer and each tent maintains one or more smoke fires. During rainy days, the reindeer spread out into the forest to graze: if the rain continues for too long, the hunters have to spend days gathering the herd. This is probably why the Evenki say that 'the horseflies and mosquitoes are our herders in the summer, they work for us'.[2] Women spend their days treating skins (smoking, scraping, or sewing), milking, and gathering berries (bog blueberries, juniper,[3] etc., cf. Vegetal cover typology). Men and children fish in the surrounding rivers, survey the herd, treat the reindeer, and train the calves to become accustomed to humans. The considerable size of the human group (in contrast to the winter, when nomads live mostly in isolation) allows the Evenki to hunt collectively: they hunt elk, red deer, and, more rarely, bears. During each hunt, even in a period of scarcity, the Evenki forbid themselves from shooting females or calves. One elk can feed approximately five tents over the course of a month.[4]

At the end of summer or at the beginning of autumn, nomads prepare for the long snow period by repairing their sledges and harnesses or making new ones, sewing clothes and shoes, gathering large reserves of berries, etc. Nevertheless, the summer is perceived by the Evenki as a moment of idleness when the long days (around 19 hours of sun) leave time to play various games and enjoy life.

During this season, activities linked to the camp and to the herd are collective: the tasks must be shared equally among all the members of the encampment. These tasks include building wooden structures around the smoke fires and the enclosures, collective hunting, sharing meat, and gathering the herd. Thus, each unit of nomadisation (i.e. a nuclear family living in a single tent throughout the year) is responsible for doing its part of the collective work. If this rule is not respected, tensions and conflicts can arise between the tents. Nevertheless, some activities remain 'individual', like milking, marking, and the medical treatment of one's own reindeer. The same is

1 The smoke is produced from the combustion of fresh larch covered by either larch branches or a block of humid *ialbuka, йалбука* moss. Around twice a week, the entire encampment helps to gather this moss.

2 The Evenki distinguish between different species of horsefly with the names *soroptun, сороптун* and *sohindo, сохиндо*: they also distinguish them in accordance with the species to which they are attracted (wild reindeer, elk, or red deer).

3 *Juniperus dahuria* Pall., *mozhzhevelnik/можжевельник* in Russian.

4 In summer, meat is cut into long pieces that are dried in the sun on a specific type of pole known as a *lokovun, локовун*. The flies often lay their larva on this meat, which is not to the taste of the nomads. Another technique is to cook the entire piece of meat, cut it into smaller chunks, and dry them under the heating stove. This dried meat is full of energy, but the Evenki like to eat it only during long trips or during periods of scarcity. Thus, the day after hunting is when the best parts of the meat are eaten: the meat is fresh and consumed in large quantities. This is the case for ritual feasts, during which one must stuff oneself with fresh meat.

true for the making of clothes and harnesses and the setting up of tents. Mutual aid can be given in the course of such activities, but there is no obligation to do so. Thus, it is crucial for the entire encampment to safeguard the harmonious balance of the relationships between members, which are based on the principle of a just allocation of game and tasks. All know that this balance is fragile; at the end of the summer, everybody is happy to leave the group for a more 'isolated' way of life, since they recognise that collective life requires a lot of effort.

The re-birth of the environment celebrated by humans, insects, birds, rainbows, and bears

Summer is when nature is reborn after nine months of snow: it 'arrays itself with thousands of colours'. This moment is long anticipated. In around mid-June, the first songs of the cuckoo announce the beginning of this period: nomads in various camps ask each other over the radio or satellite phone if they have heard the song of the cuckoo, the shaman bird. During her fieldwork, Lavrillier heard the following expression from the nomads: 'The cuckoo is the shaman bird, it is the double of the shaman. In the past, we had a lot of shamans and there were a lot of cuckoos, but now we have very few shamans and cuckoos are rare nowadays'. This bird gives other indications about events in the environment (cf. Evenki climatology) and in life. The cuckoo is thought to give a vision of the entirety of one's future if one kills it, cooks it, eats it, and puts the remains under one's pillow: the individual concerned will see their future in a dream. Very rarely do the Evenki do this: most have little interest in knowing their futures in advance.

The beginning of the summer is also marked by the young larch needles that illuminate the taiga with a bright green colour. During the summer, various flowers decorate the forest. It was during this season that the most important collective ritual of the Evenki occurred – the *ikenipke, икэнипкэ*, a round dance lasting eight days in which the shaman and all the community participated (Vasilevich 1957; Mazin 1984: 91–94; Lavrillier 1996, 2003, 2005). The elders I worked with about this ritual remember it as a meeting between many nomadic groups, with lots shamanic rituals, games, dances, songs, and joy. It was a time for weddings and economic agreements: shamans also added new elements to their costumes and drums and new shamans were elected. The last of these collective rituals happened in the 1950s.

Nevertheless, even without a shaman this period is still perceived as the 'best moment of the year', a time when one 'plays' a lot. It is also a happy moment because the children, who live all year in a boarding school, join their parents and the herd: an encampment can contain as many as 40 persons (instead of three to six during the snow period). The current Evenki nomadise along the same rivers as their ancestors did three or four generations ago. This means there are traces of old encampments and remains of shamanic ritual constructions from the 1950s not that far from the current encampments. This type of site constitutes the cultural landscape of the nomads and bears a sacred character: their positions are kept secret from alien peoples. Such places are named *nimŋakit, нимӈакит* (from *nimŋakar, нимаӈакан*, the spoken part of a shamanic ritual), *evikit, эвикит* (place of games), and *samakit, самакит* (place of shamans). The nomads know the locations of such places, but avoid walking there or taking anything from such places lest they disturb the spirits and bring upon themselves a tragic fate.

Another important event in the summer is the passage of migratory birds (*deyil, дэԑил*).[1] They are considered to open and close the doors of the summer and so their passage is looked for. At such times, nomads make sure that they have filled all the buckets with water or have turned them upside down: 'They are flying from far away, they will drink water. But, if you have empty buckets, then they will drink your blood and you will have an empty year without luck'. Sometimes, nomads throw salt in the direction of the migratory birds and ask them to come back next year.

Summer is the mating season for reindeer. Humans like to observe their mating games in large spaces: here, herbs are squashed during the mating fights. During the summer, the Evenki do not hunt bears unless they pose a danger to the reindeer. In such cases, the nomads will fire into the air in order to frighten the animal; only if there is no other choice will they fire upon it. Out of respect, women always leave berries in each bush for bears and other animals reliant on this foodstuff.

Insects are also omnipresent in the summer and give rise to games. Men, women, and children like to catch horseflies and observe their development, colour, hairs, wings, etc.[2] After this, they free the insect, saying 'Let's go and herd my reindeer!' Ladybirds of different colours are called 'shaman insects'. One can catch these insects and give them a small piece of wood, which they take and turn with their legs. The Evenki say that this is the ladybird playing a drum (the small piece of wood) and shamanising: 'Play, little ladybird!'[3] Ants (*irikte, ирикте*) are described in the following manner: 'The ants are like the Evenki: they spend their time working hard and they eat very little'.[4] One can forecast the weather with their behaviour (for instance, when

1 *Deyil, дэԑил* (pl.) derives from the root of the verb 'to fly' (*deyil-da, дэԑил-да*) and is used for birds in general. Another generic term, *chipkachan, чипкачан/chichak, чичак*, designates non-migratory birds.

2 There are numerous species of horsefly and each has its own name in Evenki. *Soroptun, соронтун* lay their eggs under the skins of reindeer. The larva (*kuikte, куиктэ*) develop there throughout the snow period and then pierce the skin, fall to the ground, and turn into flying insects. *Irgakta, иргакта* designates horseflies that do not bite: instead, they spurt their eggs into the nostrils of the reindeer. There the larva develop throughout the winter before exiting and turning into flying insects. Sometimes, they lay their eggs in the nostrils or eyes of humans – this causes terrible pain for several days, but the larva cannot develop in humans.

3 In the Even language, one of the names of the ladybird is *hamakan, намакан* (literally 'small shaman') (Cincius 1977: 314).

4 The old female shaman Matrena Kulbertinova told Lavrillier in 1994 that 'the Evenki descend from ants, while the Russians descend from pigs and the Yakuts from horses'.

they walk lazily it is a sign of rain).[1] There are also two species of coleopteran with which the Evenki play. The large coleopteran (*Cerambyx cedro* L., *iyegdy, ийэгды*[2] in Evenki) with long antenna are caught and put into people's hair in order to frighten them and make them the subject of fun. The Evenki also organise fights between these coleopterans. Nomads use their larva, which they take from under tree bark, as bait in fishing. With its long 'nose' (rostrum), the insect *Byctiscus betulae* L. (*kenggerepki, кэнггэрэпки, semerepki, сэмэрэпки* in Evenki) is the subject of much attention. The Evenki like to have a lot of such insects climbing around inside their tents: they catch them to observe them, to race them, or to put them on the noses of their children: their squinting in order to see the insect is amusing. The Evenki say that the arrival of such insects in a tent is a good sign, and one can also use their behaviour to forecast the weather. They put *sipsyki, сипсыки* grasshoppers on their backs, since they like to see how the insects try to turn themselves upright by swinging their heads from right to left: this makes a swishing sound. There is also the giant woodwasp (*Urocerus gigas* L., *beiutkan, бэйуткан* in Evenki ('little game')), which the Evenki consider to be the soul of an elk: they have to catch them and put them into a fire in order to accelerate the process of re-birth. This will mean that an elk will appear close by, which the Evenki can then hunt. The bee, *diuvuktu, дйувукту* is considered a useful animal: the nomads think they need to be stung once a year to regenerate their blood. Some Evenki consider bees to be the chiefs of insects, since they have 'richer fur'. Finally, there is the spider, *ataki, атаки,* considered to be the guardian of house fires and sometimes connected with ancestors. The appearance of a spider in a tent is a sign of a future birth or of luck in hunting; they teach the young not to kill spiders because doing so could provoke the death of a relative. In contrast, some insects are not appreciated because they are associated with the world of the dead. Such are those coleopterans that like to eat rotting carcasses and look like black cockroaches.[3] Among the other animals considered to come from the world of the dead, there are mice (*teperikan, тэпэрикан; chiŋerikan, чиӊэрикан*), which are active throughout the year, and lizards (*iyela, иҕэла*). The Evenki are really frightened by these two small animals, even though they are accustomed to dealing with large and dangerous creatures. Flies (*dylkachan, дылкачан*) and midges (*urmikte, урмиктэ*) are not appreciated because they lay their larva on fresh food.[4]

---

1 According to the Evenki classification, there are three species of ants. The first kind is small and lives in ant hills; the second, of a middling size, lives in the decomposing hearts of larch trunks; and the third, the biggest (from 1 to 2 cm in length) spends the winter sleeping in dried larch, where they delicately dig complex tunnels: this 'art' provokes respect among the Evenki. The males of this last species have wings.

2 The term comes from *iye-*, ийэ-, 'antlers' or 'branches'.

3 They are called *kovordo, ководо*, 'the frying pan', from the Russian *skovoroda, сковорода*. This name is attributed to them because of their form and colour.

4 Here again the Evenki make a distinction between 'domestic flies', which are always present in the tent and are not a danger to food, and 'external flies' that lay larva on fresh food (meat and fish).

In sum, the summer is a moment of re-birth for life in all its forms, which is a miracle for the Evenki. During their observations of the appearance of all these 'lives' in the summer, the nomads narrate the path between birth and death for each species with the same wonder, considering it a miracle (for instance, the insect that spends the winter without drinking or eating in hibernation within tree trunks or the ice). Sometimes, they joke, 'if we were like an insect, we would have fewer problems purchasing food and goods and would only see the best moment of the year'.

The first sign of the end of the summer is the torrential rain, which arrives together with thunderstorms, lightening, and rainbows. According to the Evenki, rainbows take water from the rivers to put it into the clouds: it is then poured onto the ground. Thus, rainbows are considered to be the origin points of the re-birth of the natural environment (cf. Indigenous science of climate). Torrential rain, *buya soŋollon, буҕа соҥоллон* (literally 'the sky cries') and thunderstorms are eagerly awaited by the nomads, since they regulate the life cycle of the horseflies (cf. Evenki climatology).[1] The disappearance of the horseflies is a sign of the end of summer: in around two weeks, the temperature will fall and the forest will become tawny, decorated in yellow and red colours. The time has arrived for the members of the summer camp to depart from each other.

Autumn – the promise of lucky hunting (bolo, боло): The first autumn (without snow)

In the red landscape of the autumnal taiga, the Evenki nomadise in small groups and move downstream. The camps consist of two or three tents in this season: each tent contains one nuclear family. This small association of families share herding activities collectively: this is still an important practice, but it also allows enough time for individual hunting. The small number of persons in the camp gives each hunter the opportunity to hunt in a relatively large area. This first autumn is one of the most critical periods for reindeer herding. Gathered by two-to-five individuals, the reindeer graze in order to gain weight and strength for the winter, looking for specific grasses and mushrooms (cf. Vegetal cover typology, topographic typology). Men spend entire days looking all over the forest for the reindeer. In addition, another animal is now a danger to the reindeer – bears are seeking to gain weight in order to survive the winter. To protect the reindeer against bear attacks, the Evenki have certain methods. One is inherited from the Soviet period: some armed herders will survey the herd day and night from a platform. After the collapse of Soviet authority, the Evenki recov-

---

This insect is present in myths where, for instance, a hero transforms himself into a fly after death in order to escape from a trap and then is reborn as a human (Vasilevich 1966).

1 The Evenki consider thunderstorms and lightning to have the power to destroy the spirits of those humans who have suffered bad deaths and thus disturb the living. One can see an association between 'herders' (horseflies) and the spirits of ancestors.

ered the ancestral methods of herding described by Vasilevich in the 1930s (1969: 72–80) and observed by Lavrillier in 1998. This technique is based on the strong link between calves and mothers: nomads place the females into an enclosure at the centre of the camp, meaning that the calves will graze very close to the camp and thus be secure from predators. During the night, the calves are tethered in the camp while the females graze nearby. Since the herd gathers naturally (the females, calves, and breeding and castrated males come together) during this season, keeping calves and mothers in the camp means that all the herd will be close to humans and thus safety. If a predator still prowls close to the camp, the Evenki will first try to frighten it by firing into the sky, shouting: 'Go away, go far away, we are not attacking you, so do not attack us!'. If the predator persists, men from one or several camps will hunt this bear or wolf. The activities in this period are numerous and are mostly performed individually: gathering the herd, preparing for the snow season, completing the construction and repair of sledges, putting summer equipment into storage, gathering winter clothes, preparing food reserves and goods for the next nomadisation, travelling to the village to purchase food and goods for the snow period to last until the Russian New Year, hunting wild reindeer, birds, and fish, gathering berries, milking the reindeer and freezing the milk reserves, and finishing the treatment of skins[1] and the sewing of winter clothes. This autumn without snow is often a period of scarcity because the herd requires a lot attention and hunting is difficult (the lack of snow means there are no tracks). Thus, the Evenki make reserves of milk, berries, meat, and fish. These reserves can be kept in a good state because the ground into which the food is placed is entirely frozen. The nomads constantly and closely observe the environment, animals, and the vegetal cover in order to make decisions relating to hunting or herding. For instance, they know that when the larch needles fall to the ground, it is time to fish for the last time before the installation of a strong ice layer on the rivers: they are aware that the graylings and salmons are about to go downstream for spawning (cf. Evenki climatology).

The reindeer must be castrated one month before the installation of the snow during a period with temperatures of -10°C/-15°C degrees. Only the most skilled men perform this task. Traditionally (and also today), young reindeer are castrated with teeth (which are used to cut the tail of the vas deferens within the *bursa*), while reindeer older than two years are castrated either with specific tongs from the Soviet period or via a minor surgical operation. In such cases, the herder must open the *bursa* with a knife, cut out the testicles, attach a thread to the vas deferens, and then close the *bursa*. Considered to be very tasty, the testicles are either grilled so that they can be eaten or thrown into a river. Sometimes, the surgery is not entirely complete: a reindeer may be only half castrated, and this causes a specific form of behaviour. Such 'half-male'

1 In this season (and during the second spring), the Evenki treat the skin with smoke from *ilte* (rotten larch) in a small hut called a *nutinek, нутинэк* (this term comes from *nuny, нуны* – 'the smoke' from wood) (cf. Evenki climatology).

reindeer are called *nara, нара* (a word also used to denote an infertile man). After the operation, the reindeer is kept in the encampment for 24 hours until it collects its wits; then the herders observe the reindeer so that the wound does not become infected.

Throughout this period, the discussion of men focuses on only one topic: 'When will the snow cover be installed?' Their eagerness is great: 'I want the snow to fall as quickly as possible and for the taiga to become like an 'open book', where the life of each animal is written down with tracks!' is a frequent refrain. The small society of the camp is happy to eat meat after the summer, with fish serving as an everyday food.

The second autumn – snow cover installation[1]

Around mid-September (in the 1990s) and October-November nowadays, the nomads transform their tents in preparation for the future installation of the snow cover. At the basis of the tent, a small wall 40–50cm high is made from horizontal trunks and placed in the lower part of the tent. This offers better insulation from the cold, and from the snowfall in particular – snow offers good insulation. The snow installation offers a rest to the herders, who can now find the herd more easily by following the tracks it leaves. In the 1990s, the installation of the snow cover was expected around 15 September. From the 2000s, there has been a delay of one or two months. This day was called the 'day of Simon' (*den Simon, дэн Симон*) by the Evenki. According to the elders, St. 'Simon' (after St. Nikolai) was one of the most favoured saints in Siberia, which is why this important day was named after him. St. Simeon's day is 14th September (oral information by A. Sirina). Coinciding with a crucial shift in the seasons, this saint had a specific place in the Evenki calendar. However, the Evenki today do not make a link between this date and any Orthodox saint. Thus, for the Evenki, the day of Simon is a kind of frontier in the calendar which is supposed to correspond to the installation of the snow cover and the beginning of the hunting season.

Between 15–30 September in the 1990s (and nowadays between mid-October and November), the winter slowly arrives and the snow gradually reclaims the taiga, which is covered by the glacial mist of the wintertime. The Evenki hunt sable and wild reindeer, offering the nomads fresh and fatty meat (a real joy for the entire camp after the scarcity of the autumn). This is the occasion for feeding/making offerings to the spirits: this one does by throwing meat and fat into the 'mouth' of the stove.

The other important event of this season is the mating period for domestic reindeer and wild reindeer (one or two weeks later). It is interesting to note that while there are different names for domestic reindeer (*oron, орон*) and wild reindeer (*beiun, бэйун*), the term for their mating period is the same (*sokan, сокан*). The first signs of this event are that the skin on the young antlers begins to dry, tear, and fall off, the

1 Nowadays the Evenki term for this season is forgotten. Already in the early 1990s, few elders knew about this term. Lavrillier recorded this word at the time, but could not find it in her fieldwork notebooks.

necks of the males become bigger, and the reindeer start to mimic mating fights with harsh sounds (a sort of training). The nomads like to watch this training: young boys play by mimicking such fights and holding onto the antlers of very young reindeer. The Evenki distinguish between real fights and 'training' by referring to the latter as *kuhi-da, кухи-ða*, a kind of play. Just after the mating period, the males lose their antlers. While 'training' is not dangerous for the reindeer, the real mating fights are very hazardous and can even be fatal. The Evenki mitigate the risk by sawing sharp points off the antlers and taking the collars off the reindeer to prevent strangulation (antlers can get stuck in the collars). Just before the mating period, which is supposed to begin with the first full moon of October, the nomads guide the domestic females to a territory far from the camp, bringing back only the young reindeer, the castrated males, and the sterile females.[1] During the mating period, the herders survey the herd (females and breeding reindeer). Here one can see a specific hunting technique where domestic reindeer are used as bait for the wild reindeer: attracted by a domestic female, a wild male will arrive and then be hunted by the Evenki.[2] Normal hunting methods are also used. The meat of the wild reindeer is marked by the strong and specific smell of the mating period. It is considered to be very healthy meat.

After this event, the nomads wait for the complete installation of the snow cover to a sufficient depth and the formation of a strong layer of river ice so that they can move by sledge. They take out their winter clothes and put the autumn ones away in the storage house until the spring. The nomads prepare to hunt sable, waiting for the temperature at which the sable fur will be ready for sale (-30°C/-40°C). If a killed sable is not ready for sale, its fur will be used as a hat by the hunter or his family or to decorate other items made of fur, like festival costumes (cf. Lavrillier et al. 2016).

Winter – hunting (tuye, туҕэ)

From the snowy autumn and the beginning of winter, the Evenki intensively hunt sable for around 1.5 months, during which time they will gather enough income for the entire year. During this intensive hunting, the camps are composed of a single nuclear family: the men and some of the women and pre-teens spend all day hunting and their evenings treating the skins. Herding does not require much effort: one only has

1 During this period, bears are supposed to be hibernating and thus do not pose a threat to the reindeer.

2 Breeding between wild and domestic reindeer is not really desired by the Evenki because it raises the risk of the 'wildening' the domestic herd. If a calf is born to wild and domestic reindeer, it is obvious by the colour of its fur and its corpulence. The decisions taken in relation to such calves vary. Called *beiutkan, бэйуткан* ('little wild reindeer' or 'little wild'), they are observed for the first few months of their life. If, as the Evenki say, the wild side is more marked, it is slaughtered and eaten during the next autumn. In contrast, if the 'domestic side' has won out, the calf will grow up with the herd and is expected to 'refresh / reinforce' the blood of the herd. Often, such reindeer become good racing reindeer.

to find good grazing pastures, gather the herd in the camp each morning, and choose the reindeer to be used in the hunt. The reindeer naturally keep themselves gathered together, and the Evenki chose a place with enough snow to prevent the reindeer from straying far. The quality of the snow cover is crucial for reindeer herding in this period (cf. Snow and ice typology). Early in the morning, the men and some of the women go into the forest to bring the reindeer to the camp. After eating a substantial breakfast, hunters leave the camp for the entire day, often until late in the evening. Only women and children stay in the camp. If the couple have small children, spouses take turns to hunt, babysit, and perform domestic tasks. After the hunting day, the hunter must prepare firewood for the evening and morning, cook food for the dogs and for him/herself, and cut up the killed sable and treat the skins: they will go to sleep very late. In addition, men hunt wild reindeer for food from time to time. This exhausting hunting is ended only by the harsh cold (-50°C) or when the snow becomes too deep. Hunting with dogs is extremely tiring because it requires a lot of walking and riding on very difficult roads. The hunters sweat a lot and catch colds because of the exceedingly low temperatures. Each day, the hunters come back from hunting with clothes and hats wet from sweat, icy trousers, and faces covered with frost. The hunting is so intensive that the dogs and humans lose weight very rapidly. With climate change, dogs are being used in hunts less frequently because excessively deep snow arrives with the very first snow fall: hunting with traps (less exhausting and does not require dogs) is practised almost throughout the sable hunting period.

After intensive sable hunting, the nomads switch to wild reindeer hunting. This kind of hunting is conducted during cloudy days (cf. Indigenous science of climate). The winter days are very short (from 10–11 hours to 15–16 hours): days of very harsh cold are the time for discussions about hunting experiences, horrible tales about the spirits, and funny stories. It is during this period that the nomads are most afraid of the spirits: they are prepared to travel for many kilometres even in the harsh cold in order to avoid crossing the old camps of their ancestors or uninhabited hunting houses (they are not so afraid of these old camps in the summer). It seems that the Evenki believe that the spirits active in the landscape during the winter and summer (and the night and the day) are not the same. The other preferred subject of discussion is the hibernation tactics of bears and how they hide themselves from humans: 'It would be good if we could also spend the winter in hibernation, our life would be easier'. During storms or days of very harsh cold, the Evenki stay at home because they know even the most experienced among them can get lost and freeze in such conditions. Nomadisations become less frequent than in other seasons, and the camps are separated by greater distances. During the very cold weather, the men go first to the next camp and set up a tent: thus, on the next day, the nomadising family will find an installed tent with a stove that they can light and warm themselves with after having spent five to seven hours outside in temperatures of -40°C/-50°C.[1]

During the period of wild reindeer hunting, women stretch the skin of the killed animals as soon as the carcasses are brought into the camp: they do so with a variety of different stretching techniques (with a frame or on poles). The skins of wild reindeer are very much appreciated: they are stretched, dried, and treated (with a scraper and a tool for making the skin softer) in the winter and are sewn into nomadic winter shoes in the spring and autumn. They are also used for making the reindeer skin boots (*unty, унты*) that are sold to Russians and Yakuts in villages and towns.

In the winter, domestic tasks are particularly time consuming. First of all, a huge amount of dry wood is consumed, since the tents are heated to a temperature of +18°C to +°30°C. Even if the tents are never heated during the night, in 24 hours a small wall of dry larch logs will still be consumed.[2] In addition, most of the rivers are entirely frozen: so, in order to get water for drinking, cooking, and washing, the nomads must go to the river, cut big blocks of ice out of it, and thaw them in big tubs on the stoves. This consumes the time of men, women, and children. Reindeer are used to transport ice and wood.

Winter is the season for hunting. Hunters often come back home in the dark, sometimes even close to midnight. As soon as the women and children hear the dogs barking and the swish of the sledge on the snow, they must put a lot of logs onto the stove to make things very warm for the hunters: a rich meal will be prepared to fortify them.

The first spring – the birth of the calves (nelkini, нэлкини)

This starts in March. In the 1990s, the signal for this period was a rise in temperature from the -30°C to -25°C. From the 2000s, a rise from -20°C to -15°C is expected, since winters are much warmer than in the past. Other signs are a strong luminosity and its reflection on the snow. During this period, various villages host the feast of the reindeer herders. This feast was institutionalised during the Soviet period as one of the professional feasts that replaced the saints' days of the Orthodox calendar. During the same period, the collective shamanic summer ritual was forbidden: the Evenki started to practice the forbidden summer rituals discretely during the reindeer herders' feast in the springtime. Most men, women, and children leave the camp to join the feast in the village, the central event of which is a series of reindeer races. Such

---

1 During trips in the harsh cold, the traveller's limbs always start to freeze after the first few hours, despite the quality of their clothes. It is vital to move one's fingers and toes from the very first feeling of cold. If one is really freezing, it is necessary to stand up and walk in the deep snow while dressed heavily until the entire body warms up. If one fails to respect these rules, limbs may be lost.

2 In order to extend the duration of the fire, the Evenki sometimes add dry larch: fresh larch burns more slowly but does not produce enough heat during severe cold spells.

collective meetings with reindeer races existed long before the Soviet period, but, according to the elders, they took place in the forest. Nowadays, the reindeer herders' day is considered 'traditional'. This feast is a 'total social fact' (Mauss 1969, 1978), where many important decisions (social, economic, ritual) are taken and alliances concluded. Only one or two individuals stay in the forest to survey the herd. The feast is a unique moment for nomads to meet each other after many months, if not years, of separation. Prior to 2012, when the Evenki shaman Savelii Vasilev (supposedly the last Evenki shaman) died, this feast was an occasion for collective shamanic rituals. Of course, it is also the best time to purchase enough food and goods to last until the summer (and ideally to the autumn), since very soon the rivers will melt and flood, forming frontiers that wedge the nomads between them. After the feast, the Evenki consider the spring to have started. The nomads rush back to the forest for the birth of reindeer calves.[1]

This is the mating season for the *tuksaki, туксаки,* hare, whose appearance announces the beginning of spring: in the snowy nights under the moonlight, the mating songs of hares ring out, 'tududut, tududut'. The Evenki explain that the hare is the first animal to have its mating ritual at the end of the winter: 'Soon the wood grouse will have their mating ritual, then the cuckoo will sing!'. Although the taiga is not yet ready to give up its carpet of snow and ice, this mating ritual announces the arrival of the first spring to the Evenki.[2]

Wild reindeer hunting continues until mid-April, the birthing season for domestic reindeer. The Evenki know that the same event will happen among the wild reindeer some ten days later. Thus, for sustainable hunting, they decide that 'there are no more wild reindeer' and leave the species alone until the next autumn so that they can reproduce in peace.

This is the second period of food shortage in the year (along with the first autumn): it is also the time when reindeer herding activities are most important. In addition, there is a problem with the conservation of meat: the rising temperatures and the *salgyn, салгын* air/wind (cf. Evenki climatology) dry out and blacken the meat.

Some days before calving, which the herders identify by observing the form of the paunch and the vaginal orifice, the females are guided to an isolated place (usually the same place each year); here, in around two weeks, the females will calve and take care of the new-borns. Pregnant females are not used for transport during this period. This is the time for training young reindeer (two years old) to wear packsaddles and pull sledges. During this period, women, men, and children visit the female herd every two days and bring them salt. The nomads avoid touching the calves, since doing so could cause the mother to reject them (because of the human smell). The nomads

1 About the feast and collective rituals of the Evenki, see Lavrillier 1995, 2005, 2003.

2 If the mating season of the hare (which is hunted only when it is encountered) is long awaited, the mating season of the musk deer (which is hunted) in January/February does not attract particular attention from the nomads.

force the females and calves to stay gathered and survey the rest of the herd so that they will not approach them: the infertile females, the males, and the castrated reindeer can be aggressive towards the calves. They survey the behaviour of the reindeer and the weather in order to learn when most of the births will occur (cf. Evenki climatology). Just after calving (*sonkan, сонкан*), the females lose their antlers.

During this period, the camps are composed of two or three nuclear families gathered together to deal with the calving period. The collective activities are gathering the herd and repairing the enclosure and the storage house. In contrast, each nuclear family (men, women, and children) treat and mark their own reindeer, among other activities. Each person must keep their dogs in check, since they present a real threat to the calves during this season. This species of dogs is close to wolves and, in any case, they are hungry after the long winter. In order to prevent attacks on the small reindeer, the dogs are either tethered within the encampment or have one of their legs put through their collars: this is maintained until the second autumn.[1]

During this period, men and young children go to some regions of the forest to collect birch trunks in order to build the rigid parts of the sledges (the skis, legs, and horizontal parts) for the next winter. They will leave these trunks in a precise place in the forest, lying in the snow under the *pinus pumila* (cf. Vegetal cover typology). It is surprising how easily the hunters find these pieces of wood months later in the middle of the forest without any signs. This period is ideal for parts of the skin treatment process. The body skins of the cervidae that were stretched and dried in the winter are scraped and then tanned with smoke from decomposing larch branches (*ilte, илтэ*). The embossed *chuchun, чучун* tool is used to scrape the skins and make them stronger. During this period, the Evenki also make lassos, ideally from the neck skin of a male wild reindeer killed near the autumn. In autumn, this part of the skin is particularly resistant. The skin is soaked in water (usually in rivers) and the fur is removed. Then, starting from the edge in a circular pattern, one cuts a strip as long as possible until one reaches the centre of the skin. While wet, the lasso will be plaited out of three or four of these strips. For good results, such work requires a moderate temperature and a certain degree of humidity in the air common in this season (and also in the autumn) (cf. Indigenous science of climate).

At the end of this season, the Evenki go to their storage houses to store the winter clothes and to take out their summer ones. In addition, they continue to benefit from the last snow until the very final moment, using it to facilitate the movement of food and goods via sledge to the sites of future camps.[2]

---

1 We can see how the double activity of reindeer herding and hunting with dogs seems to be contradictory and difficult to conduct simultaneously.

2 On a sledge pulled by two reindeer, the Evenki can carry an absolute maximum of 200kg: on the back of one reindeer, one can carry an absolute maximum of 50kg. Nevertheless, the riding reindeer can transport a person weighing around 75kg for ten hours. Thus, the reindeer can transport, be it by sledge or on their backs, more than their own weight (45–70kg). From the

The Evenki consider the first spring as a season in which the very weather acts upon the metabolism of animals and humans. The most noticeable sign is an irresistible somnolence, in particular just before precipitation (snow, rain) (cf. Evenki climatology). In addition, the Evenki say that during this season sleepwalkers, epileptics, pregnant women, and *olon, олон* (the Evenki concept for people who lose control of themselves when they are suddenly frightened by a sound, gesture, or an unexpected meeting) face the aggravation of their mood or troubles.[1]

In parallel, the spirits are supposed to be particularly active at this time; as such, it was in this season that many hunting and herding rituals took place. In addition, there is a particular proscription in this season against burning birch or pine as firewood (this can be tolerated in other seasons). A failure to respect this proscription is supposed to increase the number of deaths among the calves. Equally, the Evenki forbid hanging copper or bronze bells on the reindeer during this season (although such is recommended in the winter): failure to comply with the prohibition might cause the disappearance of the animal. Engaging heavily in metal forging is not recommended at this time, since it may cause many reindeer to disappear. Sewing is also forbidden during thunderstorms in the second spring and summer (cf. Evenki climatology). As an additional piece of evidence that the Evenki believe that the spirits are different (or act differently) according to the season, the Evenki in the past would only gather ochre in the springtime and make an offering of coloured fabrics at the place where they found this stone.[2]

We see here, as in other places in this book, that the Evenki associate a pragmatic understanding of their environment with an explanation related to the world of spirits and collective representations. In terms of our attempts to bridge scientific and indigenous knowledge, it is important to note that the Evenki do not lack a more practical explanation of a phenomenon (which environmental science can deal with) just because they also have a spiritual explanation. We can quote, for instance, the Evenki explanation for a type of forest fire: '*The main spirit of the natural environment* Buγa *decides when the forest must burn for its regeneration. Some fires are provoked by the forest itself*.' We can find a similar explanation in Bouchiat et al. (1999), where biologists explain that, after some years, the taiga produces a gas that provokes fires; these fires are indispensable for the regeneration of the forest.

2000s, the Evenki have complained that their reindeer are much weaker and less tough than in the past: they link this to changes in the climate and the environment.

1 These seasonal nervous or psychological troubles were observed among other Siberian peoples. Éveline Lot-Falck wrote several papers about this phenomenon among the Yakut and its relation to shamanism (Lot-Falck 1972, 1974, 1971). This phenomenon can also observed in villages and towns. For instance, indigenous students suffer from such troubles in towns during this season. In towns, it is also a period with a high suicide rate.

2 The act of decorating with colours and motifs (*onio-do, онйодо* 'draw', and *evi-da, эви-дa*, 'play') is considered to be a ritual action (as are dancing, singing, shamanising, and the performance of rituals) (Lavrillier 2005).

The second spring (without snow) (Neŋneni, нйэӊнэни)

This is the period between the melting of the snow and the song of the cuckoo. The snow disappears and a dark brown taiga emerges, burned out by eight or nine months of frost. The camps are joined by new units of nomadisation to prepare for nomadising in the summer. The second spring is principally occupied with herding activities. The reindeer have been weakened and made thin by the long winter: they rush in groups of between two and five to find fresh grazing pastures rich in protein (like the *nirgakte, ниргактэ* grass, cf. Vegetal cover typology). From this moment, the herders walk all around the forest daily with saddles and lassos as they look for the reindeer one by one. The first riding reindeer found is caught and saddled: the search for the rest of the herd goes on until the end of the day. The reindeer are guided to the camp and kept in the enclosure: at night, they are set free. In this season, the nomads set up their camps on the edges of tussock fields (*kever, кэвэр*), where the herb *nirgakte, ниргактэ* (a delicacy for reindeer) grows (cf. Topographic typology and vegetal cover typology). The Evenki maintain these fields by burning this grass while there is still a little bit of snow left (which reduces the risk of forest fires): such actions help the grass to grow quicker.

The females and calves are now in the company of the breeding males, the castrated males, and the sterile females. The herd is led daily to the enclosure in the camp, where it stays until the evening.

Bears come out of their dens and prowl close to the herds.[1] The Evenki use the same technique as in the autumn to protect the herd: they keep the herd in the camp by tethering up the females or calves. However, the calves must be surveyed at all times because they are not used to the tethers and can strangle themselves by getting tangled up. During this season, the reindeer are a sorry sight: the males and females are without antlers and are sleepy because of the rising temperatures. They have lost half of their winter coats, giving them the impression of being moth eaten and skinny. In addition, the larva of horseflies, injected during the previous summer and developing under their skins throughout the winter, are rendered visible by the pustules they form. To avoid infection, the herders spend some hours pushing the larva out of the poor reindeer with their fingers. As this is a boring task, it is often given to children or elders.[2]

Close to the end of May, the antlers of the males and females grow: in the summer, they are covered by grey velour. Prior to the beginning of the 2000s, the nomads used

---

1 During every walk in search of the reindeer, the nomads look for bears and their tracks: however, they do not seek to hunt them, only to frighten them away by firing into the air so that they will not bother either the herd or humans.

2 The larva of horseflies usually pierce the skin and fall to the ground in order to further develop, but sometimes they take a wrong route and pierce the ribcage and the lungs: if this happens, the reindeer dies.

to cut off the young antlers of the reindeer for sale slightly before the summer. In the 1970s and 80s, the state farms sent a helicopter with a special team to gather young antlers *en masse* for the Chinese and Korean markets; they did not give anything to the Evenki.[1]

The presence of calves in the enclosure is the occasion for several tasks and games. Women cut the initials of the family who own the reindeer into the fur of the animal; women and men cut marks of ownership on the ears according to a specific code. In addition, they pass a piece of coloured fabric (*sekan, сэкан*) through the ears of the reindeer. This last marking is a ritual intended to help the calf grow without problems. As the calves enter the enclosures, women and children join them. Human babies are presented to the reindeer babies: children try to catch the small reindeer to stroke them, while women and men catch the calves to kiss them on the muzzle and gently stroke them so that they become accustomed to humans. The enclosure is full of the sounds of reindeer calls, human voices, and laughter. During this very joyous moment, the nomads say: 'The babies, be they humans, reindeer, dogs, cats, or whatever, understand each other: they come from the same world, they speak the same language!'

After the melting of the river ice, there is mass flooding. This wedges the nomads into one camp while they wait for the water level to fall. Of course, there is no hunting or fishing during this period, which means a food shortage. From calving onwards, the nomads eat very little: rare fish, grouse, partridge, and some berries saved in the snow from last summer. Then comes the mating season of the wood grouse (*oroki, ороки*). At sunrise or sunset, the 'song' of the wood grouse fills the forest with a sharp, hammering 'toctoctoc, toctoctoc'. The nomads (women, men, children, and elders) hunt the grouse as if going to a spectacle: 'They dance so beautifully that very often one forgets to fire!' This hunt lasts around ten days, during which the nomads shoot only the males after fecundation. In order to say that they will hunt wood grouse, the Evenki say 'we will toctoc'. There is a game attached to this hunt: in order to put a real or potential son-in-law to the test, he is asked to break the bones of a wood grouse using only two fingers of one hand.

---

1 These antlers are made of a bone that is still cartilaginous and full of blood. In Russian, they are called *panty, панты* and in Evenki *nimekte, нимэктэ*: in English, we use the term 'young antlers'. They are appreciated in Chinese medicine for millenars: today, Chinese and Koreans use them as an aphrodisiac. The Russian pharmaceutical industry uses a reinforcing treatment called *pantokrin, пантокрин*. The Evenki use them traditionally two or three times in the summer by burning a small piece of the young antler on a camp fire and then eating it half raw. They consider that it reinforces their organism with energy. Prior to the beginning of the 2000s, the Evenki sold the antlers of their reindeer *en masse* to Soviet state farms and, after the collapse, to local merchants. This market dried up because the cross-frontier sale of antlers was prohibited. Even though they lost an important source of income, today the herders are happy not to cut off huge numbers of antlers because they know doing so disturbed the hormonal balance of the animals.

The sun starts to warm people after the long, cold winter: everybody sits in front of their tents to complete all of the technical tasks required for the summer: making and repairing harnesses and packsaddle bags, packing the winter clothes and taking out summer ones, curving the sledge skis, etc. Working in the sun and singing, it soon will be summer, the time for many games.

From this moment, the nomads watch the state of the larch buds and for the signs of the appearance of light green larch needles. If the buds show a small red point at the end, this is a sign that soon *muchukte, мучуктэ*[1] needles will emerge from the buds (cf. Vegetal cover typology) to illuminate the forest with a green light. This is when some fish species spawn. All of this shows that the summer is approaching and that soon the shaman bird, the cuckoo, will sing.

A game occupies both the children and the adults: catching *merepki, мэрэпки* squirrels (*Tamias sibiricus* Laxm.). Despite this very busy period and considerable workloads, nomads will spend a lot of time and energy trying to catch the squirrel. Each year one can observe the same enthusiasm when people gather around a tree in which a squirrel is sitting and try to devise a new system to catch it. Each family will attempt to capture a squirrel and spend a lot of time on this. If the animal is captured, one will make a small wooden cage for it and feed it: after a few days, it is freed. It seems as though the capture of the squirrel is supposed to bring some luck in hunting. This is perhaps linked to the fact that the squirrel is considered to be the worker of *amaka, амака*; *amikan, амикан* ('grandfather'), or *koŋnoren, коӈнорэн* ('brown') bears: bears eat the food reserves of squirrels and are considered to be the ancestors of humans.[2]

When the larch branches become green, they are spread over the floors of the tents and disseminate a nice fresh smell. The taiga smells fresh: indeed, the fields under ledum trees smell so strong when this tree flowers that one's head spins and there is a risk of losing consciousness. Then, one camp asks the others by radio or satellite phone, 'has the cuckoo sung in your area?' Soon a long caravan with packsaddles will move upstream towards the source of the river and the mountains or towards the north to relive once again the best moment of the year – the beginning of the summer, the re-birth of the natural environment.

Through the co-produced documentations of Evenki typologies, we will see in subsequent chapters how the traditional ecological knowledge of the Evenki is much more complex and detailed than in the ethnographic description above or the accounts produced via the anthropological approach (Lavrillier 2005).

---

1 This term is interesting, since it also designates the small larch used in all ritual constructions. In the ancient ritual *ikenipke, икэнипкэ*, it symbolises a mythical invisible river *eŋdiekit, эӈдйэкит*, along which the recyclable soul *omi, оми* circulates between the poles of death and life. This term comes from the root *muchu*, meaning 'to come back' or 'to return', which is used in both ritual and daily contexts.

2 These two animals are omnipresent in myths linked to the appearance of humans and hunting (Vasilevich 1936, 1966, Romanova and Myreeva 1971).

2.2 Natural landscape
by A. Lavrillier

This chapter aims to present two important typologies of Evenki TEK indispensable for how reindeer herders understand the changes they have observed. We decided to assemble them under the appellation of 'natural landscape', in contrast to another chapter focusing on the 'human-made landscape' (e.g. non-natural landscape). Although such a chapter was prepared, it was too large to publish here. In addition, we also had to leave the typology of fauna aside for a later publication. We decided to concentrate here on two typologies essential for the understanding of changes in the snow cover and climate (cf. following chapters): the topographic typology and the vegetal cover typology.

In this encyclopaedia-like part (2.2.), we preferred to integrate the analysis from the anthropologico-TEK co-production directly after the explanatory texts attached to each entry, along with additional information from previous ethnographic research (in italics). The explanatory tri-lingual texts attached to each entry are given just as the Evenki herder-hunters put it: they were mostly composed by S. Gabyshev, the other Evenki co-researchers (Vasilii Gabyshev, Albert Kolesov, Oleg Iakovlev, etc.), and members of the two-dozen nomadic families interviewed for the project. (cf. Preface)

2.2.1 Topographic typology
by S. Gabyshev and A. Lavrillier

Kever / Кэвэр

A tussock field containing the grazing grass *Eriophorum vaginatum*

Кочкарник – марь с кочками, и с травой «пушица»

Ideal for setting up the camp during earliest spring because the snow melts quickly in *kever*; very soon after, the *nirgakta* grass *(Eriophorum vaginatum,* tussock cotton-grass) for which reindeer look starts growing. Thus, it is easier to find reindeer and keep them close to the camp. If this grazing grass does not grow properly, it will be burned in the autumn (or spring) so that it will grow quicker later. Wild reindeer, which arrive from the small mountains during the summer, also graze in *kever*: One can lie in wait for them. *Kever* is bad for transport by sledge because of the tussocks. (cf. Vegetal cover typology: nirgakte)

Таду нйэӈнэ илгимадйэми айа бивки. Кэвэрду имакунди иманна унивки, иматмарит йувки чука, иматмарит йувки ниргактэ. Орор тара айамамат оӈкодйовкил, таду ороpду айакикин оӈкодйоми. Таду оронмо айатмар бакадйами. Ниргактэ эрэкин балдырэ, нуӈанмам которонэвкил – чука йудан, боло-у нуӈанмам которонэвкил – нйэӈнэ иматмарит йудан. Кэвэрилтыки бэйур нйанэ эвувкил нйэӈнэ, таду нйан карабдйаннэнны. Сыргакитту элйэ эрукун сыргат ӈэнэдйэми – чумникаккун.

Весной хорошо там стоять, потому что на *кэвэр* быстро тает снег и там в первую очередь выходит трава *ниргактэ* (пушица), которую олени обожают, там оленям хорошо кормоваться. В таких местах оленей легко найти. Если трава *ниргактэ* там хорошо не растет, ее иногда сжигают осенью (или весной), чтобы эта трава быстрее росла. На *кэвэр* дикие олени спускаются в начале лета с горы и их там иногда караулят. Для поездок на нартах *кэвэр* плохое место из-за кочек.

Avlan / Авлан

A flat land without trees close to a river

Равнина без леса возле реки

Good for summer camps because it is windy here. There are high levels of visibility (for both reindeer and people), which makes it easier to survey the herd and calves or to see predators from a distance. Domestic animals, seeing the predators immediately, can run away quickly. In *avlan*, the solid and dry ground types (*maŋarne* and *keteme*) allow for the establishment of a good camp. This landscape is very suitable for making nomadic roads because it is flat. It is also excellent for hunting because game are highly visible, which gives one time to aim. It is a wonderful location for gathering berries. The flat and hard qualities of the ground type make it easy to travel. In the past, this was the preferred place for collective shamanic rituals, festivals, and games. This word often appears in ritual songs.[1] This type of landscape is rarely found in the taiga. (cf. diagram Reindeer grazing versus topography) Дйуҕаниду айакикин таду илгимадйами, таду тар адындйэвки, соӊукикин. Айамамат ичэвувки орор горокунду ичэвувкил, бакавувкил имакунди. Горокундук орорбо, эӊнэкарбэ, бэйӊарбэ ичэвкил, имакунди олонэвкил. Таду тар маӊарнэкун, кэтэмэкикин, айакикин таду уринчэдйэда. һоктодйоми талы со айа. Горокундук бэйуна ичэдиӊас, нйурмадиӊас. Таду кэтэкун диктэ балдывки. Таду талы тар нулгидйами айакикин, кэтэмэкикин. Нонон таду саманидйавкил-да, эвидйэвкил-да бичал. Таргачин тар умукокор авлар бивкил.

Летом здесь хорошо разбивать лагерь, ветер поддувает, прохладно. Видимость хорошая, оленей видно издалека, их легко найти. Издалека видно и оленей, и оленят и хищников. Увидев хищников, домашние животные успевают убежать. Там два типа твердой земли, которые хороши для установки стоянки (*маӊарнэ* и *кэтэмэ*), там хорошо стоять. В таких местах очень удобно делать дорогу – место ровное. Там издалека увидишь дикого оленя, успеешь сделать выстрел Там много ягод растет. По таким местам хорошо кочевать – твердая земля. В старину там всегда проводили камлания, игры, праздники. Такие места попадаются редко. (см. схемы, с. 152–158)

Emker / Эмкэр	Steep river bank Крутой берег реки с обрывом

This is a steep river bank: it can prevent flooding. Very often, the Evenki place their camp at the top of the *emker* because the hard and dry ground (*keteme*) is ideal for camp installation. It is also very windy (a defence against horse flies, flies, and other insects). This location is a favourite among children because they can play with the sand. This is an ideal place for taking larch roots, since they are curved into a V-shape, which allows one to make them into a packsaddle.

Тара бира дйапкадун ойодун эмкэр бивки. Тар эмкэр ойолын му эвки уруру. Таргачирду уринчэдйэми нйан айакикин, кэтэмэл бивкил. Бивкил нйанэ адынмар таду айакикин бидйэми. Тадук ирйактэни тэкэнмэ гадйавкил эмэҕэнмэ одави.

Это крутой берег с обрывом. Этот крутой берег держит воду во время наводнений. В таких местах тоже хорошо стоять из-за твердой сухой земли (кэтэмэ) а также летом из-за постоянных ветров. Оттуда берут корни лиственниц для изготовления вьючных седел.

1 See also Lavrillier and Lecomte 2002; Lavrillier and Matic 2013.

Emkerkur / Эмкэркур

High, steep river bank

Высокий крутой берег реки с обрывом

A high, steep river bank that prevents flooding. Very often, the Evenki place their camp at the top of the *emker* because the hard and dry ground *(keteme)* is ideal for camp installation. It is also very windy, which is useful against horse flies, flies, and other insects, as well as for drying meat.

Тара һэгдыкун эмкэр. Тар ойодун нйан уринчэвкил, талы тар му этан окинда мударэкин уруру. Таду тар окинда кэтэмэкикин бивки. Нйан адындйакикичэвки таду.

Высокий крутой берег реки с обрывом. Там тоже стоят, воды во время наводнения точно никогда не будет. Там всегда бывает твердая и сухая земля *(кэтэмэ)*, что хорошо для стоянки, а для летних стоянок хорошо, что дует постоянный ветер.

Amnunna / Амнунна

Wide and flat river basin

Широкий ровный бассейн реки

This is a crucial landscape type for the Evenki. Ideal for **autumn camps** because sivak grass (*Equisetum arvense L.*, similar to horsetail) grows in great quantities in *amnunna*; by consuming this grass, the domestic reindeer are able to gain weight.

Closer to the spring, wild reindeer come to the *amnunna* from the small mountains to graze. Such places are useful for transport because the grass and bush do not disturb trips by sledge or riding reindeer. *Amunna* is a good landscape type for reindeer because there is a lot of lichen and other pastures (including *oktalyk* bushes) around the edges. Reindeer graze close to the edge of *amnunna* and than come towards the centre for the *sivak* grass, especially during the autumn. Around the *amnunna's* edges, there is always hard and dry ground *(keteme)*: a lot of lichen grows in this type of ground, and it is not marshy. It is also a suitable location for installing a camp because the winds protect people and reindeer from insects. In the **winter**, it is also a good place for camp because the water flows out from under the ice *(ulan)*: people can drink it directly and thus do not have to thaw the ice. Domestic reindeer can lick the salty ice. During **springtime and summer,** it is possible to stay there because of the presence of *sivak* grass, from which the reindeer can gain weight after / before the long winter; however, this can only be done on large *amnunna* so as to prevent the reindeer from exhausting the grass. People gather berries on the *amnunna*: the very cold water also means the presence of a lot of fish.

Мунду Эвэнкилду тар со нада. Боло илгимадйами таду со айама, таду кэтэкун сивак балдывки, орор таду айамамат сивакдйэвкил, туӡэрдивэр умуксэйэвэрэ. Бэйур тала тар эвдйэвкил. Нульгидйами талы айакикин, һоктодйоми. Нйукучукокон чука балдывки амнунна дйапкалдутын оӊкококун окинда бидиӊан. Орор маӊарнэлду оӊкодйовкил. Кэтэмэлду бидйэвкил дйапкалын. Дйуӡа кэӡа эвувкил или ӊариллэкин. Лэва талы ачэн. Туӡэ нйан таду айа илгимадйами, уландйэвки, окинда му бидиӊан. Орор имадйэвкил дйукава. Талы нулгиктэдйэми нйан айа. Дйуӡаниду нйан таду илгимавкил һэкдыһинча амнуннаду, оӊко эдатын манавра. Таду адындйэвки. Таду кэтэкун диктэ балдывки. Кэтэкун олло бивки эмур мудукин.

Для нас Эвенков такие места очень нужны. Осенью там очень хорошо делать стоянки, потому что там растет много травы *сивак* (хвощ полевой – *Equisetum arvense L.*) и олени хорошо поправляются к зиме, набирая жир. Ближе к весне дикие олени спускаются с горы на *амуннах*. Через такие места очень хорошо кочевать или дорогу делать. Мелкая трава вдоль берега *амнунны* растет, всегда много оленьего корма в *амнуннах*. Олени сами по себе кушают ягель по берегам *амнунны*, а осенью *сивак* кушают в центре *амнунна*. Олени на сухом месте кормятся по берегам. Летом вечером или к утру олени спускаются с горы на *амнунна*. Там болота типа *лэва* нету. Зимой там тоже хорошо стоять – из-подо льда выходит вода *(улан)* – всегда есть питьевая вода (не надо таять лед). Олени облизывают соленый лед на *амнуннах*. Кочевать через такие места тоже хорошо. Летом тоже стоят, но на больших *амнунна*, чтобы олени корм не кончили. Там ветер дует, мошки и комары почти не бывают и есть много ягоды. Там много рыбы из-за холодной воды.

hegdykun amnunna / һэгдыкун амнунна

Big large and flat river basin

Очень широкий ровный бассейн реки

In such a landscape type, one can keep a large reindeer herd, since they will not exhaust the *sivak* grass. It is good to stay there in the autumn so that the reindeer can be kept close to the camp (grazing on the *amnunna*). This means there is no need to pen them in close to the encampment in order to protect them from predators, a requirement elsewhere. In such places, there are also a lot of fish, which swim down stream during the autumn.

Таргачирду кэтэ Орочил уринчэдйэвкил, сивак эвки манамналчара. Боло айакикин болодйоми энэ уйучэрэ. Таду нйан со айа олло нйан эйаргидйэвкил тала, оллококун бивки.

В таких местах многие Орочоны ставят свои стоянки. Трава *сивак* не кончается там. Осенью хорошо стоять, потому, что олени всегда рядом со стоянкой кормуются на *амуннах*, значит не надо ночью их держать на привязи (как обычно делают, чтобы их держать на стоянке ближе к человеку и подальше от хищников). Там тоже очень хорошо из-за рыбы, которая спускается по течению.

Niukuchukokon amnunna / Нйукучукокон амнунна

Small, wide, and flat river basin

Маленький широкий ровный бассейн реки

Such locations are usually around 150 metres wide. These places are suitable for nomadising and setting up camp. There is some *sivak* grass, usually enough for a stay of one or two days: any longer may exhaust the grazing grounds. Fish and berries are also present.

Таргачирду нулгиктэдйэми, айа уринчэдйэда. Сивак нйан кэтэкан бивки, дйулэлэду-илалладу уринчэвун айа, эдатын оӈкойэ манавра. Оллочи нйан бивки, диктэкакун нйан бивки.

Примерная ширина 150 м. По таким местам хорошо кочевать и хорошо разбивать лагерь. Трава *сивак* там тоже есть, но немного, поэтому делают только двух-трехдневную стоянку, чтобы не кончили корм. Там и рыба тоже есть и много ягоды.

Oio / Ойо
The summit of narrow, high mountains
Вершина горы

Some *oio* are used for passing to another river valley (during nomadisation, hunting, and other movements), in particular during the summer because the strong winds defend against the horse flies that usually disturb the reindeer during nomadisation. In spring, one can camp there for around two days if drinking water is closeby. It is impossible to stay there in the winter because the winds are too strong. People gather *Pinus pumila* cones (further ‘cones’) here, but the most important activity is wild reindeer hunting: the wind prevents the animal from gaining the hunter’s scent if he/she is correctly placed down wind from the animal. Reindeer often graze on *oio* due to the large quantities of lichen. You can find permanently fresh lichen there (the strong wind breaks the small extremities of the lichen branches, which means that fresh ones are always growing in their place). The reindeer can graze to their heart’s content on these young lichen branches (cf. diagram Reindeer grazing versus topography) and thus are often found on summit *oio*.

Таргачирдули нулгидйэми нйан айа, аландйами уӈту бирала, ойоду адынмар. Нйэӈнэ, му бидйэми айа таду илгимадйами дйулэйэ. Туӻэниду эвкил таду илгимарэ. Таду болгиктэ балдэвкил, оӈко кэтэкун, нйаӈта кэтэкун. Дйолочи биһин. Бэйуктэдйэми нйан айа. Ойоду адындйэрэкин оӈко ойовон адын анавки тар эӻиркикин оӈко балдыдйавки. Тавар тара орор айавувкил, ойолдунэн оӈкодйовкил.

Через такие вершины хорошо кочевать, переваливать на другую реку. Особенно летом хорошо, ведь на вершинах всегда дуют ветры, спасают от оводов и обеспечивают спокойную кочевку. Весной там можно стоять дня два, если есть питьевая вода. Зимой там не стоят из-за сильных ветров. Там много кедровых стлаников *(Pinus Pumila Pall.)* и шишек, много оленьих пастбищ и много камней. Там охотиться тоже хорошо. Корм оленей всегда хороший на вершинах – ветер сдувает кончики ягеля и поэтому всегда свежее пастбище бывает. Олени любят такой ягель и на вершинах часто кормятся. (см. схемы, с. 152–158)

Alakit / Алакит	A nomadic road passing into another river valley Перевал (дорога) на другую речку

In such a landscape type, one can easily nomadise or drive the heard. These roads are sometimes formed from natural paths and sometimes from human intervention in the form of cutting down bushes and trees. Some exist for several generations, while others are brand new. In the latter case, one must make marks on the trees by cutting out a piece of bark *(ilken)*, so people will not get lost and will use this new nomadic road, thereby inscribing it on the landscape (i.e. frequent passage will mark the vegetation, giving the road an enduring presence). If the word *alakit* is used to name a place, it means that there is a pass to the next river system.
Thus, place names relate information about the possible uses of land. This may belong to the typology of 'human-made landscapes', since humans, even though they are nomads, are clearly transforming and organising the landscape.

Талы айакикин нулгидйэми, ororбо илбадйами айа. Талы окинда икэн бидиҥан, бэйэ манын талы алакитйа овки, һоктол талы окинда бивкил. Тар алакикилдулы гороптылдук окинда. Ноктодйоми сукэт алдыдйаннэнны, һокто сапдан – илкэн гунэвкил, эдатын кайдйэрэ бэйэл, һоктолы ҥэнэдйэксодатын.

Там хорошо кочевать или оленей гнать. Там всегда бывает *икэн* (наклон между горами), человек сам может найти дорогу, чтобы перевалить, тропы там всегда есть. Если *алакит* говорят об одном месте, то это значит, что всегда там есть дорога. Этими дорогами пользуются из поколения в поколение, но можно самим тоже проложить там дорогу, топором делая зарубки на деревьях вдоль тропы *(илкэн)*, чтоб люди знали дорогу, не терялись и по ней ездили.

Iken / Икэн
Slopes between mountains created by a small river
Наклон между горами, созданный родником

In *iken*, wild reindeer graze well, making it a good place to hunt them. When travelling through such places, one must be careful not to frighten them. It is possible to pass through all *iken* easily and thereby quickly move into the valley of the next tributary. Domestic reindeer also use such paths and can run along them. It is an excellent location for driving the herd and spending the night during a nomadisation.

Икэрду бэйур айат оӈкодйовкил, тала ӈэнэдйэми бэйурэ сэрэнчэдйэнэ ӈэнэдйэннэнны, бэйуна бултаӈнэнны. Экуты-да экурдули аланда айа, иматмарит уӈту бирала аландиӈас. Орор нйан тала урудйэвкил. Экундули уӈту бирала аландйавкил, таду бэйэ анӈадйэвки.

На *икэн* дикие олени хорошо кормятся, если ехать через такие места, то надо быть осторожным, чтобы не спугнуть диких оленей и поохотиться на них. Куда и откуда бы не ехал, через *икэн* можно легко и быстро перевалить на другую реку. Олени тоже через такие места уходят и можно их гнать через *икэн*. Где легко перевалить, там по дороге можно заночевать.

Solokit / Солокит	A nomadic road following a river to its source
	Дорога идущая вверх по ручью

If a place is known as a *solokit*, it means that one will always find a way along a small river to the pass. These small rivers are excellent places for setting up a camp; however, they must be short-term encampments (two or three days at maximum), since the small quantities of lichen can be quickly exhausted by the reindeer (eating and trampling). If such occurs, they may flee far away to the next river system. Such a landscape type is also used by wild reindeer and is thus a suitable location for hunting.

Солокит бирэкин тала һокто окинда бидиӈан, солокит биракан долын бивки. Бэйэ солокуки талы ӈэнэдйэвки биракандули. Таду уринчэда илалэйэ-дйулэйэ бидйэми нйан айа. Биракар эчэл һэгдыл бирэ, горойо эпкил бирэ оӈко имакунди эдан манавра. Манаврэкин оӊко орор горолдйодйоҕоро, уӊту бирала аландйаҕарэ. Таргачирдулы бэйур солокитчирдулы. Талы бэйуктэдйэми нйан айа.

Если место *солокит* называется, значит там всегда есть дорога к перевалу вдоль маленькой реки вверх. Там тоже хорошо стоять, но недолго, два-три дня, чтобы быстро не кончить олений корм. Если там корм кончится, олени далеко уйдут и перевалят на другую большую речную систему. По таким местам дикие олени переходят реки и там хорошо охотиться.

Iiergekit / Ийэргэкит	A nomadic road down a river Дорога идущая вниз по ручью

If the word *iiergekit* is used to denote a place, one will find a way down a small river towards a bigger river. At the place where the small river *(iiergekit)* and the bigger river meet, there are often camps on the *amunna* (large river basin), since such locations are good for setting up encampments. Such paths are used by both wild and domestic reindeer.
We can see that the place names inform the possible uses of a landscape type and that landscape types are very often analysed by the Evenki in a dynamic way: this helps them to conceptualise movements from one landscape type to another in the form of a chain. In other words, one landscape type seems to be attached to another specific landscape type, which in turn is also attached to a particular landscape type, and so on. Thus, if you have landscape type X, you expect to find landscape type Z after it, and so on; this is a way to know in advance about a landscape you have never seen before. We can say that this is knowledge about the 'logic' of a landscape system, not just knowledge about the landscape or places.[1] (cf. naldy)

Талы нйан арбун һокточи бивки аланми бираканма ийэргидйаӊнанны, тар дйапкалын һокто бивки. Ийэргэкит биракан нйан нйукучукокон бивки, ийэргидйаӊнанны даптулан, даптудун уриӊнэнны, таду амнунна айа уринчэдйэми бивки. Амнунна эчинда нйан айа уринчэдйэми. һэгды бирала эмэӊнэнны ийэргэкитли.

Если *ийэргэкит* называется, значит всегда есть дорога вниз по маленькой реке на большую реку. Там тоже склон с дорогой. Чтобы перевалить, спускаешься

1 In the 1990s, I associated this logic, and the ability of the Evenki to think continually in terms of paths and roads during hunting and herding activities, with the fact that most of them are excellent chess players (Lavrillier 1995, 2005). The relationship between the ways of thinking used in hunting and herding and skills in chess has been studied among another Evenki regional group by Davydov (Davydov 2015).

по течению маленькой реки по берегу, на котором есть дорога, ведущая к устью *(дапту)* и там разбивают лагерь, потому что там *амнунна* (широкий бассейн реки) есть. Там хорошо стоять. Через *ийэргэкит* всегда приезжают на большую реку. (см. налды, с. 80)

Gonyvun / Гонывун	The face of a mountain Сторона горы

Bohoγo / Боһоҕо	A sunny mountain face Солнечная сторона горы

Antaγa / Антаҕа	A mountain face in shadow Не солнечная сторона горы

Birakan daptun hegdy birala / Биракан даптун һэгды бирала
The confluence between a small stream and a big river
Устье ручья, впадающего в большую реку

A small stream converging with a big river. It is very comfortable to stay next to such a confluence. It is easy to fish in the big river and the location is close to the routes of the animal makchika: they are very numerous in such places. By following the small stream to its conflucence with a big river, a human can always find other humans, either directly or thanks to the many tracks along the big rivers. Big rivers are like motorways for the Evenki. There are a lot of willow trees *(siekte)*, with which one can make the linking parts of a sledge (the rest is made of birch). Further up the small stream, it is possible to allow the reindeer to graze: they will not go far from there. In such places, it is also possible to install traps for the sable that constantly pass through the small stream. (cf. daptu)

In some ways, this landscape type reflects the social network, since it helps individuals to find other humans in an expansive and sparsely populated territory. Through the example of daptu, *we see that topographic landscape types are cognitively linked by information about every possible ecosystem service they can offer, from vegetal tools to the presence of hunted animals to ways of passage (or lack thereof), as we will shortly see.*

Биракан даптун һэгды бирала – тары тар биракан даптувки һэгдыкун бирала. Таду уринчэдйэми сома айа. Таду тар олломотчоми нйан со айа, макчикал аҕактадйами айа. Макчика кэтэкун бивки. Таргачирду, биракун һэгдыкунду, бэйэ уҥту бэйэйэ бакадиҥан – удйан окинда бивки. Таду сйэктэкакун бивки, сыргала одйаҥнэнны. Бираканма манман солокуки ороробо тинэҥнэнны – эвкил горолло. Таду капкарба наннэнны, андаһи окинда ҥэнэдиҥан.
Нйан тар даптудуҕачин.

Это маленькая река, которая впадает в большую реку. На таком устье очень хорошо стоять и рыбачить. Там также хорошо перекрывать дорогу животным *макчика.* Там их очень много. В таких местах, где ручей впадает в большую реку, человек всегда других людей найдет: или самих, или их следы – всегда бывают следы, потому что большие реки, как автострады для эвенков. Там очень много тальников *(сйэктэ)*, из которых делают бечевки и ими соединяют элементы нарт из березы. Вверх по маленькой реке можно отпустить оленей одних кормоваться, они не уходят далеко. Там еще можно ставить капканы на соболей – они всегда по мелким рекам проходят. (см. дапту, с. 74)

Amut / Амут	Lake Озеро

In such places, there are two types of marsh (*leva* and *nire*). Due to this, it is impossible to set up camp there. In such a landscape type, one can lie in wait in order to hunt the elk that swim in the lake during the summer. People fish for pike, crucian carp, minnow, grayling, and *Brachymystax Lenok (P.).* In such places, one can hunt geese and ducks.

Тарҕачирду лэвал бивкил, нирэл. Эвкил тарҕачирду уринчэрэ. Таду тар коҥнокувэ карабдйавкил, коҥнокур амутту элбэскачйэвкил. Олломотчовкил таду, согдоннол дйавассэдйэннэнны, галйарба дйавассэннэнны, адылдун нйэручар, майгулэ. Таду эҥнэнны окинда илгимарэ – лэвадукин, нирэлдукин. Дэҕичарэ боло бултактэннэнны.

В таких местах есть болота типа *лэва* и мари *(нирэ)*. Из-за этого там невозможно стоять. Там летом лосей караулят, которые купаются на озерах. Там и рыбачат, ловят щук, карасей, гольянов, хариусов иногда, ленков. Там и охотятся на уток и гусей осенью.

Ily / Илы
A piece of rock sticking out of a mountain
Часть скалы, торчащая с горы

One never stays in such places and reindeer never walk there. It is only possible to walk around the edge, close to the routes taken by *makchika*. *Makchika* hide there and dogs rest nearby. Such landscape types are used only for hunting, sometimes even for red deer. In the past, people came here to gather ochre for decorating the packsaddle. In some *ily*, one can find sacred places (sometimes with rock paintings) where one must make an offering (bullets, matches, cigarettes, coloured tissues) to the spirits. If one has nothing, then one must cut off a young larch branch and plant it close to the sacred place. Every hunter respects these habits.

Талы эңнэнны окинда нулгирэ. Орор нйан эвкил таду ңэнэрэ. Талы тар гиркуктадйаминун или дйапкалын айа макчикал аӡактадйада. Макчикар таду дыкурэкачэвкил, нинакир таргачирду макчикарба илэвувкил. Тала тар эмэңнэнны булдаңнэнны – кумака адылдун таргачирду нан илэвки, нинакир илэвканывкил. Нонон таду тар дйололбо гаңкитын, эмэӡэрбэ онодйэммэн. Илылду адылдун улганникичил бивкил, таргачирду наннэнны ирйактэкачар токтонноннны, наннэнны адылдун.

Олени там не ходят. Там никогда люди не стоят. Там только ходят по краю, чтобы закрыть дорогу животным *макчика*. Они там прячутся и в таких местах собаки их ставят. Туда приходят только охотиться. Благородного оленя можно там собаками ставить иногда. В прошлом люди брали оттуда камни охры для украшения вьючного седла. В некоторых *илы* находятся священные места для угощения духов (иногда с наскальными рисунками), там оставляют угощения (пули, спички, сигареты, ткань яркого цвета) или, если ничего нет с собой, то надо рубить молодую лиственницу и воткнуть ее возле священного места.

Kadar / Кадар	Cliff
	Скалы

photo: see *ily*; фото см. *илы*.

This is a cliff on the edge of a mountain. In such a landscape type, one can hunt *makchika* with dogs. There are also sacred places where one must offer bullets or put a young larch branch in the ground. After performing such an offering, one is put into a good mood. *Some Evenki and Even reindeer herders use the cliffs as a natural enclosure for keeping the reindeer herd gathered in one specific area.*

Дяпкалдутын, урэлдули бивкил кадарил. Таргачирду макчикал илэвкил, нинакир илэвканывкил. Улганикичил нйан бивкил, таргачирду надйэӊнэнны патрокарэ, ирактэкачарэ токтоннонны, тыка нэкэми айа овки мевандодун. Таргачинду ичэчэӊнэнны или ӊэнэдйэми нйан айакикин.

По краям, вдоль горы, *кадар* бывают. На таких местах собаки ставят животное *макчика*. Там тоже места угощения бывают, там надо положить патроны или маленькую лиственницу воткнуть в землю как угощение. Когда угощаешь, хорошо на душе станет. И оттуда очень удобно смотреть куда можно ехать или идти. *Некоторые эвенкийские и эвенкские оленеводы употребляют скалы как натуральный загон, чтобы олени далеко не уходили.*

Murki / Мурки

Pingo ground

Пинго, бугор пучения

If such places are sizeable, then one can set up a camp for a short time in the summer. It is comfortable to place a tent in such a location because of the dry and hard ground *(keteme)*. It is always windy, which is good for reindeer. Within the *murki*, one finds permafrost: it is kept very fresh by the blowing of the wind. Sometimes, one digs a hole within *murki* in order to prevent fresh meat from rotting. Hunters also use *murki* to hide from animals.

Тары таду нйан мурки ойодун уриннйэннэнны, эӊнэнны горойо бирэ, палаткйэ таду айа надйэми, кэтэмэкикин бивки. Адындйэвки окинда, орордун айа. Мурки эрэдун дйука бивки, соӊутмар адындйэрэкин. Адылдун тала тар чокынналны, тала тар уллэвэ наннэнны, мурки эрэдун айат, эдан мунурэ.

На вершинах таких бугров пучений можно даже поставить лагерь ненадолго, там хорошо палатку ставить, там твердая и сухая земля кэтэмэ. Там всегда ветер дует, что хорошо оленям. Внутри мурки лед с землей есть (мерзлота) и прохладно, когда ветер дует. Иногда там роешь ямку, чтобы мясо хранить внутри в холоде, чтоб не гнило. Когда охотятся, за ним *(мурки)* прячутся.

Daptu / Дапту	The confluence of rivers Устье реки

At the confluence of small or big rivers *(daptu)*, it is always good to set up camp. The confluence means a big river is nearby. *Sivak* grass *(Equisetum arvense L.)* horsetail is present for reindeer. It is also a good place for fishing. *Makchika* are found here. You can also find a nomadic neighbour whom you can ask where other people are, where they are staying, what they are doing, and you can travel from confluence to another. There are also a lot of willows *(siekte)*, which are useful for repairing the linkages in sledges. Further up the small rivers, one can allow reindeer to graze, since they will not wander off. It is also possible to install traps for the sable that always pass through small rivers. It is always windy along big rivers (see also Birakan daptun hegdy birala).

Бира-ҕу, биракан-ҕу даптадун – тара тар гунэвкил: дапту – айакикин уринчэда. Даптун окинда тар һэгды бирала бира даптувки. Таду сивак бивки, олломочоми айакикин, макчикакакун бивки. Таду бэйэйэда бакадиңас, нийавэл нимакийа анңуктаммэн экуравал, экудыдук уңтул даптулан эмэча элйэ тала ңэнэктэңнэнны. Таду сйэктакакун бивки – сыргалба авйэстаңнэнны. Бираҕудун даптадун бираканҕадун солокуки ороpбо тиндйэми айа, эвкил горолло. Таду капкарэ наннэнны. Адын окинда ңэнэдиңан.

На устье маленькой или большой реки – называется *дапту* – очень хорошо стоять. Если устье, значит всегда будет большая река. Там трава хвощ *сивак (Equisetum arvense L.)* бывает для оленей, там хорошо рыбачить, и животное *макчика* есть тоже. Там и людей найдешь, соседа чтобы спросить, где другие люди, куда едут, ездишь к одному и к другому устью. Там много тальников, из которых делают бечевки *(авйэс)* для ремонта нарт. По маленькой реке отпускаешь оленей вверх по течению – далеко не уходят. Там и капканы на соболей ставишь. Ветер всегда бывает по большой реке. (см. Биракан даптун һэгды бирала, с. 70).

Edekit / Эдэкит	River crossing Место для перехода реки на ту сторону, брод

Humans can always cross big rivers in these locations. The water level is low and there is not much of a current; thus, one can cross to the other bank. This makes these locations suitable for driving herds across a river.

һэгдыкун биралдулы бэйэ окинда талы эдэдйэвки. Таду арбакукан, окинда бивки эйан, дйапкалы эдэӊнэнны. Талы ororби илбаннэнны айакикин, баргыскаки ӊэнэдйэми со айа.

На больших реках человек всегда там перейдет на другую сторону. Там уровень воды низкий и нет сильных течений. Через *эдэкит* оленей гонишь через реку на ту сторону.

Bargila / Баргила	Any landscape on the other side of the river Любой ландшафт на той стороне реки

The term *bargila* indicates the opposite side of a river. Such locations make good camping sites. This term is one of the four used for orientation: our side of the river *(dyski)*, the opposite side of the river *(bargila, bargyski)*, up the river *(solokit)*, and down the river *(eiaki)*.[1]

Тары тар, бирава эдйэми – баргила гунэвкил. Уринчэдйэми нйан айа, айа бидиӊан.

Это, если реку переходить, указывают *баргила* т.е. противоположный берег реки. Там хорошо стоять. Это один из четырех ориентиров: на нашу сторону реки *(дыски)*, на ту сторону реки *(баргыски, баргила)*, вниз по течению *(эйаки)*, вверх по течению *(солоки)*.

Mitkula, mitkidan, mitkidadun / Миткула, миткидан, миткидадун	The landscape on our bank of the river (term of orientation) Любой ландшафт на нашей стороне реки

This term means 'on our side of the river'. Socially and in terms of landscape knowledge, it is a very important concept because the nomads do not know either the people or the landscape (as they say, 'other lands – other peoples') across big rivers like the

1 Ref. to Lavrillier 2005–2006, 2010.

Olëkma: they thus that act like spatial and social frontiers.
We should also stress the importance of the noun mit *(the root of this expression).* Mit *can mean 'I' and 'we' at the same time, which, in contrast to the ordinary 'we'* (bu) *in Evenki, designates the speaker together with 'his' people (relatives, allies, friends with whom he/she nomadises); in other words,* mit *defines the ego as a social group.*

Миткула, к нашему берегу. In the direction of our (or ego) bank of the river.
Миткидан, наша сторона. Our (or ego) side of the river.
Миткидадун, на нашей стороне. Социально-географически важное значение – за большие реки – это как большая граница, за чертой которой не знают людей и территории (например за рекой Олекма) той стороны.

Eiakur / Эйакур	Strong rapids
	Сильные пороги

When nomadising there, one will not be able to cross the river. There are a lot of fish in such a place, like *Huchen taimen (P.), Brachymystax lenok (P.),* and big graylings. In such places, one will not set up camp because the reindeer calves sink, even the big ones. Since one cannot cross the river, one will have to stay put. The strong streams are very noisy and prevent one from hearing either predators or domestic reindeer coming to the camp. It is not possible to hunt in such a place. The stones on the ground are very slippery.

Талы нулгидйэми этанны эдэрэ. Таду олло кэтэкун бивки, дйэлил, майгул, нйэручар һэгдыкур, бургукур бивкил. Таргачинду эӈнэнны уринчэрэ, таргачинду эӈнэкар чэпэдйэвкил, һэгдыкур нйан чэпэвкил, баргыски этанны окинда эдэрэ. Чаваркун эйакунын – бэйӈал этанны долдырэ, этанны долдыматтэ он орор эмэрэ. Дйолол тыгдарэкин балдакикур.

Если туда будешь кочевать, то не сможешь перейти реку. Там много рыбы: таймень, ленки, и хариусы там большие и жирные. В таких местах не устанавливаешь лагерь, там оленята могут утонуть, даже большие. На ту сторону реки никак не перейдешь (значит без движения останешься). Сильные течения очень шумные – и хищников не услышишь и своих оленей, как на стоянку приходят, не услышишь. И там не поохотишься. Камни скользкие каждый раз после дождя.

Bira / Бира

River

Река

In these locations, it is always possible to go down river: thus, one nomadises along these big rivers. One travels up river to purchase goods, meet people, or ask something. It is always comfortable to travel along rivers; one can travel in many directions, driving the heard as you go. Such places provide excellent hunting grounds for wild reindeer or sable. Along the rivers, one will always see some tracks. It is also possible to fish here, including with nets.

Let us note that the perception of landscape here is centred on the observer (ego), because, in the case of the observer who has given the interview which is translated, the village where nomads buy food and goods is situated up the main river system.

For the Evenki, rivers are crucial: they function as the main roads and thus offer opportunities to find and meet other people; equally, it is always possible to set up a camp. For the Evenki, river systems are a scheme for orientating themselves, remembering kinship, and maintaining their social networks (Lavrillier 2005–2006).

Талы эйандйэвки окинда таргачирдулы. һэгдыл биралдули нулгидйэӊнэнны, солокуки дйэпгайави ганнадйэнны, бэйэлбэ бакалдыдйаӊнэнны, анӊуктэдйаӊнэнны. Нулгидйэми окинда айа бидиӊан, илйакат нулгидиӊас, илйакат ороpби илбадиӊас. Бэйуктэдйэми, соболдйэлэми нйан айакикин. Таргачирду удйалэ окинда ичэдиӊан. Олломотчоми нйан айа, адылдйами нйан айа.

Туда всегда можно спускаться по течению. Вдоль больших рек кочуют, вверх по течению ездишь за продуктами, людей можно встретить в пути, чтоб что-нибудь спросить. Вдоль рек всегда хорошо будет кочевать, куда бы не кочевали, куда бы оленей не вел. Там хорошо охотиться на диких оленей или на соболей. Вдоль рек всегда можно увидеть следы. Для рыбалки тоже хорошо: и удочкой и чтобы сетки ставить.

Birakan, birakakan / Биракан, Биракакан	Small river Ручеек

These are very small rivers, along which one can hunt. It is not possible to set up your camp for a long time, since the reindeer will overgraze. On very small rivers, it is impossible to stay put except in some rare instances.

Эчэ олус һэгды биракан бирэ. Таргачирду нйан бэйуктэдйэӊнэнны. Эӊнэнны горойо уринчэрэ, оӊко эдан манавра. Олус нйукучукон биракан, таргачинду эӊнэнны уринчэрэ, адылдун уриӊчэннэнны.

Это река совсем небольшая. Вдоль таких ручейков охотиться хорошо. Там нельзя долго стоять, чтобы корм на пастбище не кончить. На очень маленьких речках не стоят или в редких случаях.

Liopike / Лйопикэ	A very small stream which is barely noticeable Маленький ручей вместе с деревьями, ели заметный ручей

Liopike designates a barely noticeable small stream alongside which a line of trees grows *(teŋke)*. If one starts to follow this tree line, a very small river appears. One either walks or skis along such mini-streams. Reindeer can rarely cross to the next river system in such places. Bilberries and mushrooms grow well in these locations. Wild reindeer find plently to graze on and thus spend the night.
Once again, part of a river system is used as orientation in order to find one's way home after hunting. Very few hunters know this word. (cf. diagram Liopike)

Лйопикэвэ гунэвкил тар эӈикан саврэ. Тарканма ойолдулы тэӈкэ эвучэдйэрэн таргачинма гунэвкил: лйопикэ. Тэӈкэвэ бододйоһиндиӊас талы бивки биракан одиһинивки, нйукучукокон, эһи саврэ. Таргачирдулы гиркуктаксоннэнны киӊначчэннэнны-у. Орор адылдун умнакар аландиӈарэ. Черника кэтэкун балдывки, дэвуӊнактакакун. Бэйур айат оӊкодйовкил, анӊачэпкил ӈэнэдйэрил.

Называют *лйопикэ* еле заметный мелкий ручеек, вверху которого растет полоса деревьев *(тэӊкэ)*. Если идти вдоль этой полосы, после него начнется маленькая речка – маленькая, еле заметная. Вдоль таких или пешком ходят, или на лыжах. Олени иногда через такие речки переваливают. Там растет хорошо черника и грибы. Дикие олени при миграции там ночуют и кормуются. (см. схему, с. 115)

Oŋokto ureni /
Оӊокто урэни

The foot of a gently sloping mountain

Конец горы с пологами

This term literally means 'the nose of the mountain'. It is comfortable to climb the mountain or to go down into the valley from these locations. The reindeer graze well and it is quite warm when the sun is shining. It is also a useful place through which to drive the herd. Such places make excellent camp sites, since the reindeer can graze here, and the snow is soft *(duiukun)* and shallow. It is also a good point to observe the roads: if one climbs to the top, one can see everything around.
We can note the anthropomorphisation or zoomorphism of the landscape in this name, since it means the 'nose of the mountain'. The Evenki term 'nose' in this name is used for both humans and animals. Such a place offers several ecosystem services: orientation, herding (good grazing and good snow quality for accessing pastures), and the potential to build human camps (it is suitable for herding and travelling, since it is a kind of crossroads between big rivers and mountainous terrain).

Талы айакикин туктыдйада бивки, эвдйэми нйан айакикин. Орорду оӊкодйоми нйан айакикин, дылачадйэрэкин нйаматмар бивки. Илбадйэми талы нйан айакикин ороpбо. Оӊокто урэду эрэдун уринчэдйэми айа орор таду айамамат оӊкодйовкил, таду иманна окинда арбакун бивки, дуйукун иманна. Таргачирдули сачйадйанны илйа урудо һоктоли. Ойолон туктыми оӊоктолы ичэччэми нйэскуки айакикин, экунада ичэдиӊас.

Называется буквально «нос горы». Там хорошо подниматься на гору или спускаться с горы в долину. Там олени хорошо кормятся, и когда есть солнце, там теплее, чем в других местах. Тоже удобно оленей гнать через такие места. Внизу таких мест хорошо стоять, потому что там олени хорошо кормятся и снег там всегда мелкий и мягкий *(дуйукун)* будет. Оттуда тоже удобно смотреть, куда дорога ведет. Если залезть наверх – это хорошая точка наблюдения, можно все увидеть.

Deran / Дэран	Beginning of a river, around two or three kilometers from the source
	Начало реки, от истока 2–3 километра

Deran is the beginning of a river. In such locations, one also can set up camp, but only for one night, on the way downstream. When the sun rises, the lead hunter goes in search of various animals to hunt. One cannot spend a day there because the reindeer will run away downstream. In places where humans have never set up a camp before, there is always game. This is why a single lead hunter is sent ahead along the road: doing so means that the game will not be frightened by the noise made by people.

Дэран бидйэрэн иду бира овки. Таду нан уриндйэвкил умулоконду ңэнэдйэнэ эдйэливи эйаргирэ. Ңариллэкин, дйулгу бэйэ элэнэлэ бэйңалэ арытчанэ ңэнэдйэвкил. Эңнэнны инэттэ, орор эдатын эйаргирэ, анңачиђа эйаргиммэн айа урикитла.

Дэран – это начало реки. В этих местах тоже разбивают лагерь, только на одну ночь, проезжая вниз по реке. Когда рассветает, первый охотник выезжает на розыск всяких зверей. Там не днюют, чтобы олени не ушли вниз по реке. После ночевки кочуют вниз по реке на хорошую стоянку.
В новых местах всегда дичь бывает, поэтому на охоту сперва только один охотник едет вперед, чтобы было меньше шороха от людей и не напугать зверей.

Naldy / Налды	A small tributary, the length of which is equal to the part of the main river situated between the confluence with this tributary and the source of the main river (up river). This does not depend on the position of the observer (ego).
	Небольшой приток реки, длина которого одинакова с длиной части главной реки: от устья данного притока до истока главной реки (вверх по течению другой реки) – где бы не стоял говорящий.

This is a tributary of a small river. The length of this tributary is equal to the part of the main river situated between the confluence with the tributary and the source of the main river (up the main river). This does not depend on the positions of the observer or the tributary. Along such tributaries, one can look for nomadic roads. This is a useful landscape type when one wants to climb mountains further upstream in order to pass into another river system. One can easily pass into another valley either through the source of the main river or that of the *naldy* tributary itself. If one walks along the *naldy* upstream, one will come to the source of the other river over the pass. If one walks up the the river itself, one will come to the *naldy* of the river over the pass. In such places, one can remain for several days in comfort. At the confluence on the side of a *naldy* or between a *naldy* and the main river, there is always a large river basin *(amnunna)*, where it advantageous to camp either for a night or several days. This is because one can let the herd go up river (be it the main river or *naldy*): the reindeer will graze well there. From such places, one can go in various directions to hunt wild reindeer. (cf. diagram Naldy)
Each river has a naldy *tributary, as the Evenki explain. We should stress the mathematical and geometrical aspects of the* naldy *concept. It conceptualises the lengths between the confluence of the main river and its source and between the confluence and the source of* naldy. *This highlights a profound knowledge of the local geomorphological system, since it is possible to deduce that, in case of presence of a tributary* naldy, *one will find the precious landscape type* amnunna *(large river basin).* (cf. amnunna, diagram Naldy). *The Evenki system of landscape knowledge is knowledge of the logic of the landscape. A particular landscape type is followed by either one landscape type or by a limited range of types. In other words, upon seeing a specific landscape type, the Evenki can deduce which landscape types must follow, and he/she knows the possible uses of each landscape type.* (cf. *Iiergekit*)

Тар бирани налдын биракан бивки. Бирани налдын урэдйэвки биранун манын ңоныммэтпи, бира ңоными, налды ңонымин уратчэвкил дэралдулавэр. Экудыкат бираду бидиңан налды. Таргачиндулы һоктойо ичэччэми айа, ңэнэңэтпи-у. Таргачирдулы алакичилтыки ңэнэдйэми айа солодйоми. Дэрандукин аландйами нйан айакикин, экудыкат налдылы – биралы-да, налдылы-да. Таргачирду анңакчэми айа, даптулдутын окинда амнуннал бивкил мардулитын. Налдылы амнунна бидиңан, биралы амнунна бидиңан даптудын. Даптулдутын уринчэдйэми айакикин, ororбо тиндйэми солокуки биралы-да, налдылы-да айа бидиңан. Талы экудылакат бэйуктэнадиңас. Налдылы нйан айакикин бэйуктэдйэми.

Приток реки и небольшая река. Длина этого притока одинаковая с длиной части главной реки от устья данного притока до истока главной реки (вверх по течению). И это – где бы не стоял этот приток, и где бы не находилась стоянка, говорящего это слово. По таким притокам удобно смотреть дорогу или ехать. Там также удобно ехать к перевалам, чтобы перейти на другую речную систему. От его истока удобно переваливать, в общем то, можно перевалить пройдя вверх по *налды* или по самой реке. Если перевалить через *налды* можно попасть на другую реку находящуюся за перевалом. А если идти по самой реке, то можно перевалить на *налды* другой реке за перевалом. В таких местах ночевать хорошо, ведь по устьям со стороны налды там всегда широкий бассейн реки есть *(амнунна)*, между главной рекой и *налды* есть всегда *амнунна*. На устьях хорошо ночевать, там и оленей можно спокойно отпускать вверх по реке (или по *налды* притоку или по самой реке), хорошо будет. Оттуда можно хоть куда ехать, на охоту на диких оленей. По самому *налды* тоже хорошо охотиться на диких оленей. (см. схему, с. 112)

Urigdan / Уригдан	A quiet stretch of river without a significant current where fish live and people catch them
	Тихий плес реки без течения, место, где рыбачат

This is a very good place for installing fishing nets because there is no current. Even if there is significant flooding, one can cross the river on a *mereke* boat in order to transport food and goods or to drive the herd to the opposite side of the river. In the absence of a current, both domestic and wild reindeer are able to cross easily.

Таду тар адылдйами со айа, эйанэ ачэн. һэкдыкун мудакун бирэкин мэрэкэт эдэбдйэми айа дйэпгалбэ-да, орорбо-да баргыскаки эдэбдйэӈнэнны. Талы эйан ачэн, талы тар орор айамамат эдэдйэвкил эйандукин ачэнду. Бэйур талы нйан эдэвачэвкил эйанэ ачэрдулы баргыскаки.

Очень хорошо там сетки ставить, потому что нет течения. Даже при сильном наводнении можно его переплыть на эвенкийской лодке мэрэкэ, чтоб за продуктами съездить или оленей перевезти на ту сторону. Там течения нет, и олени хорошо переходят реку из-за отсутствия течения. Дикие олени там тоже хорошо переправляются на ту сторону.

Tykis / Chykis (only in some places) Тыкис / Чыкис (олекминские УН) /	Waterfall
	Водопад

In the Aldan, no such landscape exists: only the nomads of the Olëkma river system know the term.

Тара мутту Алданду ачэн.

У нас в Алданской стороне нет такого ландшафта.

Iukte / Йуктэ

The source of a river, a spring

Исток реки, родник

These are located at the origins of a river. One can stay one or two nights upon setting up camp. The wild reindeer used to drink water from such places. One finds small *nire* marshs. Wild reindeer often come down from the mountains to such places because they can always find *chuka (long grass)* on which to graze. This is a good place for observing the directions in which one can go. From here, one can always go down stream and find a big river.

We have seen several times that the discourses explaining several terms of topographical typology evoke the use of the landscape to 'find one's own road', 'to see where to go further', or 'to reach a main river'. It seems to denote the fact that the Evenki do not always know the details of the places they cross or where they are exactly: the indigenous people's knowledge of the landscape allows them to use some types of landscape to reach an axis along which they can find spaces they know better and thus can find their way home. In addition, this knowledge is so systematic that it can be applied to all southeastern Siberian landscape systems. This probably explains why the Evenki were able to adapt to conditions across the vast region of Siberia and why they were the preferred guides during the discovery of Siberia even when they were outside their usual nomadisation spaces. In addition, some Evenki regularly (roughly every five to ten years) 'explore new lands' in order to find future potential replacements for their annual nomadic routes or hunting areas (Lavrillier 2005–2006).

Йуктэ тар дэрандун бивки. Анҥанами тала умулйэду-дйулйэду таргачинду уриҥдйэннэнны. Бэйур таргачирду муйа умнадйэвкил. Таргачирду нирэкар бивкил. Бэйур тала эвдйэвкил, чука окинда балдыдйадиҥан таргачинду, бэйур чукалыдйэвкил. Ичэччэҥнэнны талы или ҥэнэдйэми нйан айакикин, йуктэдук эйакуки урунэ, онда бирала эвдиҕас.

Это в самом начале реки. Чтобы ночевать там, можно разбить лагерь на один-два дня. Дикие олени оттуда воду часто пьют. Там иногда бывают небольшие мари *(нирэ)*. Дикие олени часто спускаются с горы на такие места. Там всегда растет длинная трава, и там дикие олени кормятся. Там тоже можно хорошо осмотреться, куда дальше ехать. От йуктэ, по течению если ехать, всегда можно выйти на большую реку (и найти свою дорогу домой).

kapchan / kapchakun
Капчан / капчакун

A mountain surrounded on both sides by high and abrupt peaks

Гора, окруженная высокими горами с обеих сторон, где горы крутые по краям

One finds a *kapchan* in places surrounded by high and abrupt peaks. It is a bad place for setting up camp. The reindeer can graze on top of it, but they are often distant from their herders: it is also very uncomfortable to drive them though such a landscape type. It is also a very poor place for wild reindeer hunting because of abrupt declines and inclines. *Makchika* are present but one can hunt them only on foot. In such a landscape type, the snow is always shallow.
This topographic peculiarity related to the quality of the snow cover is important for the Evenki, especially during years when the snow cover is of a bad quality (too deep or with ice, for instance). In such cases, herders know that they can bring the herd to graze close to kapchan *places, where better snow conditions always prevail. This represents a specific set of ecosystem services which are dependent on the topographic landscape; i.e. some particular reliefs offer good snow cover qualities (shallow and soft snow) every year, even during years with bad snow conditions. This is essential for reindeer herding. This is a good example for proving that indigenous knowledge is a key tool when adapting to extreme weather events.*

Манын капчакун, капчан-у бирэкин таду урэкур бивкил, эврикур. Уринчэдйэми таргачинду эруми. Орор таду ойодолон оӈконовкил, горолэвкил, илбадйэми нуӈарбатын эруми талы. Бэйуктэдйэми нйан уруми эврикур, туктэрикур. Таргачирду макчика кэтэкун бивкил, гиркуктадйаминун макчикарба аӈактадйаӈнэнны. Таргачирду иманна арбакукан окинда бивки.

Сам *капчан* или большой *капчан* есть, когда вокруг есть большие крутые горы. Там очень плохо стоять. Олени на вершинах таких мест кормятся, далеко уходят, и очень неудобно их гнать через такие места. Также плохо охотиться на диких оленей из-за резких спусков и подъемов. Там очень много животных *макчика*, пешком только можно им дорогу перекрыть. Там снег всегда мелкий.

Bira chokchokodioron /
Бира чокчокодйорон

River bend

Поворот реки

A river bend is a very good place for setting up camp. At the bend of the river, the mountain's slopes are not abrupt. On such slopes, reindeer are able to graze easily; wild reindeer also come down from the mountains. A river bend is where a river turns; it offers a comfortable location for staying in one place.

In the descriptions of several landscape types, it seems that there are two important realms. One realm is situated at high altitudes: 1. the mountains on which the wild reindeer or elk live and sometimes come down from; and 2. the cliff and rocks where makchika live and are hunted. The other realm is the valley or large river basin where domestic reindeer and human circulate and can find each other thanks to tracks. It seems that both realms can coexist thanks to landscape knowledge and a balanced distribution of the two kinds of inhabitants in the natural environment (humans and domestic reindeer on the one hand, and wildlife on the other). What is interesting is that this suggests a conception of a three-dimensional symbolical distribution of beings in the environment: the 'higher' world (mountains and cliffs) is the wild realm, while the 'lower' realm (valleys and plains) is considered to be the domestic and human world. However, this does not reflect reality.

Each landscape type is conceived very systematically through its potential uses (or lack thereof) for the Evenki economies (reindeer herding, hunting, fishing, gathering) and movement (e.g., ensuring movement by giving the possibility to see further in the landscape and its ability to sustain a campsite).

Бира чокчокодйорон – таргачинду нйан уриндйэми айа. Чокчоколдутын урэл эчэл эврикур бирэ, нэптэкуһинча эллэӈэ бивкил, орор айат огнкодйовкил, бэйур талы эдэвачэвкил, эвдйэвкил таргачирдулы. Бира чокчйокодйорин тар бира манын чокчйокодйовки, тар чокчйокоду уринчэвкил.

Поворот реки – это тоже хорошее место для стоянки. На поворотах горы крутыми не бывают, пологие бывают и там олени хорошо кормятся, дикие олени спускаются с гор по таким местам. Поворот реки – это, когда сама река поворачивает и на этом повороте устанавливают стоянку.

Nepteke urel / Нэптэкэ урэл	Gently sloping mountains or high mountains with a flat top
	Пологие – ровные низкие горы, или высокие горы с ровной вершиной

It is very comfortable to nomadise and set up camp on such flat tops. It is also a very good place for sable hunting, because there are either no stones or very few, which allows for one to move and carry things easily. In addition, there are always good pastures for domestic reindeer. In the summer, there is always fresh air, which the reindeer like. After New Year feasts, there is always deep snow, after which it is not possible to stay there any longer: the reindeer can only draw poor sustenance in deep snow conditions. Before the New Year, however, if the snow is not too deep, it is a very good landscape type for sable hunting with dogs and for herding. In such places, there are many wild reindeer.

Нэптэкэл урэлдули – таргачирду нулгиктэдйэми нйан айакикин, уринчэдйэми таргачирду айакикин. Андаһидйами со айа, дйоло ачэн бивки, умукокор. Оӊко идукат таргачинду бивки. Таргачир ойоду, таргачинду тар соӊутмар бивки дйуӡа. Туӡэниду новый год амардукин, иманна суӊтакун бивки. Таргачирду эвкил уринчэрэ, орор эрут оӊковкил иманна суӊтадукин. Таргачинду андаһидйами со айа бивки иманна арбадукит Новый год дйулэдукин нинакиксодит. Бэйукакун бивкил таргачирду.

Вдоль гор с ровной вершиной тоже хорошо кочевать и стоять. Там охотиться на соболей очень удобно, камней нету или мало, и не мешают передвижению. Там и корм для оленей всегда есть. На вершинах летом всегда прохладно. Зимой, после Нового года, снег очень глубокий там становится. И тогда там никак нельзя стоять, потому что олени плохо кормятся на глубоком снегу. Но, до Нового Года там очень хорошо на соболей охотиться с собаками. В таких местах много диких оленей.

Teŋke / Тэӈкэ

A dense forest on a river bank

Густой лес по берегам реки

Teŋke is where there are dense patches of forest alongside big and small rivers. It is a very good place for an encampment during the snow period because of the lack of wind, the shallowness of the snow, and the presence of *sivak* grass *(Equisetum arvense L.)*. It is also a good place for domestic reindeer. In such landscape types, there is high-quality wood for making skiis: it cleaves along the grain at the first chop. This is also a suitable location for preparing firewood: the trees are tall and thus provide enough wood for a long stay. Domestic reindeer graze easily on *sivak* grass and live very well there. Wild reindeer also eat the *sivak* grass. Birch and willow trees are also present, allowing for one to build and repair sledges. Even Russians install their hunting houses within the *teŋke*.

Тэӈкэ тар манын ирйаккур бивкил, аһиӈкур бивкил бирадули. Таргачирду уринчэдйэми айакикин, адын эвки адыннэ, иманна арбакукан бивки, окинда сивак бидиӈан. Ороду нйан айа. Таргачирду колтоҕодйоми киӈна айакикин, муӈнакикинди колторговкил. Молыдйами нйан айакикин, мо ӈонымкур, горокунэ элэкчэвки. Орор таду сивакдйэвкил, айамамат бидйэвкил. Бэйур нйан таргачирду бидйэвкил сивакдйэвкил. Чалбан таду нйан биһин, сйэктал нйан биһин упкатын биһин.

Тэӈкэ это густой лиственничный лес и ельник на берегу реки. Там разбивать лагерь в снежный период очень хорошо, ветер там не дует, снег мелким бывает, всегда трава *сивак* бывает. Оленям тоже хорошо. Там очень хорошо раскалывать ели для изделия лыж (ровно раскалывается), за один раз раскалывается ель по нити дерева. Это тоже очень хорошее место для приготовления дров для печки, дрова длинные и надолго хватает. Олени кушают траву *сивак* и очень хорошо там обитают. Дикие олени тоже кушают траву *сивак*. Там тоже бывают березы и тальники, все там есть для изготовления и ремонта нарт. Даже русские делают свои бараки в *тэӈкэ*.

Teŋkekun / Тэӈкэкун
Very dense forest on a river bank
Очень густой лес (чаща) по берегам реки

cf. Teŋke

Ure / Урэ
An average mountain
Средняя гора

On such average mountains, there are *kapchan* mounts and *elleŋe* slopes. The snow is not deep, so the reindeer can graze well. In the winter, it is also possible to set up camp at the foot of the mountain because there is no wind. People camp on such mountains in sunny periods because of the warmth. There are rivers at the edge of the mountains where one can fish successfully. There are also a lot of berries and mushroom. Such a landscape type is good for hunting wild reindeer, wild big game, and sable. The wild reindeer also eat to their heart's content on such average mountains.

Таду эллэӈэ биһин, капчар нйан бивкил. Урэлду иманна арбакукан бивки, орор айамамат оӊковкил. Уринчэдйэми урэлду нйан айакикин эрэдун, эвки адынэ. Адылдулын дылачавулын уринчэдйэӈнэнны, дылачадйэвки нйамадйанэ. Урэл дйапкалдулын бирал биһи, олломочоми нйан айа. Диктэкакун, дэвуӊнактакакун нйан айамамат балдывки. Бэйуктэдйэми, бултадйэми айакикин. Бэйур нйан айавувкил урэду оӊкотодйовэр.

Там есть склон типа *эллэӈэ*, и гора типа *капчан*. В горах снег неглубоким бывает, олени хорошо кормятся. Зимой стоять в горах, внизу гор, очень хорошо, там ветра нет. По некоторым горам на солнечной стороне, люди стоят, потому что там тепло, при ясной погоде. По краям гор, есть реки, где хорошо рыбачить. Там много ягод и грибов. В таких местах хорошо охотиться и на диких оленей и на любого зверя и соболей. Дикие олени тоже любят кормоваться в таких горах.

Urekachan / Урэкачан

A small mountain

Маленькая гора

These are small mountains with *elleŋe* slopes. On such small mountains, the snow is not too deep, which allows the reindeer to graze freely. The edge of the mountain is a beneficial place to set up a winter camp since there is no wind. On the sunny side of such mountains, people can camp there because it becomes warm during good weather. If there is a river nearby, one can usually fish there. There are both berries and mushroom in relative abundance. Wild reindeer hunting can take place because they graze in such a landscape type.

Тар эчэ һэгды урэ бирэ, эллэӈэ биһин. Урэкачанду иманнан арба бивки, орор айат оӈковкил. Уринчйэдйэми нйан айа, урэ дйапкадун эвки адынэ. Адылдули дылачавулын уринчйэдйэӈнэнни, дылача нйамадйавки. Бира бирэкин олломочйоми нйан айа. Диктэкакун, дэвуӈнактакакун нйан балдывкил. Бултадйэми нйан айакикин, бэйур нйан таду бивкил урэкачанду.

Это небольшая гора, со склоном типа *эллэӈэ*. В таких небольших горах снег мелким бывает, поэтому олени там хорошо кормятся. Там также хорошо стоять, по краям таких гор ветер не дует. На некоторых таких горах на солнечной стороне солнце сильно греет при хорошей погоде. Если там река есть, можно рыбачить и много ягод и грибов. И охотиться хорошо, ведь дикие олени тоже бывают в небольших горах.

Iang / Йаӈ	Tall and wide mountains. These are not necessarily mountains with pointed summits: they are just tall mountains where trees do not grow, dry out, or are broken by the wind. Большая высокая гора. Не обязательно острая гора, просто высокая, где деревья не растут или высыхают и ломаются от ветра.

Tall mountains provide good places to set up encampments. If staying there in the summer, reindeer will remain safe from hoof infection. The reindeer are always able to put on weight in such locations because there are a lot of rich grazing areas. There are also lakes and *sivak (Equisetum arvense L.)* and *nirgakta (Eriophorum vaginatum)* grasses: the freshness of air helps assure good health for the reindeer. People stay there until August. In the winter, nobody stays because of the very strong and harsh winter winds. There are a lot of wild reindeer available to hunt, something which the high levels of visibility makes easy. In the winter, one hunts wild reindeer on skiis. It is also comfortable to nomadise along the smaller mountains because there may be inclines. One can travel by reindeer sledge along the small inclines *(ukty)* on the surface of the snow, which has become asphalt-like thanks to the permanent winds. There are no fish, but a lot of berries, cones, and *nirgakta* grass.
This landscape type is typical in terms of the Evenki knowledge system and use of microclimates. The Evenki use microclimates they know perfectly 'to benefit from the best part of each season for the reindeer'. (For more details on microclimates, cf. Evenki climatology, Evenki calendar)

Йаӈилду уринчэдйэвкил, уринчйэдйэми со айа таргачирду. Дйуҕа уринчйэдйэвкил орор эвкил буҕлэрэ. Окинда бургукур бивкил, оӈко кэтэкакун. Амутыл бивки, сивак нйан биhин, ниргактэ биhи. Ороpду таду айакикин, соӈукикин бивки. Таду тар августала бэйэл бидйэвкил. Туҕэ таду этанны уринчэрэ, адыкун бивки туҕэнилду. Талы тар бэйукакун бивки, бултактэдйэми айакикин, горокуртыки ичэӈнэнны. Туҕэ бэйуктэдйэӈнэнны киӈналдит ӈэнэктэӈнэнны, бэйурэ арытнэ. Урэлдули айакикин ӈэнэдйэми, туктыкур бивки, нулгидйэми нйан айакикин, уктыкурдули иманна ойолын ӈэнэӈнэнны орордит. Таду олло ачэн, диктэл бивкил, нйаӈтакакун нйан бивки, ниргактэ нйан кэтэкур балдывкил.

В высоких горах устанавливают стоянки, в таких местах очень хорошо стоять. Летом, олени не болеют гноением копыт. Всегда жирными бывают, потому что там много пастбищ. Там и озера бывают, и травы *сивак* и *ниргаката*, и оленям очень хорошо также из-за прохладного воздуха. Там люди стоят до августа. Зимой в таких местах не стоят из-за сильных ветров. Там много диких оленей бывает и хорошо на них охотиться, издалека их видно. Зимой там на диких оленей хорошо охотиться на лыжах. По мелким горам тоже хорошо ехать, там всевозможные подъемы бывают, кочевать в таких местах хорошо. По мелким подъемам *укты* можно подняться на оленях по снежному насту, образовавшемуся от постоянных ветров в таком ландшафте. Там рыбы нет, но ягоды есть, также много шишек и травы *ниргактэ*. (см. с. 43–59, 175–177)

Boltoko, boltokokon, boltokokun / Болтоко, болтококон, болтококун

A small semi-circular mountain

Гора маленькая полукруглая

Boltoko means a semi-circular mountain. *Bolotokon* refers to a small semi-circular mountain. *Boltokokun* indicates a big semi-circular mountain. In such landscape types, wild reindeer graze, and it is possible to hunt them. Upon climbing to the top, one can see wild reindeer.

Тар болтоко – урэкачан бивки мукчэрин. Болтококон – нйукучукокон урэкачан нйан мукчэри. Болтококун – һэгдыкун урэкачан нйан мукчэрин. Таргачирду бэйур нйан айат оӊкодйовкил, бэйуктэдйэми ойолон туктыми бэйурэ ичэдиӊас, айакикин нйанэ.

Болтоко – это полукруглая гора. *Болтококон урэ* – это маленькая полукруглая гора горбатая. *Болтококун* – это большая гора полукруглая горбатая. В таких ландшафтах, дикие олени тоже хорошо кормуются, на вершинах можно охотиться и увидеть их, если туда подняться.

Luŋtuku / Луӈтуку

Very pointy high mountain

Очень острая и высокая гора

Luŋtuku designates a pointed high mountain. It is impossible to climb to the very top: reindeer neither graze nor roam there. One only climbs such a mountain to get a better view and see all the surrounding paths that one might take. The slopes are too abrupt, making it impossible to even hunt *makchika*. Humans also cannot hunt there because of the dense forest and shrubbery *(sehi)*. On the other hand, Russians cannot hunt there and exhaust the game: this leaves more animals for the Evenki.
Let us note that this is one of the term-concepts for landscape types also used as a place name (like solokit *and so on). As the term denotes a very noticeable topographic feature, it is used for orientation.*
This is one of the rare moments where a non-native population is noted in the landscape description: they are represented as predators exhausting the game of the Evenki. Nevertheless, the taiga is an area where the Evenki and Russian mostly help and respect each other because, as they explain, 'nature makes them equal'.

Луӊтуку – сулуӊэһинча бивки, гугдакун тар урэ. Тала ойолон онда этанны туктырэ, орор тала эвкил оӊконоро. Таргачикунтыки туктыми ичэччэминун илйавэл ӊэнэми ичэдиӊас. Эллэӊэкун таду олус эврикур, макчикайэ этанны аҕактарэ. Бэйэ этан бултарэ, сэһимил дйапкалдулын. Нйан айа, бэйэл Лучал этарэ бултадйэрэ, орор таду эвкил ӊэнэктэрэ.

Луӊтуку называют острую и высокую гору. На самую вершину такой горы никак невозможно подняться и олени там не кормятся и не ходят. Если залезть туда, так это только для того, чтобы посмотреть куда можно проехать. Склоны слишком крутые, даже не сможешь погонять животного *макчика*. Человек там не поохотится также из-за очень густого леса с кустами *(сэһи)*. С другой стороны хорошо, что русские там не добывают, хоть там не кончают всех зверей, значит запас зверей остается эвенкам).

Sulin / Сулин
1. A pointed high mountain, like *iaŋ*, 2. Any pointed or sharp item.
1. Острая и высокая, возвышенность одинаковый с *йаӊ*, 2. Любой предмет острый.

Sulin is a high and pointed mountain. In such places, it is impossible to travel: the inclines are too abrupt and reindeer will not walk there. One only climbs them until one reaches an observation point. *Makchika* are present, but it is impossible to hunt them.

Сулин – урэ. Таргачиндулы этанны онда ңэнэрэ, эврикурин сокур, орор нйан таду эвкил ңэнэктэрэ. Ичэччэминун ойолон туктыми айа. Олус эврикун, макчикал бивкил таду, бэйэ этан таду бултарэ, орор таду эвкил оңкоро.

Сулин это гора высокая с острой вершиной. На таких горах никак невозможно проехать, склоны слишком крутые и олени тоже там не пройдут. Можно подняться туда только ради осмотра ландшафта. Склоны такие крутые, что только животные *макчика* там ходят. Люди там не охотятся и олени там не пасутся.

Lumburin / Лумбурин

High mountain with a rounded form

Высокая гора с округленной формой *(йаң)*

It is really uncomfortable to establish a campsite in such a landscape type. There are very strong winds and the snow is very deep: it is difficult to walk on. The wild reindeer graze here safely since no human will ever be able to climb the abrupt inclines. In such a landscape type, reindeer can sometimes graze freely. Berries do not grow there and it is impossible to gather mushrooms. One can find *Pinus pumila* cones.

Нулгиктэдйэми таргачимилду эруми. Адыйачэвки нйанэ, суңтамил иманнамилин бивкил, туктыдйами эруми. Бэйур таду айат оңкодйовкил, бэйэл тала эвкил туктываттэ уктыкур. Лумбурилтыки нйан айа орор адылдун тала туктывкил. Диктэ эвки балдырэ, дэвуңнактэйэ эңнэнны тэврэ, нйаңтаксолэ тэвдиңас.

В таких местах очень плохо стоять. И сильные ветра дуют, и снег очень глубокий, и подниматься туда очень тяжело. Дикие олени хорошо там кормятся, люди туда вообще не будут подниматься по крутым склонам, так что они могут спокойно кормоваться. На такой ландшафт иногда олени поднимаются, и им там хорошо бывает. Ягоды там не растут, там грибы есть, но их не собираешь, там только можно собрать шишки.

Iaŋkichan / Йаӈкичан	
Plateau	
Плато	

This is a good landscape for passing from one next river system to the next because there are continual strong winds that always render snow roads visible. As the harsh cold is installed, the snow roads freeze and harden (thanks to the winds) and will endure throughout winter (the winds clean the snow from the road). During the summer, nobody travels in such a landscape type. Domestic reindeer never go there, although wild ones can always be found in such places.

Таргачирдулы туҕэнирду аландйами айа адындйэксовки, һокто окинда ичэбдиҕан. Ӈэнэрэкис иӈинэлдиӈан, тар һоктос доӈкотодйоӈон, туҕэнивэ ичэбдиҕан. Дйуҕа эвкил таргачирдулы ӈэнэрэ, орор эвкил тара уруру. Бэйун талы окинда ӈэнэдиӈан.

Там хорошо переваливать в холодные зимы, ведь там ветер постоянно дует и дорога всегда видна бывает. Ехать удобно когда сильные морозы наступают, ведь там всю зиму дорогу видно бывает (из-за постоянных холодных ветров дорога [т.е. протоптанный один раз снег] примерзает на всю зиму и ветер не дает снегу заваливать дорогу, в отличии от других ландшафтов). Летом в таких местах не ездят, олени туда не пойдут. Дикие олени всегда там будут.

Keteme / Кэтэмэ	Solid ground good for establishing camps and nomadic roads
	Твердая земля, хорошая для стоянки или дорог

Keteme is highly suitable for setting up camp: good reindeer pastures and mushrooms are present. It is also a very good landscape type for travelling. In *keteme*, the ground is very hard. (cf. avlan, emker, amnunna, murki, maŋarne, ai, nepteke, elleŋe)

Тары кэтэмэлдули уринчэдйэми со айа, оӈко окинда талы бивки, дэвуӈнактэ нйан бивки. Ӈэнэктэдйэми нйан айакикин. Кэтэмэду дуннэ маӈакикин бивки.

На такой твердой земле очень хорошо устанавливать стоянку, там всегда есть оленье пастбище и грибы. Ехать по таким местам тоже хорошо. *Кэтэмэ* – земля сильно твердой бывает. (см. авлан, амнунна, мурки, маӊарнэ, ай, нэптэкэ, эллэӊэ, с. 61, 63–64, 73–74, 95, 97, 106–107)

Maŋarnekun / Маӊарнэкун	Very solid and dry ground, good for camp sites and nomadic roads Очень твердая и сухая земля для стоянки

Almost the same as *keteme*, but harder and drier.
Кэтэмэӡэчин дуннэ, маӊатмар, олготмор.
То же самое что *кэтэмэ*, только более твердая и сухая земля.

Ai / Ай	Flat land without trees *(avlan)* and covered by moss *(ialbuka)*, where cranberries *(Oxycoccus)* grow Равнина без деревьев *(авлан)* покрытый ковром мха *(йалбука)*, где клюква растет

In the *ai*, one finds *ialbuka* moss cover. In such a landscape type, the moss is used for smoke fires (it produces good smoke for protecting the reindeer against mosquitoes and horse flies). One can also gather cloudberries or cranberries, which are medically beneficial for heart illnesses. Reindeer can graze on a lichen-like plant in the form of micro tubes (reindeer adore this plant and walk on the *ai* to eat it). Reindeer will eat as much of this as they can very quickly. In such a landscape type, one can hunt snow partridges and wood grouse *(Tetrao urogallus)*. During the summer, people set up camp on the edges of this landscape type (where there is *keteme*) so they do not have to walk too far to obtain *ialbuka* for the smoke fire. During the summer, Evenki keep their campfires lit all day long: they burn *ialbuka* and fresh larch wood to produce smoke and thus protect the herd from mosquitoes and horse flies. This means reindeer do not need any further inducement to come into the camp, especially during good summer weather. In such a landscape type, it is possible to nomadise. One can find this landscape type on river banks or at the edges of small mountains.

Айилдуды йалбукал бивки. Айилду йалбукайэ самӊирду гадйаӊнэнны, нуныдйадан айат самӊин. Таду тар тэвуктэ тэвлэдйэннэнны мйэвандук айа, иӊамуктэл балдывкил. Орор таду оӊкоктодйовкил. Оӊкоӡочинма, нйан урэкан бивки, тара айавувкил. Бултактэдйэми таду элакирэ бултактэдиҥас, орокирэ бултактэдиҥас. Дйуђанилду эҥнэнны горокуртыки уринэ, ай дйапкадун нйан адылдун уриһиннйэҥнэнны, йалбукайэ гаммэн. Таргачирдулы нулгидйами нйан айапчэ ҥэнэвувки. Таргачир бивкил бирал дйапкалдулитын-у, урэкачар дйапкалдулитын-у.

По *ай*, есть ковер мха типа *йалбука*. На *ай* берут ковер мха для дымокуров, чтобы они хорошо дымили для оленей против комаров и оводов. Там еще клюкву собирают, которая является хорошим лекарством от болезни сердца, и морошка растет. Олени там кормятся. Там есть трубчатое растение, похожее на ягель (олени обожают такое и ходят по *ай*) и олени хорошо наедаются этим. Там охотятся на куропатку и на дикого глухаря. Летом стоят недалеко от таких мест, но на краю (где сухая земля *кэтэмэ*), чтобы далеко не ходить за ковром мха *йалбука* для дымокуров. Летом эвенки держат весь день костры покрытыми ковром мха *йалбука* и свежей лиственницей, чтобы дымить на оленье стадо и защищать его от комаров и паутов. Таким образом, олени сами приходят на стоянку, как только солнце встает при хорошей погоде. Таким образом, оленей и держат подальше от хищников. В таком ландшафте, кочевать тоже хорошо. Такой ландшафт бывает по берегам рек, или по краям небольших гор.

Iɣak / Иђак (UN)
A stony place (a mountain or flat ground)
Каменистое место (и гора и ровный ландшафт)

Aŋkaiku / Аҥкайку (UN)	A stony mountain
	Каменистая гора

One never nomadises in such a landscape type because of the huge stones placed on top of each other *(iɣak)*. During the summer, reindeer do not walk in such areas. Even for humans, it is very uncomfortable to travel here. If it rains, it is impossible to walk because of the slippery stones. Sometimes reindeer roam in such locations during the winter. These places can be useful for observing where one can establish a camp site. It is very difficult to hunt wild reindeer here.

Таргачирдулы бэйэ эвки нулгирэ, талы иҕаккун, тар дйолоксокун, дйолокур һэгдыкур бивкил. Орон талы этан ңэнэрэ дйуҕаниду. Гиркуктадйаминун бэйэ нэһилэ гиркудиңан. Тыгдэрэкин эңнэнны талы ңэнэрэ, дйолол балдакикур. Туҕэниду орор адылдун урувкил. Ичэччэми нйан айа, таргачирдулы иду уринда можно, ойолон туктыми. Таду бэйуктэдйэми эруми.

Вдоль таких мест не кочуют, там очень большие камни сложены друг на друге *(иҕак)*. Олени там не ходят летом. Даже человек, если ходит, то пешком там. Как только дождь идет, там никак невозможно ходить из-за очень скользких камней. Зимой олени там иногда ходят. Такое место является хорошей точкой для обзора местности, чтобы узнать где можно стоять. Там очень неудобно охотиться на диких оленей.

Nepteke / Нэптэкэ	Even or flat (top of a mountain or any ground), any flat/even landscape Ровный (вершина горы *ойо*, земли, камни), любой ровный ландшафт

It is comfortable to hunt sable in this landscape type. Nomadising is quite easy in such a flat landscape, especially since one can find *keteme* (solid and dry ground) and *kever* (tussock fields). It is comfortable to drive the herd (cf. diagrams at the end of this chapter).

Таргачирдулы андаһидйами айакикин. Нулгидйами айакикин нэптэкэлдули, нэптэкэлдули окинда кэтэмэл бивкил, кэвэрил нйан нэптэкэлду бивкил. Таргачирдули илбадйами айа.

По такому ландшафту хорошо охотиться на соболей. Кочевать тоже очень хорошо по ровному месту, так как всегда там твердая земля *(кэтэмэ)* и можно там найти кочкарник *(кэвэр)*. По таким местам удобно гнать оленьи стада. (см. схемы, с. 111–122)

Nepteke keteme / Нэмтэкэ кэтэмэ	Very dry; such flat, solid, and dry soil is excellent for setting up tents Земля сухая и ровная, идеальная для стоянки

This ground is very dry, flat, and solid, which makes it an excellent location for establishing a camp site.

Таду абдун олгокикин бивки, уринчэдйэми айакикин, нэптэкэкикин, кэтэмэкикин.

Там место для стоянки всегда очень сухое бывает и стоять там очень хорошо: ровная, твердая и сухая почва.

Nepteketmer / Нэптэкэтмэр
Low mountains, almost flat
Горы низкие, почти ровные

Such a landscape type is a good place for nomadising. The inclines of a *nepteketmer* are not too abrupt: they rise quite gently. It is a suitable place for hunting wild reindeer. Along such landscape types, one can find wide and flat river basins *(amunna)*. Berries can be gathered here.

Таргачирду нйан нулгидйами айа. Нэптэкэтмэр эчэ олус нэптэкэ бирэ, нэhилэ эври бивки, бэйуктэдйэми таду нйан айакикин. Адылдули таргачирду амуннал бивкил, диктэкакун адылдулин бивки.

Там удобно кочевать. *Нэптэкэтмэр* являются не слишком ровным, с легким, еле заметным подъемом, и охотиться на диких оленей там очень хорошо. По некоторым из этих ландшафтов есть ровные бассейны реки *(амунна)* и на некоторых есть ягоды.

Nire / Нирэ	Marsh with grass
	Марь, травянистое болото

Chuka (long grass) grows in such a landscape type. One can also find *leva* swamps nearby (these swamps are covered in a carpet of grass and soil which floats on top of a large quantity of water). During the summer, wild reindeer come to this place

from the mountains either in the evening or at the crack of dawn: hunters lie in wait at this location. It is also a comfortable place for nomadising during the summer and winter. Sometimes, one can find lakes in such a landscape type. The *chuka* contains a lot of sap, making this a good grazing spot for elk and wild reindeer. One never sets up a camp there because it would frighten away the elk and wild reindeer. Thus, one must stay very far away from such places. This type of landscape is in the so-called *kanu* zone. (cf. Leva, diagrams at the end of this chapter.)
On some of the *nire*, there are roots that reindeer love to eat in the spring. Herders have a strong preference for setting up their camps in such places (in Amudiγa, for instance). These roots are around 15–20 cm in length and carrot-like, although they are white in colour.
The Kanu *zone corresponds to several zones in the taiga that the Evenki leave free (as much as possible) from human tracks in order to preserve big game populations. The Evenki go there exclusively for hunting, during which they keep their camps far away from the area.* Kanu *zones are the diametric opposite of zones with camps and pastures, which are known as* beiechi dunne *('land with humans'). This behaviour can be considered a technique of management or possibly the 'construction of the wild landscape and fauna, similar to agriculture or even herding' (Lavrillier 2005–2006, 2008, 2011).*

Таду тар чука балдыдйавки, нирэлдули лэва бивки. Бэйур тара тар эвдйэвкил дйуӡа, кэӡа эвувкил, ңарилдйэрэкин эвувкил, таду тар карабдйаңнэнны, бултаңнэнны. Нулгиңнэнны нйан таргачирдулы айадулдулы. Туӡэ нйан ңэнэңнэнны. Адылдун, таргачирду амутыл бивкил. Таду айакикин чука бивки бэйур-да айа, коңнокур-да айа. Таргачирду эңнэрэп уринчэрэ, бэйур эдатын уруру, коңнокур урудйуӡорэ эңнэнны уринчэрэ. Городу уринчэдйэңнэнны таргачир эммэн бэйурбэ олонмуканэ. Таргачин кануду бивки.

Там длинная трава *чука* растет, вдоль мари *нирэ* иногда бывает там болото *лэва* (похож на ковер травы и земли лежащий на большом объеме воды). Дикие олени туда спускаются с горы летом, вечером спускаются или как только рассветет и там их караулишь, чтобы на них охотиться. Это тоже очень хорошие места для кочевки (летом). И зимой тоже по таким ландшафтам ездят. Иногда в таких местах есть озера. Там очень сочная длинная трава *чука* и, следовательно, это хорошее место для лосей и диких оленей. Там не устанавливают стоянки, потому что это мешало бы приходу диких зверей. Дикие олени не приходили бы, и лоси ушли бы навсегда оттуда, поэтому там не стоят. Очень далеко от таких мест надо стоянку устанавливать, чтобы не напугать диких оленей. Такой ландшафт находится в эвенкийской зоне *кану*. Эвенки описывают такие зоны как «невидимые для глаз человека» или «внутри густого леса *сэhи*». *Зона* кану *является зоной тайги, которую эвенки оставляют как можно больше без присутствия и следов человека, для того чтобы там поддерживать популяцию дичи. Эвенки туда приезжают только на охоту. Такую территорию называют* кану. *Это можно считать техникой организации и даже «конструкции» ландшафта и связанной с ним фауны, наподобие техники сельского хозяйства или даже «животноводства», только в дикой природе*) (Лаврилье 2010). (см. лэва, с. 100, см. схемы, с. 111–122)

Leva / Лэва	Marsh, swamp; more precisely, those parts where one can sink or where a carpet-like vegetal cover floats on top of the water
	Болота, топи, те части болота, в которых можно утонуть или как ковер

Sometimes, this landscape type becomes a *kudu* (a salt marsh constituted by a mixture of sand and earth). Elk, wild reindeer, and red deer create this landscape (cf. Kudu) and regularly visit it. Domestic reindeer do not visit, however. It is crucial to stay very far away from such areas in order not to frighten away the game. These landscape types are also in the *kanu* zone (a remote and wild non-human zone reserved for hunting) (cf. Nire, kudu).

Таргачинду адылдун кудул овкил, тыталакур овкил. Таргачирду тар коңнокур кудул одйавкил, кумакал, бэйур адылдун эмэвкил. Орор тала эвкил нэнэктэрэ, таргачирдук горокунду уринчэда нада, эдатын уруру. Таргачин нйан кануду бивки.

Некоторые из этих ландшафтов, становятся *куду* (наподобие земного-песочного солонца). Это лоси создают эти солонцы *куду*, но и также благородные олени и дикие олени (см. куду). Олени в такие места не ходят. От таких мест обязательно надо очень далеко стоянку устанавливать, а то люди будут спугивать дичь, которая с концом уйдет. Такой ландшафт тоже находится в зонах *кану* (см. с. 98–99, 101–102).

Debge, debg / Дэбгэ, дэбг	Soft and dry soil with a *ialbuka* moss cover in the forest
	Мягкая сухая земля с ковром мха *йалбука*, находится в лесу

During the winter, one can set up tents in this landscape. The *ialbuka* moss cover and the soft soil facilitate the placement of the wooden structure of the tent. If staying in such a landscape type during winter, the heating stove will be placed into the ground so that it is lower than the edges of the tent: the warmth of the fire slowly melts the ground under the stove and the heat will spread throughout the tent thanks to the upward circulation of the warm air. Such a landscape type exists in spruce forests. There will always be *sivak* grass and sometimes *Lonicera caerulea* berries. The landscape makes transportation easy. Wild reindeer graze there. In such a landscape type, there are many opportunities to hunt sables because they find a lot of wild mice.

Туӈэнун адылду палаткалви таду наӈнэнны. Тэлбэл айакикинди илэвкил йалбукалдула айамамат илэвувки. Таду тар уринчэдйэми печка эргиски урудйэвки, нйамадйавки мандун дйапкалыдулын, оӈчука овки. Таргачир бивкил иду тар аһиӈил бивкил. Сивак окинда бидиӈан, утумуктэ адылдун балдывки. Оӈко нйан бивки. Таргачирдулы ӈэнэдйэми нйан айакикин. Бэйур нйан бивкил, оӈкодулин. Андаһил туӈэ аӈактадйами айакикин, таргачирду тэпэрэкарэ бакадйавкил андаһил.

Зимой иногда можно ставить палатку на таком. На ковре мха *йалбука*, хорошо стоять и структура палатки хорошо втыкается в эту мягкую почву. Если там стоять (зимой), печка будет (пока от теплоты печки замерший ковер мха тает день за днем) в землю вниз опускаться слегка (ниже постельных мест), и тепло пойдет вверх по всей палатке (раз теплый воздух вверх идет), приятно бывает. Такие места бывают, где есть ельник. Там всегда будет хвощ полевой *(сивак)* и жимолость иногда растет. Там также олений корм бывает. И это очень удобно для езды. Дикие олени там ходят, кормятся. Соболи там хорошо охотятся зимой, там они находят много мышей.

Kudu / Куду

An area of naturally salt marsh made from a mixture of sand and soil

Зона природного солонца со смесью земли с песком

Elk, red deer, roe deer, and domestic reindeer lick the *kudu*. This is made from a mixture of salt sand and soil *(tytala)*: it functions as a grazing space. This means that hunters will lie in wait close by in order to shoot at game.

During the winter, the salt content of the rivers means that they do not wholy freeze over; thus, the Evenki are very careful when travelling in such places, since they do not

want to fall through the ice. This term is also used as a place name for several locations. (cf. Leva, Nire)

Таргатирду коӊноку кудудйэвкил, кумакал кудудйэвкил, гипчар кудудйэвкил. Таргачирду тар тыталакун турукэчи овки, тара дйэпивкил овки таргачинду карабдйаӊнэнны.

На таких ландшафтах, лоси ходят по солонцу и облизывают соль там, также как и благородные олени и косули или домашние олени. Это сделано из соленой смеси песка с землей *(тытала)*, которая является кормом, там и караулишь дичь. (см. лэва, нирэ, с. 100, 98–99)

Tytala / Тытала	A naturally salty mixture of sand and soil Природный соленая смесь земли с песком

A salty mixture of soil and sand *(tytala)* can be found in natural salt marshs *(kudu)*. The elk create this landscape type when they repeatedly walk over such ground to feed on the salt. The landscape type is made of sand, soil, and mineral salts. *Tytala* often appears in *leva* marshes.
Some animals, especially elk, are considered to master or transform the landscape by their behaviour and create new landscape types (cf. also kudu). During the interviews conducted by the authors of this book, several Evenki nomads stressed that with the decline in the domestic reindeer population, the landscape has changed. Indeed, they argue that a herd of 500-1500 heads of reindeer is required to reduce the over growth of the vegetal cover. Some Evenki nomads link the significant decline in the reindeer population (herds now consist of 20–600 heads) to the transformation of their landscape into very bushy and dense vegetation. Such a landscape is considered to complicate herding and hunting, as well as transport. It confirms the expression of several nomads: 'the Evenki and their reindeer create the taiga; if there were no Evenki, the taiga would not exist as such'. First, these points of view contradict the idea (widespread in general anthropology) that nomadic people do not transform their landscape, or do so very little. Second, it also brings the anthropological idea that only humans can model landscapes into dispute. In addition, this transformation of the vegetal cover that every Evenki (be they villagers or nomads) has noticed is attributed to climate change. (cf. Leva, Kudu)

Тытала – кудуду тар таду бивки, коӊнокур таптавкил турукэчи бивки тары. Тукала дуннэнун овки тар тытала турукэнун. Тытала – тукала дуннэнун овки, лэвалду.

Соленая смесь земли с песком *(тытала)* находятся в зоне природного солонца *(куду)*. Это лоси создают такое, когда постоянно ходят там, это соленая почва. Это песок вместе с землей и солью становится *тытала*. *Тытала* появляется часто на болотах типа *лэва*. (см. лева, куду, с. 100, 101–102)

Kerain / Кэраин

A small hill with a flat, long top

Маленькая горка с плоской и длинной вершиной

Kerain is a small hill (1 to 10 metres high). On the top, there is a small *ukty* (the end of an incline which abruptly turns into a flat summit). It is very useful to set up camp in such a landscape type, especially during winters when the snow is too deep elsewhere.

In such places, reindeer graze well and do not seek to leave. Indeed, winds move the snow away, freeing up the plants for grazing (in the curved part of the *ukty*); thus, both wild and domestic reindeer can eat well here. It is easy to set up a tent because of the relatively shallow snow. In February and March, one can stay for two to three days in a camp established alongside a road for transporting food and goods from the village to the storage houses placed next to the annual nomadic route. In such a case, one will not stay too long. There is always food for the reindeer. It is forbidden to camp there during the summer because thunderstorms can kill reindeer and humans. *Kerain* attracts electric storms. (cf. diagrams Kever, Vegetal cover typology: Reindeer grazing versus topography)

Throughout this set of topographic landscape types, we see how precise and detailed the collection of concepts that define the different types of mountains, hills, inclines, and slopes are: indeed, these concepts are so precise that they are difficult to translate or even describe. This differentiation between inclines, slopes, hills, and mountains is of a crucial importance in determining the depth and quality of the snow, as shown in most of the descriptions provided. Kerain, *for instance, presents an alternative in those years when the snow is too deep: it always offers herders a place with accessible pasture for their reindeer. This represents a specific and indirect use of the topographic landscape, which is is allowed only by knowledge of the interactions between different ele-*

ments of the landscape (topography, precipitation, micro-climates, and vegetal cover). In addition, please note that there are several kinds of 'deep snow'. There is deep snow that threatens herding and the reindeers' ability to dig through the snow cover in order to graze. The Evenki avoid staying in landscape types with 'snow conditions that are too deep' (e.g. river banks, deran, naldy, arbun solgu, iken, iken, *cf. Snow and ice typology). There is another kind of 'deep snow' used to keep reindeer gathered in one area: the snow is used as 'a herder' and as a protective barrier against predators, since predators cannot enter the deep snow. (cf. diagrams Snow as a herder, Ukte, Kever, and Lavrillier and Gabyshev 2017)*
Кэраин тар урэкачан нйукучуһинчакан бивки. Тар ойолдун уктыкар бивкил. Тара тар кэраинду уринчэдйэми нйан туӡэ айакикин: иманна суӈтакун уӈкту дуннэду бирэкин, орор кэраинду айат оӈкодйовкил. Таду тар адын анавки иманнавэ, таргачинду оӈкодйовкил орор-да, бэйур-да. Палаткэйэ ойодун тэлэӈнэнны имакунди иманна арбадукин. Орор таду айат оӈкодйовкил иманна арбадукин. Таргачирду мартаду, февральду илгимадйаӈнанны, таһаӡаспэ урубдйоми, дйэпгалбэ урубдйоми. Таргачирду илэӈнэнны, дйулэйэ- илалэйэ илгимаӈнанны, эӈнэнны горойо бирэ. Таду оӈко окинда бидиӈан. Дйуӡа таргачирду эӈнэнны уринчэрэ ойолду, ойо кэраинду – агды иктэвки оронмо-да, бэйэлвэ-да вадиӈан, кэраин танэвки агдыва.

Кэраин это маленькая горка (1–10м высоты). На вершине есть маленький *укты* (конец подъема, который резко становится ровной поверхностью) и в таких ландшафтах, очень хорошо устанавливать лагерь зимой, особенно когда в других местах снег сильно глубокий. В *кэраин* олени хорошо кормовались и далеко не уходили. И там ветер сдувает снег (на уровне сгиб горы – *укты*) и именно там олени, (и домашние и дикие) отлично кормятся. Рядом можно легко поставить палатку в неглубоком снегу. В феврале или марте, когда перевозят продукты с поселка и развозят по амбарам вдоль годовой кочевой дороги, удобно по таким *кэраин* останавливаться на два-три дня. Там долго не стоят. Там всегда бывает корм для оленей. Летом на вершине нельзя стоять, поскольку там часто бьет гроза и может убить и оленей и людей. *Кэраин* притягивает к себе грозу. (см. укты, кэвэр, с. 105, 60; см. схемы, с. 346–347, 114, 111)

Ukty / Укты	The end of an incline which abruptly turns into a flat summit. On the tops or flat land without trees *(avlan)*, the wind creates either a hard layer of snow or blows away the snow, almost exposing the ground. Конец подъема, который резко переходит на ровную поверхность. На вершинах или от долины без деревьев *авлан*. Там ветер создает твердый наст снега *(чуйур)* или сдувает снег почти до земли.

Ukty is an area where strong winds blow away the snow. Sometimes, such winds leave almost no snow, while on other occasions they can create a hard layer of snow *(chuiur)*. It is very comfortable to hunt wild reindeer on skiis on the hard layer of snow. It is easy to nomadise along the hard snow. By observing the traces made on the snow by the wind, one can easily identify the north (most strong winds come from the north) (cf. diagram Ukty, other diagrams at the end of this chapter).

Укты – Тар иду тар адынмупча иманна адылдулин арбакукан бивки иманна адынмупча, маӊакун иманна. Ойолдулытын киӊначчэми айакикин – маӊакун иманна. Чуйурилдули ойололитын киӊналди ӊэнэӊнэнны, бэйур нйан таргачирду оӊкоктодйовкил. Нулгидйэми нйан таргачирдули айа маӊакун уктыду.

Укты – это где снег сдувается ветром: иногда снега почти не остается, иногда снег твердый (снег типа *чуйур*). Наверху на лыжах идти очень удобно – там снег твердый. По снегу типа *чуйур* хорошо на лыжах по вершинам охотиться на диких оленей, которые тоже по таким местам кормятся. Кочевать тоже хорошо по твердому снегу. Изучая линии, нарисованные ветром на снегу на *укты*, можно сразу определить направление севера – оттуда дует большинство сильных ветров. (см. схемы, с. 114–122)

Tuktyri / Туктыри	Uphill slope Подъем в гору

Tuktyri is an uphill slope which rises along the mountain until the top. It is comfortable to nomadise along such inclines when one desires to reach the top. Along such inclines, it is possible to close the routes taken by wild reindeer. If *tuktyri* is used as a place name, it means that there is a road or path to the top.

Гунэвкил тар туктыдйами урэвэ ойолон. Таргачирдули нйан нулгиктэдйэӊнэнны туктыдйэнны, ойолон иста. Талы бэйуна аӈадйами нйан туктырилдули нйан айа бэйунэ аӈадйами.

Называют *туктыри,* дорогу, которая поднимается по горе к вершине. По таким местам хорошо бывает кочевать, поднимаясь вверх, чтобы достичь вершины. На таких подъемах хорошо перекрывать дорогу диким оленям. Если топоним – *туктыри*, значит там обязательно возможен подъем по горе.

Evri / Эври
A decline
Спуск

Evri is a decline from a mountain into a river basin. One can find nomadic roads and descend towards a small river. There are also wild reindeer. One hunts wild reindeer along the declines.

Эври гунэвкил талы бэйэ эвдйэвки нйэскаки бирала эвуӈнэнны. Адылдулин таргачирдулы һокто бивки, нйан эвуӈнэнны бирала. Бэйурэ нйан таду, бэйуктэдйэӈнэнны эбдйэнэ эврили.

Эври – это где человек спускается с горы вниз в бассейн реки. Вдоль некоторых из них есть дорога вдоль реки и можно также на мелкую реку спускаться по таким *эври*. Там тоже бывают дикие олени, по таким и охотишься на диких оленей, спускаясь по спуску.

Elleŋe / Эллэӈэ
The slopes of hills and mountains with dry soil ground
Склон сбоку горы и горки, с твердой сухой почвой

Along the slopes of mountains and hills, the snow is really dry and soft (the snow type *duiukun*); the snow is not too deep, making it a very good place for reindeer grazing. Wild reindeer also graze well here. In such a landscape type, there is always solid and dry ground (*keteme*). On some of the *elleŋe* slopes, there are *Pinus pumila* trees, pine forests, or burned forests.
As shown in several of the descriptions of topographic landscape types and in the chapter devoted to the typology of snow and ice, the concept of elleŋe *is crucial for determining the quality and depth of snow, as well as the quality of reindeer pastures.*

Эллэӊнилдули иманнан дуйукукан бивки, иманнан арба бивки, арбакукан бивки, орор оӊкокичилдулитын айакикин бивки. Бэйур нйан оӊкодйовкил таргачирду айат. Таргачирду кэтэмэл окинда бидиӊан, болгиктэл балдывкил адылдун. Адылдун дйагдаҕил бивкил, адылдун буҕарил бивкил.

По склонам горы или горки снег мягким бывает (тип снега *дуйукун*) и неглубоким, это очень хорошие места корма для оленей. Дикие олени там тоже очень хорошо кормятся. Там всегда сухая и твердая почва (*кэтэмэ*) будет. В некоторых *эллэӊэ*, кедровый стланик растет, в других есть сосняк, а иногда горелое место бывает.

Daldy / Далды	A place in a river where one can build a dam to catch fish Место в реке где можно перегораживать рыбу

Daldy is a place in a river where one can catch fish by building a dam. There must be a weak current. In such a place, one dams the river so that the fish will not go along the stream. This can be done with stones or with willows. On the middle dam, one can install a fishing basket. *Among the eastern Evenki, this has not been practised for around 20 years, but the Even fishermen of Kamchatka still do this every summer.*

Далды тар далкит. Таргачирдулы эйакар нйукучукокор бивкил, таргачирдулы далыӊнэнны олло эдан ийэргирэ, дйололди далыӊнэнны, адылдун сйэкталди далыӊнэнны.

Далды это место перегораживания реки. Там течение должно быть небольшим. Там и перегораживают реку, чтоб рыба не ушла по реке и его ловить, можно камнями или тальниками, посередине мордушку ставят. *Сейчас так не делают у восточных эвенков уже лет 20, но у эвенов рыбаков Камчатки еще делают так каждое лето.*

Oŋkuchakso / Оңкучаксо	Places with many pits (of different sizes) alongside some rivers: here, underground water flows.
	По некоторым рекам, места с ямами (разного размера) в земле, под низ которого течет подземная вода. Редко бывает.

It is very dangerous to cross through such a landscape type. Reindeer can stumble, drop the packsaddle, or be injured. Even humans can get wounded in such a landscape. One must not cross through such places. Water flows under the *oŋkuchakso* and the nearby ground. There are a lot of dense bushes, so one cannot see anything under one's feet. (cf. diagram Oŋkuchakso)

Оңкучаксо таргачирдулы эдэдйэми эрукун. Орон талы оңкучакла тудиӈан, инива буридиңан-у вавдиңан-у, манын-да бэйэ тунадиңан эруми. Талы тар эңнэнны эдэватчэрэ, эруми талы бивки. Оңкучакилдулы эрэйадун му эйандйэвки, манын дуннэ. Октальікми бивки, эвки оңкучакми ичэврэ.

Через такой ландшафт очень плохо переходить. Проходя там, олень может оступиться, уронить вьюк или пораниться. Человек может сам себе ногу поранить там. Там не переходят, это очень плохое место, под *оңкучаксо* вода течет, под самой землей. Там много густых кустов и под ногами вообще ничего не видно. (см. схему, с. 113)

Iŋa / Иңа	The stony bank of a river
	Каменистый берег реки

This is a river bank made up of small stones. It is always comfortable to cross the river at such places. The current is never too strong.

Тар бирал дйапкалдулитын дйолокор бивкил нйукучукокор. Биранил дйололын, талы тар эдэкичил адылдулын бивкил, эдэдйэми айакикин таду нйан бивки. Талы эйан арбакукан бивки.

Это берег реки бывают с маленькими камнями. По таким камням всегда очень удобно переходить реку. Там течение небольшим бывает.

Arbun / Арбун	The foot of a mountain downstream or upstream of a river confluence
	Подножие горы на устье рек

(cf. Arbun edighu, arbun solgu and diagram / с. 109, см. схемы, с. 116–118)

Arbun ediγu / Арбун эдиҕу	The foot of a mountain downstream from the confluence of small and big rivers / The foot of a mountain at the confluence of small and big rivers, situated downstream from the big river Склон горы, подножием упирающийся в устье реки по течению главной реки вниз, подножие горы вниз по течению главной большой реки

Arbun ediγu denotes the foot of a mountain downstream from the confluence of small and big rivers: in other words, they are found at the confluences of small and big rivers, situated downstream from the big river. In such places, the sun always warms the ground, meaning that the snow melts quicker during the spring: it is thus a pleasant place to camp. The wild reindeer also spend some time on the ground where the snow melts *(iliakaγil)*: hunters search for them there. It is also possible to bring your herd here so they can graze. (cf. diagram Arbun solgu, arbun ediγu)

Арбун эдиҕу – тар урэр манапдйарил, бира даптулын бивкил. Тар урэкундули эдиҕулин. Окинда дылача дылачадйэвки, талы имакунди иманна унивки, талы тар уринчэдйэми нйан со айа. Бэйур нйан бивкил. Илйакамэлчэвки таду. Илйакаҕилдули бэйурэ арычиңнанны, ороро онодйоннэнны.

Арбун эдиҕу это подножие горы на устье реки, которая стоит вниз по течению главной большой реки. Всегда там солнце греет и снег быстро тает; там ставить лагерь очень хорошо. Дикие олени там тоже бывают, стоят на той части земли, где снег растаял весной *(илйакаҕил)*, и там их ищешь и туда можно оленей пригнать, чтобы кормовались. (см. схемы, с. 116–118)

Arbun solgu, Arbun gonne / Арбун солгу, арбун гоннэ	The foot of a mountain upstream from the confluence of small and big rivers. / The foot of a mountain at the confluence of small and big rivers, situated upstream from the big river Склон горы, подножием упирающийся в устье реки по течению главной реки вверх / Подножие горы, которое упирается в устье реки по течению главной реки вверх / То подножие горы вверх по течению главной большой реки

Arbun solgu refers to the foot of a mountain upstream from the confluence of small and big rivers: in other words, they are found at the confluences of small and big rivers, situated downstream from the big river. During the spring, the snow always melts slowly in such places. This is a place where one can safely leave the herd to graze without fear of loosing them (i.e. you can follow their tracks in the snow). In such places, there are rich pastures. (cf. diagram Arbun solgu, arbun ediγu)

Арбун солгулын, нйан даптудун урэ манапдйари, оӈоктодун. Тара тар «арбун» гунэвкил. Талы тар иманна эвки унмэлчэрэ. Тала тар орорбо нйан тинэӈнэнны бакаммэн ороробо. Талы иманначи бивки. Талы нан орор оӈкококун бивки.

Арбун солгу – это подножие горы, упирающееся на устье реки, которая стоит по течению главной реки вверх. Там снег никогда быстро не тает и можно оленей отпустить весной, чтобы их не терять (ведь можно их найти по следам в снегу). Там много оленьего корма. (см. схемы, с. 116–118)

Kever (tussock field) is related to ***kerain*** (a small hill with a flat top) (indigenous geomorphology)
Кэвэр (кочкарник) связан с *кэраин* (горка с плоской вершиной) (коренная геоморфология)

Kerain does not allow water to flow away: the water stagnates in the flat land and forms *kever*, the best place for spring camps / *Кэраин* не дает воде вытекать – вода застаивается на ровном месте в *кэвэр* – лучшее место для весенней стоянки.

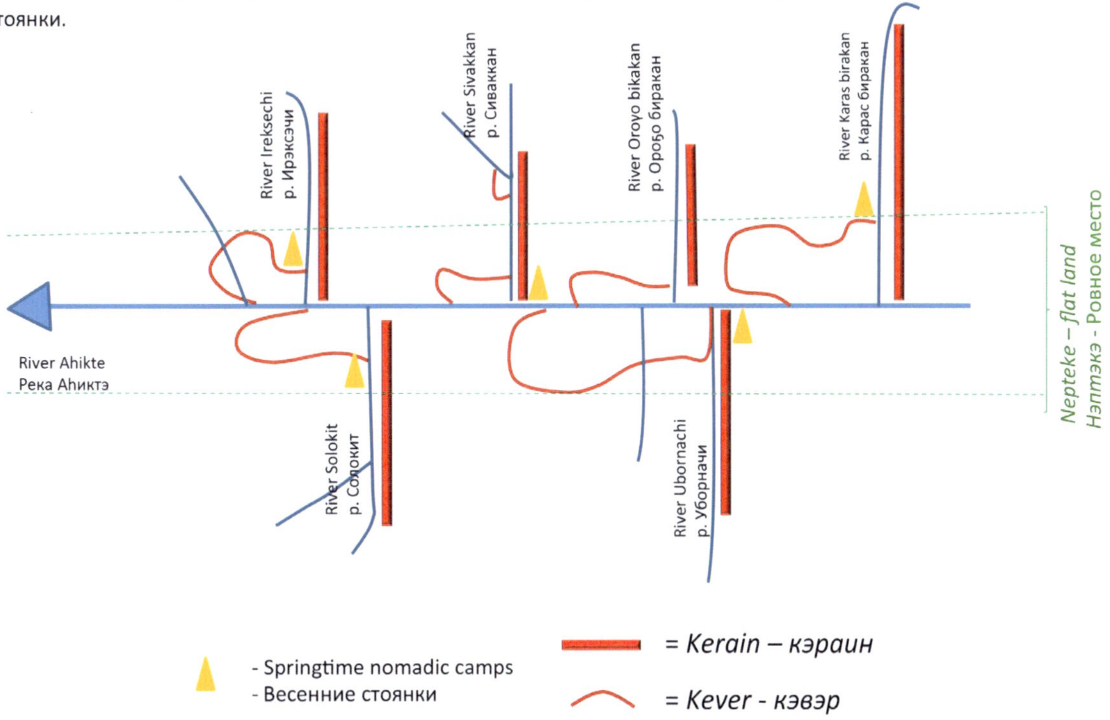

Cf. Vegetal cover typology.

Naldy – A small tributary, the length of which is equal to the part of the main river between the confluence with this tributary and the source of the main river (upstream). This does not depend on the position of the observer.

Налды – Небольшой приток реки, длина которого одинакова с длиной части главной реки: от устья данного притока до истока главной реки (вверх по течению другой реки) – где бы не стоял говорящий.

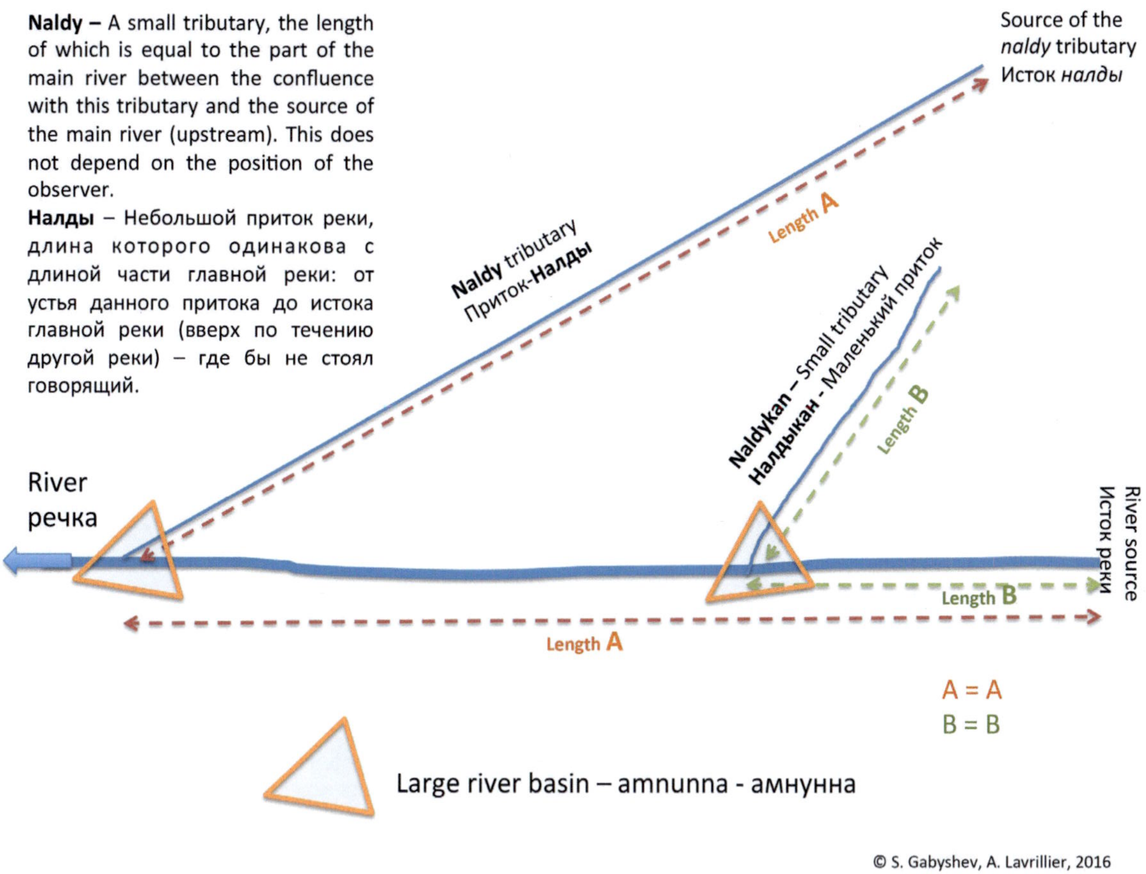

Oŋkuchakso - Оӈкучаксо

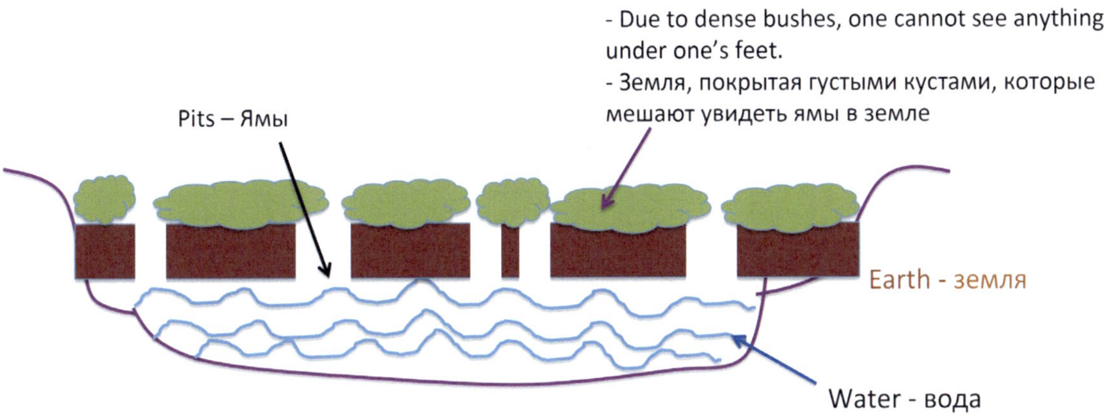

Along some rivers, there are places with many pits of different sizes: underground water flows here / По некоторым рекам места с ямами (разного размера) в земле, покрытые густыми кустами, под которыми течёт подземная вода. Редко бывает.

Ukty – the end of an incline which abruptly turns into a flat summit.
Укты - Конец подъема, который резко становится ровной поверхностью.

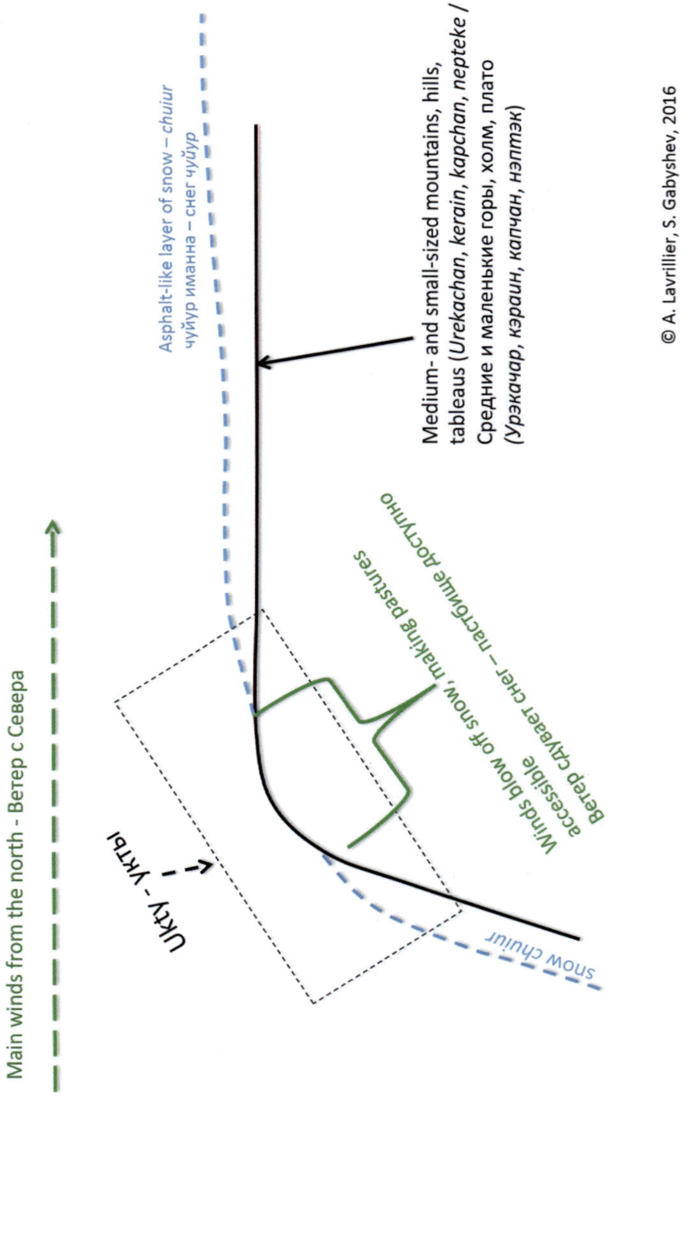

Liopikel – very small streams/ лйопикэл / маленький ручей
Small river / биракан / речка
01/03/2014 04:50 PM

At the confluence of a small river with a big river, the main winds come approximately from the north (because big rivers mainly flow from south to north). So, when looking at the small river from the point of view of the confluence (like in this picture), there are two types of mountain or hill feet: the *arbun ediyu,* situated downstream of the big river, and the *arbun solgu,* situated upstream of the big river.

На устье большой реки основные ветра идут, примерно, с Севера (большинство больших рек текут с Юга на Север). Значит, если смотреть от устья главной реки на маленькую реку (как здесь на фотографии), есть два типа подножия гор или горок: *арбун эдиҕу* – вниз по главной реке и *арбун солгу* – *вверх по главной реке.*

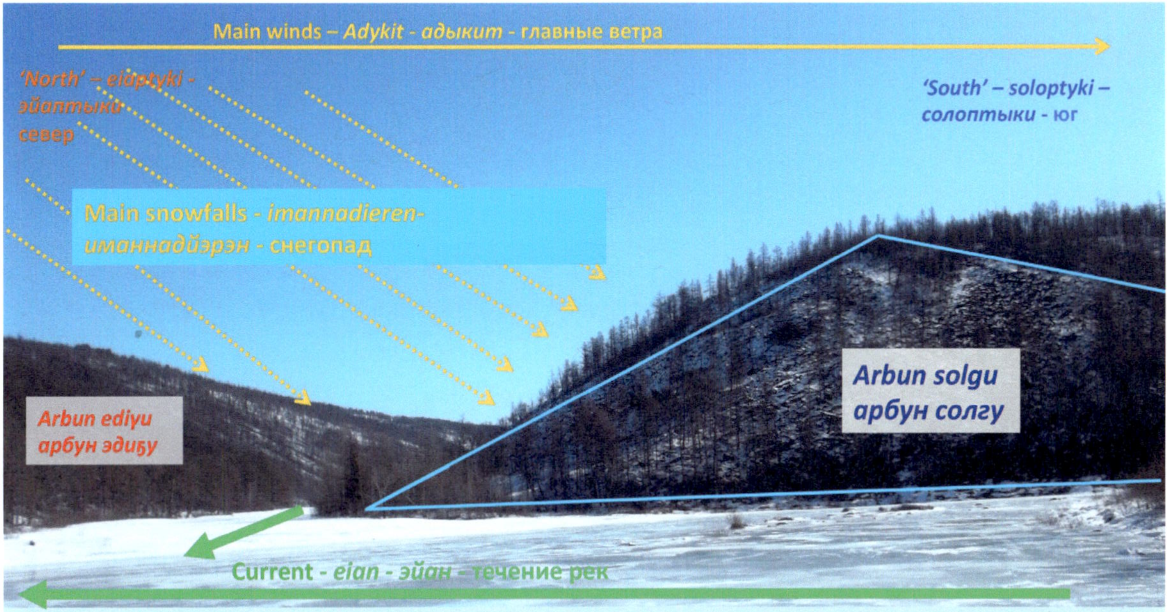

The *arbun solgu* mountain or hill foot receives more snow than the foot of the other mountain or hill (*arbun ediyu*) because of the northern origin of the main winds: however, as shown in the third slide, the *arbun ediyu* foot receives more sun. The Evenki deduce several factors from the position of mountain feet: the depth and quality of the snow, the quality and quantity of the vegetal cover (including that for grazing), and, consequently, the presence of wild and domestic animals (*cf. next slide*).

На подножие горы *арбун солгу* из-за того, что большинство ветров идут с Севера, снега выпадает и накапливается больше чем на подножии другой горы (*арбун эдиҕу*). Как видно на третьей диаграмме *арбун эдиҕу* зато получает больше солнца. По этим признакам эвенки делают различие по глубине и качеству снега, и также по качеству растительного покрова между подножиями этих гор, которыми они пользуются в своей повседневной жизни (*см. след. диаграмма*).

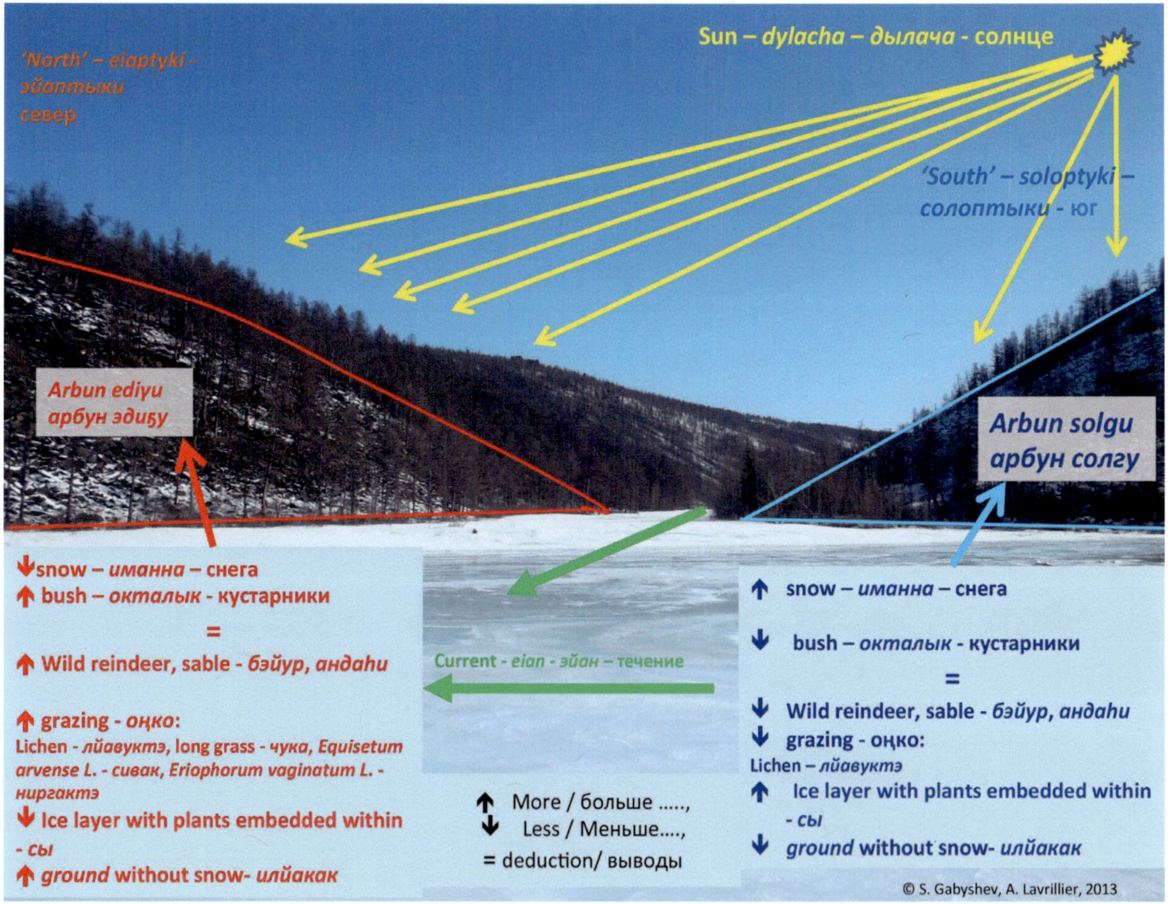
Sun – dylacha – дылача - солнце
'North' – eiaptyki - эйаптыки север
'South' – soloptyki – солоптыки - юг
Arbun ediyu арбун эдиђу
Arbun solgu арбун солгу
↓snow – иманна – снега
↑ bush – окталык - кустарники
=
↑ Wild reindeer, sable - бэйур, андаhи
↑ grazing - оңко:
Lichen - лйавуктэ, long grass - чука, Equisetum arvense L. - сивак, Eriophorum vaginatum L. - ниргактэ
↓ Ice layer with plants embedded within - сы
↑ ground without snow- илйакак
Current - eian - эйан – течение
↑ More / больше .....,
↓ Less / Меньше....,
= deduction/ выводы
↑ snow – иманна – снега
↓ bush – окталык - кустарники
=
↓ Wild reindeer, sable - бэйур, андаhи
↓ grazing - оңко:
Lichen – лйавуктэ
↑ Ice layer with plants embedded within - сы
↓ ground without snow- илйакак
© S. Gabyshev, A. Lavrillier, 2013

Reading landscapes (1)

Reading landscapes (2)

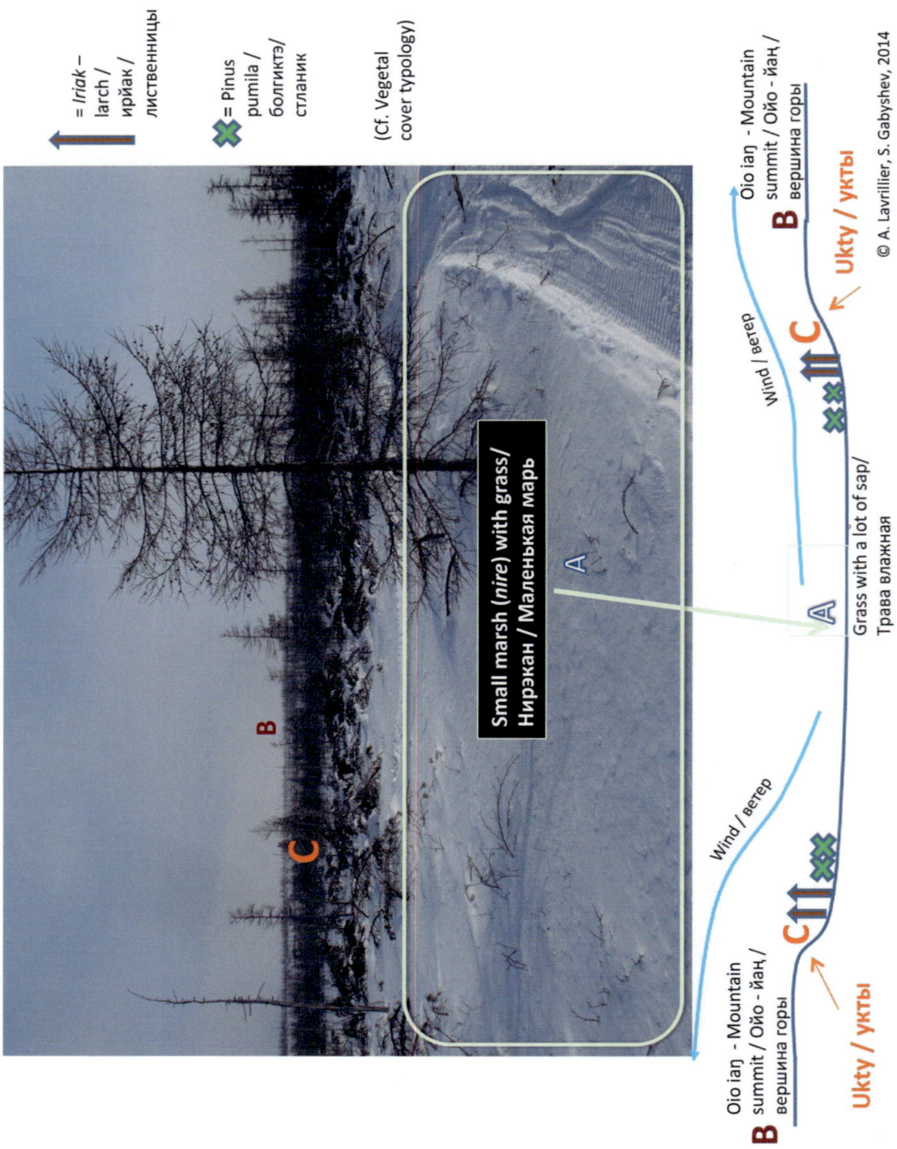

Reading landscapes (3)

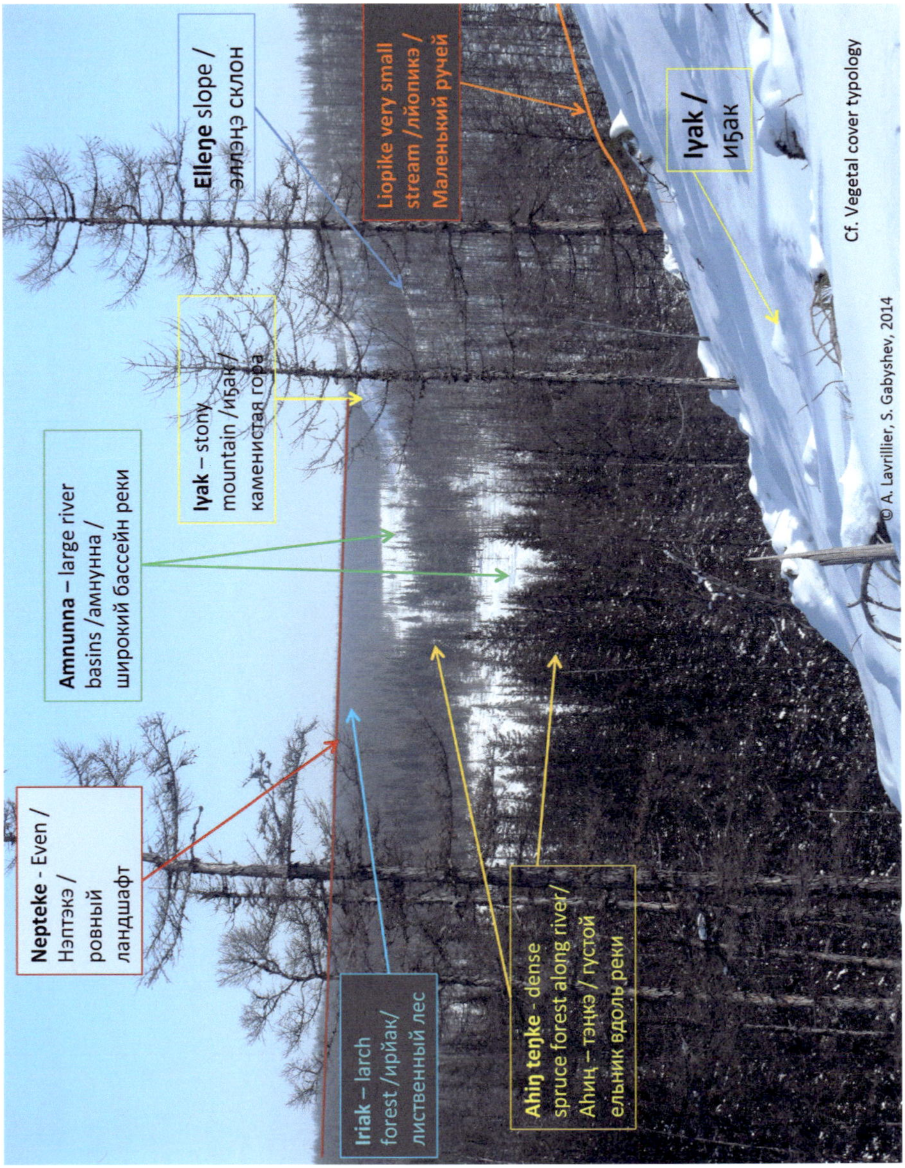

Reading landscapes (4)

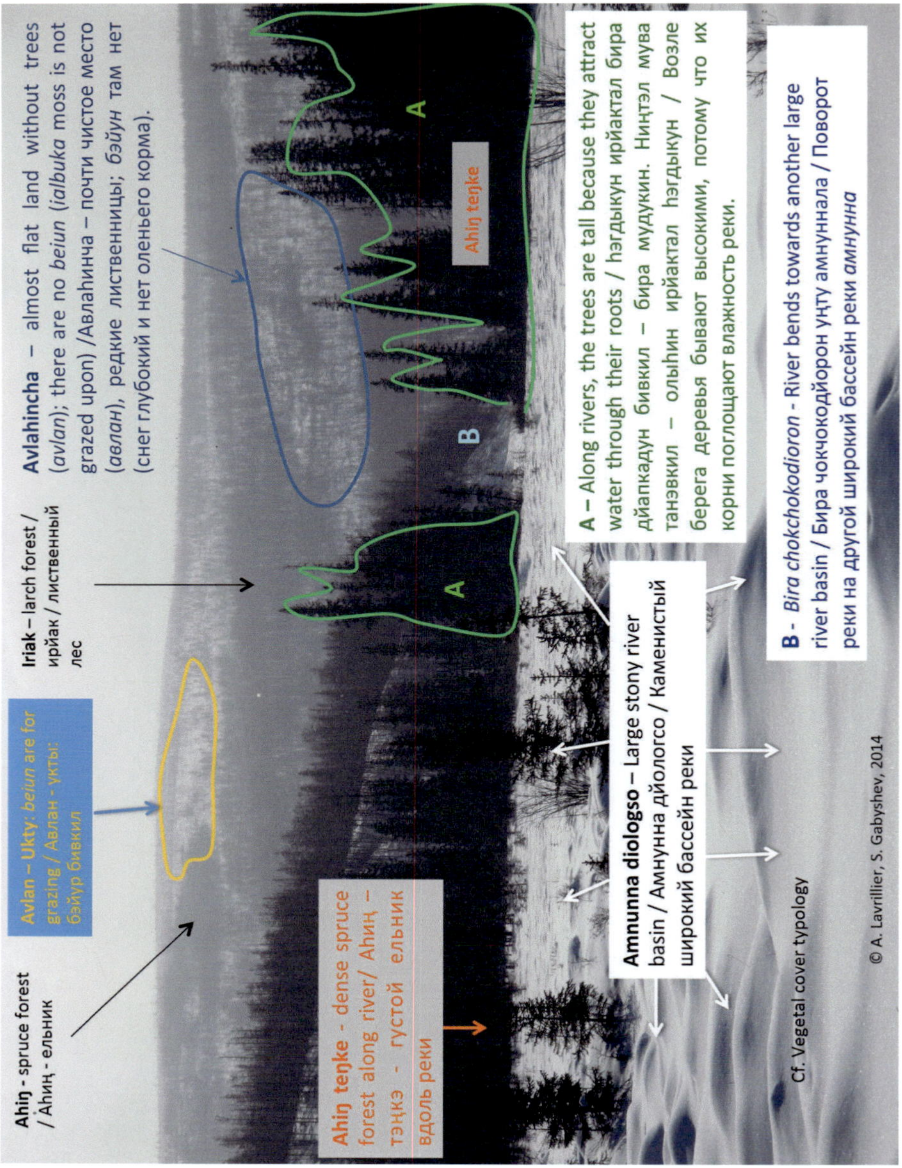

2.2.2 Vegetal cover
by S. Gabyshev and A. Lavrillier

Diagdag / Дйагдаг

Pine forest

Сосняк

In a pine forest, the vegetal cover on the ground is extremely light: there are no bushes. Thus, it is a very suitable place to look for wild reindeer and hunt them. It is also easy to hunt sable. In such a forest, one can always find wood grouse (also capercaillie, *Tetrao urogallus*) to hunt. Pine trees are very easy to chop apart so that one can make boards for sledge construction. In such forests, there are a lot of berries, pastures for reindeer, and mushrooms.

Дйагдаҕилдулы олус эвки балдырэ, окталык эвки бирэ, талы тар бэйуктэдйэми айа, арытчада бэйурэ айа, андаһидйами талы нйан со айа. Орокил таду окинда бивкил, таду бултадйэми айа. Дйагдал айакикинди йолторговкил таду тар, адарайил айакикин одйами. Таду тар диктэкакун бивки, оӊко нйан кэтэкакун бивки, дэвуӊнактэкакун.

В сосняках мало растительного покрова, нету кустов, например, и удобно там охотится на диких оленей, легко там дикого оленя искать и на соболей тоже там хорошо охотиться. В сосняках всегда бывают глухари и там охотимся на них. Сосна очень легко раскалывается, из него доски хорошо делать для нарт. В сосняках очень много ягод, также, как и корма для оленей, и грибов.

Chalbuk / Чалбук

Birch forest

Березняк

Elk eat birch, so one can find and hunt them in a birch forest. One can also find hazel grouse and *makchika*, which use the forest to hide. In such a forest, one takes wood from the trees in order to build skis for sledges: one must carve the wood into a curved shape. Also, one takes long and thick sticks from these forests in order to make the long horizontal parts *(battyk)* of sledges. Equally, the wood can be used to make axe handles, knives, and *cherkan*, a wooden trap for sable. Birch wood is the strongest wood in the region and is excellent for making almost anything. Birch bark can also be used to make fires, immek paddlesacks, *guievun* (baskets for gathering bog blueberries), *mereke*, and small traditional Evenki bags. The reindeer like to eat the large quantities of mushrooms that grow in such woods; however, neither berries nor lichen can be found in these places. It is possible to find *chaga* mushrooms *(Inonotus obliquus)*, which can be boiled and then drank to treat stomach illnesses and to purify the kidneys.

Чалбуҕилдулы коӈнокур чалбукилба дйэпчэвкил, коӈнокурбэ бултадйэӈнэнны, нинакир илэвувкир коӈнокурбэ. Почйокар талы нйан бивкил, макчикал таду тар дыкурэкачэвкил. Таду тар сиӈакилэ гаӈнэнны, матаӈнэнны, сыргалави оммэн, баттыква гаӈнэнны. Эҕирэ сукэлду одйаӈнэнны. Эҕиӈатэлэ, котолду, чарканэ оӈнанны чалбандук. Чалбан маӈатмар уӈтул ирйакталдук, айакикинди экунада одиӈас. Талуйэ тадуккар гаӈнэнны, илавунэ талуйэ оӈнанны иммэкилэ одйаӈнэнны, гуйэвуннэ оӈнанны, мэрэкэйэ оӈнанны, авсакарэ, упкатылва одйаӈнэнны тавадук. Орор таду тар дэвуӈнактэдйэвкил, дэвуӈнактэ кэтэкун, таду оӈко ачэн, диктэ ачэн. Туҕэ чаҕа балдывки, уйупчэннэрэм чаҕэвэ умдйаммэн урду айакикин, экундуда, боhоктоло айат силкивки мандус.

В березняках лоси питаются березой и там можно охотиться на них с собаками. Рябчики тоже бывают и животные *макчика* там прячутся. В таких лесах надо брать сырье для лыж нарт. Их потом загибают в нужную загнутую форму, чтобы сделать полозья для нарт. Также берутся палки для горизонтальной верхней структуры нарт *баттык*. В таких лесах также берут сырье для ручек топоров, ножей и для деревянных ловушек на соболей *чэркан*. Береза – это самые крепкие дрова среди всех деревьев, что у нас растет, и годится для любого изделия. Оттуда же берется береста, которая употребляется для разжигания огня, а также для изготовления вьючных сумок *иммэк*, корзинки для сбора голубики *гуйэвуни*, лодки *мэрэкэ*, и также сумочки. Можно делать много разного из бересты. В таких лесах олени кушают многочисленные грибы. Зато там нету ни ягод, ни корма для оленей. Там собираем чагу, которую кипятим, чтоб пить. Это очень полезно для желудка и для всего организма, в том числе для почек – хорошо прочищает.

Iriak / Ирйак

Larch forest

Лиственный лес

It is possible to manufacture many things from larch, such as tent poles, reindeer enclosures, high platforms for meat storage, and staffs. This wood is also very easy to build with because one can remove the bark without difficulty. In addition, larch is used for the construction of the tall storage houses for food and goods. It is the main type of firewood, because it burns well when dry. One can also gather dried branches from the ground to use as firewood for the campfire outside the tent. One can even manufacture a table from such wood. It is possible to make a tripod for the campfire and for drying meat out; indeed, when it is fresh, it does not burn easily, meaning one can put it close to a fire. If there is no birch, larch can be used to make sledges. Larch is also used to make reindeer hobbles that prevent them from going very far. In such forests, there are always a lot of grazing pastures, berries, and mushrooms: this means a great many sables will be present.

Ирйаҕилду тарылду палатканатту оӊнанны, курэйэ оӊнанны, дэлкэнэ одйаӊнэнны, алдыдйаӊнэнны со айа. Сайбэйэ оӊнанны, мойа гаӊнэнны, олгокирэ айакикин дэгдэвкил. Упкатйэ одйаӊнэнны, остолэла оӊнанны. Гарала тавуӊнэнны илачавунави гулувунду. Гадйаӊнэнны гарала, айакикинди дэгдэдйэвкил. Сонанэ оӊнанны, уллэлйэ локовура, колболо оӊнанны, эдатын чипкачар дйэпчарэ уллэвэс. Адылдун чалбан ачэн бирэкин, сыргала оӊнанны. Чаӊгайэлэ таду одйаӊнэнны орор эдатын горолдйоро. Оӊко окинда бидиӊан, диктэ нйан таду бивки, дэвуӊнактэ нйан бэйур дйэпивкил, андаһил талы бивкил.

Из лиственницы готовят структуру палатки, изгороди для оленей, платформы для вещей и мяса, и легко очищается от коры. Из этого дерева и делают воздушные амбары, где продукты и вещи хранятся. Дрова для печки тоже готовятся из него, потому что сухим оно очень хорошо горит. Также сухие ветки собираем для разжигания костра, очень хорошо разгорается. Можно также и стол сделать, ну и все можно делать. Можно также делать треножник для костра, чтобы мясо сушить, также как и воздушные платформы для мяса, чтоб птички и мышки не кушали мяса. Если нет березы, можно делать нарты из лиственниц. Из них делают путы *чаӊгай*, чтобы олени далеко не уходили. В таких лесах олений корм всегда есть, также как и ягоды, грибы и соболи.

Siektak / Сйэктак
Место, где много тальников
Willow grove

In such places, one looks for elks and red deer because they like to eat willow. There are always small tributaries that join larger rivers in these locations: these are useful for orientating oneself and discovering where it is possible to nomadise. Willows are used to provide the vegetal links for sledges and wooden reindeer collars: these are attached to the *chaŋgai* hobbles that prevent the reindeer from straying too far. Sometimes one can make a *tagan* saddle from willow: these allow for small children to safety ride on reindeer in a sitting position. Water can always be found under the snow here, which is very useful during the winter since it allows one to obtain drinking water without breaking and melting ice.

Таргачирду коӈнокурэ арычиӈнанны, кумакалвэ, сйэктэлбэ дйэпчэвкил; кумакал, коӈнокур сйэктэлбэ дйэпчэвкил. Талы окинда биракан бидиӈан, бира-ҕу бидиӈан. Талы ичэдиӈас окинда или нулгида. Сйэктадук авйэстаӈнэнны сыргалави. Чаӈгайиптэлэ оӈнанны орорду, чйаӈгаил локовурэ, эдатын городйоро. Адылдун таганэ оӈнанны куӈакарду. Талы тар окинда му бидиӈан иманна эрэдун.

В таких местах ищут лосей, благородных оленей, которые кушают тальники. Там всегда бывает речка и рядом с ней река, что поможет ориентироваться при кочевках. Из тальника мы делаем бечевки для связки элементов наших нарт, но и также деревянные ошейники, на которые привязываем путы *чайӈгай*, чтобы олени далеко не ушли. Иногда можно делать рамку *таган*, чтобы ребенка посадить на верхового оленя. И там всегда вода будет под снегом, это удобно зимой и лед не надо долбить и таять.

Sivak / Сивак

Horsetail *(Equisetum arvense L.)*

Хвощ полевой

Horsetail is found in amnunna river basins, spruce forests, and on river banks. Reindeer will eat their fill from such a grass type and thus gain weight for the coming cold season. Reindeer go by themselves toward grazing with such a grass type. (cf. Topographic typology: amnunna, teŋke, daptu, iaŋ, debge, cf. diagram Reindeer grazing versus topography)

Сивак амунналду, аhиңилду бивки, бирал дйапкалдулитын. Орор тала айамамат айавувкил, бургудйувкил айакикинди. Туӡэрдивэр бургувкил, горокуна тарачэвкил, айат ңэнэвкил орор.

Хвощ полевой находится в широких бассейнах рек амнунна, в ельниках и по берегам рек. Олени хорошо кормуются на таких пастбищах и хорошо поправляются. Они там набирают жир, чтобы выдержать зиму, и долго упитанными бывают (быстро не худеют) от такой травы. Олени хорошо идут на такие места. (см. амнунна, тэңкэ, дапту, йаң, дэбгэ, с. 63–64, 87, 74, 90, 100–101, см. схемы, с. 152–158)

Chukalyk / Чукалык

A place with a lot of long grass

Место, где много длинной травой

One can find a lot of long grass, *nire* marshes, and *kever* tussock fields. At the beginning of the spring (just as the snow is melting) and during the summer, reindeer can be found grazing on the long grass. Such grass is also found on river banks. Wild reindeer and elk come down from the mountains to the *nire* marshes to graze on this grass.

Таду чукал балдывкил. Чукалыкил балдывкил нирэлду, кэвэрилду нйан балдывкил. Окинда талы тар чука айакикин бидиңан. Орорду айат дйэпивкил чукава. Дйуӡаниду кэвэрилду, нйэңнэллэкин иманна унэкин дйэпчэвкил чукава айакикиндит. Бирал дйапкалдулитын нйан балдывкил чукал. Нирэлтыки бэйур эвувкил, коңнокур эвувкил.

Там много травы растет. Такие места бывают в болотах типа *нирэ*, но также в кочкарниках *кэвэр*. Там всегда хорошая сочная трава бывает. Летом на кочкарниках, а также, как только весной начинает таять снег, олени хорошо пасутся там и кушают эту длинную траву. Длинная трава также растет по берегам реки и на болотах типа *нирэ*, куда дикие олени и лоси спускаются, чтобы ее покушать.

Ahiŋ / Аhиӊ

Spruce forest

Ельник

Elk and wild reindeer use spruce forests to hide. It is very fresh in such forests, especially during the summer. From spruce, one can manufacture traditional skis. In the winter, it is very pleasant to lay spruce branches on tent floors like a kind of carpet (normally, larch is used for this): they offer good insulation from the cold ground. In spruce forests, there are a lot of bird nests. One must not use spruce for firewood since it produces a very unhealthy kind smoke which can cause inflammation and infection.

Аhиӊилду дыкинчэдйэвкил коӊнокур, бэйур. Талы тар аhиӊдодун соӊутмар. Аhиктэлдук киӊналиви оӊнанны. Туӡэ бирэкин сэкдйэми айакикин аhиктэлдук. Талы тар кэтэкун чипкачар умуктаӡилтын.

В ельниках лоси прячутся, также как и дикие олени. Внутри ельника прохладно бывает, особенно летом. Из ели делают лыжи. Зимой хорошо стелить пол палатки ветвями ели – создается хорошая изоляция. В ельниках много птичьих гнезд бывает. Ель не сжигают из-за дыма, который вызывает инфекцию ран и воспаление суставов.

Bokoto / Бокото

1 Fungus growing on tree bark (larch, spruce, pine, or cedar) Polyporaceae;
2 cones growing on larch branches

1 Гриб на дереве или на елке, лиственнице, сосне или кедре,
2 Шишка на ветви лиственниц

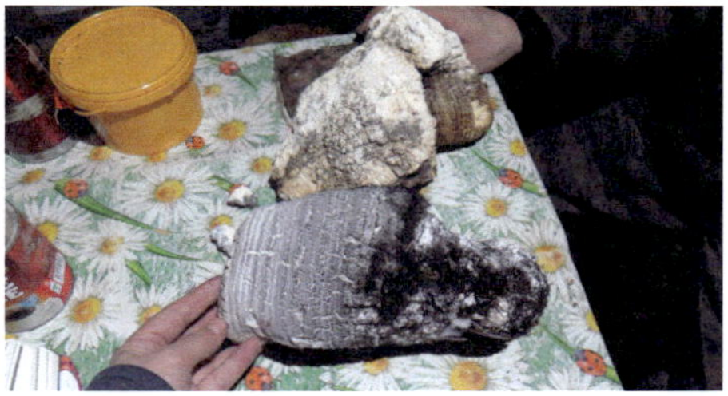

When the winter comes, *bokoto* (a fungus-like plant) starts growing on larch bark. One can boil it and drink it, since the resulting liquid can treat many stomach illnesses. As such, *bokoto* is a form of medicinal treatment.
Please note that the Evenki consider bokoto *to be distinct from the mushrooms species that grow on the ground. According to the nomads, this fungus-like sprout appears on the outside of the tree because the winter cold affects the circulation of the sap. It is considered to be a concentrated form of the tree's properties. Nomads still gather* bokoto *and use it as a medical treatment because of its efficacity.*

Туҕэ оракин, бокото ирйактэлдук йувки, тара тар умдйаҥнэнны айакикин урду бивки. Бэҕэвэ одйаҥнэнны мандуви бокотодук.

Когда наступает зима, гриб *бокото* произрастает на лиственнице, его кипятят и пьют от расстройства желудка, хорошо помогает. Так сами себе готовим лекарство из такого гриба – *бокото*.
Эвенки считают эти бокото *не грибами, как те которые растут на земле, а как отростки, которые выходят зимой из-за изменения циркуляции сева деревьев с холодом, в которых концентрируются все полезные вещества. До сих пор люди собирают эти* бокото *и используют в традиционном лечении.*

Oktalyk / Окталык

Bush; dwarf birch

Кустарник; карликовая береза

When the summer starts *(daldy)*, the reindeer eat the young leaves on bushes and dwarf birches. Along treeless flat lands close to a river *(avlan)*, such bushes are small because of the wind, while in the spruce forest bushes are high, dense, and numerous, which makes travelling very difficult in such locations. Wild reindeer and elk eat the young leaves on such bushes: in the winter, elk eat the bush itself. One can make very beautiful knife handles from the roots of such bushes. (cf. diagram Vegetal cover varia)

Далдырэкин, октакар абданналбан орор дйэпдйэвкил. Авларду октакар нйукучутмар бивки. Аhиҥилду октакар hэгдыкур бивкил, кэтэкун бивки, ҥэнэдйэми эруми талы. Бэйур, коӈнокур нйан дйэпивкил октакар абаданналбатын. Коӈнокур дйэпдйэвкил туҕэ найан манман. Экундутын-кана, тэкэндукин оӈнанны эҕирэ котолду, айакикин овки.

Как только устанавливается лето *(далды)*, олени кушают молодые листья кустов (в т.ч. карликовых берез). На равнине без леса, возле реки *(авлан)*, такие кусты маленькими бывают из-за ветра. А в ельниках кусты очень большие и многочисленные, и туда проехать очень тяжело из-за этого. Дикие олени и лоси тоже кушают молодые листья этих кустов. А зимой лоси кушают сами кусты. Из корней этих кустов можно делать ручки для ножей очень красивые. (см. схемы, с. 148–151)

Bolgik, bolkikun / Болгик, болгикун	
A place where there are a lot of *pinus pumila (bolgikte)*	
Место, где много длинной травой	

Niangta / Нйаӈта	
Cone	
шишка	

Koktychal / Коктычал	Empty cone
	пустая шишка

Bolgik (places with a lot of *pinus pumila*) appear on slopes made up of dry soil *(elleŋe)* and on the summits of mountains. In such places, a lot of cones grow which one can gather. Sable and bears eat these cones, which means that one can hunt sable in such locations if there is a good cone harvest (such good harvests happen every several years or so). During the winter, it is very difficult to cross such places: large *pinus pumila* branches are hidden by the snow and present obstacles to movement. When crossing such places in the summer, one must be very careful because bears can appear unexpectedly. The arcs placed on the front of sledges, reindeer collars, and the frames of *immek* packsaddles are made from these trees. *Pinus pumila* can save lives if there is a need to spend a night outside during the winter. In such conditions, one builds a camp fire on either sand or a large flat stone. Once the fire has heated the surface, it can be extinguished: then one places *pinus pumila* branches on the warm surface as bedding. *Bolgiks* are very warm, even during the harshest cold weather. (cf. diagram Reindeer grazing versus topography)

Let us stress the science of surviving in extreme weather conditions by using even the smallest elements of the surrounding environment.

Болгик – таргачин тар эллэӈэмилдули, ойолдули балдыдйавкил, талы нйаӈтакакунма тавуӈнэнны, андаhил тара нйаӈталба дйэпдйэвкил, амикар дйэпдйэвкил айат. Талы тар бултадйэӈнэнны андаhилба, нйаӈта бирэкин. Туҕэниду эруми бивки болгиктэмил нйэла овкил иманна эрэлан, ӈэнэдйэми эруми. Дйуҕаниду сэрэнчэчэдэ нада ӈэнэдйэми амикакурдук. Болгиктэдук овкил камургарэ сыргалду, чаӈгайиптэрэ нйан оӈнанны. Туҕэни тулинду аӈначалми, тукаладу-у, каптарэн дйоло ойодун-ӈу гулувунэ илаӈнанны. Тадук дйолог окулэкин, гулувунмэ синэнны. Тадук болгиктэ сэҕинэнны ойодун аhиммэн – нйамакикин бивки.

Болгик (место где много *болгиктэ*) бывает на склоне горы с твердой сухой почвой *(эллэӈэ)*, а также на вершинах. Там растет много шишек, которые можно собирать. Соболи тоже едят шишки, также как и медведи. Следовательно, там можно охотиться на соболей, если урожайный год и шишек много там есть (только раз через 3–4 года). Зимой туда проехать очень тяжело из-за того, что внизу под снегом находятся толстые ветви стланика, которые лежат и препятствуют передвижению. Летом, если туда проехать, то надо быть очень осторожным из-за медведей, которые часто там находятся и могут внезапно появиться. Из стланика делают переднюю полукруглую дугу для нарт, также как и ошейники для оленей, чтобы привязывать путы, или еще делают твердую рамку для вьючных сумок *иммэк*. Также стланик может спасти жизнь зимой. Если зимой приходится ночевать на улице, то можно на песке или на большом плоском камне разжечь большой костер. Потом, как только камень или песок нагреются, можно потушить костер, постелить ветви стланика и ночевать на них – очень тепло бывает, даже при морозе. (см. схемы, с. 152–158)

Taktykan / Тактыкан

Cedar
Кедр

Cones grow on the top of cedar trees. These trees are sometimes found in bands close to larch groves and, more rarely, spruce trees. The large cones that these trees produce are food sources for sables, wood grouse *(Tetrao urogallus)*, and spotted nutcrackers, which means these animals are plentiful in such locations.

Тактыкан ойодун нйаӈтал балдыдйавкил, таргачир манак эвкил балдырэ, аhиӈилду балдывкил, адылдун ирйаҕилду, дйагдаҕилду балдывкил тэӈкэвлэҕэчин. Талы кэтэкун нйаӈта, нйаӈталын hэгдыкур. Андаhил дйэпдйэвкил тариӈан тар, оӈолол нйан дйэпдйэвкил. Талы андаhил кэтэ бивки, орокил дйэпдйэвкил, кэтэ бивки.

На верхушке кедров растут большие шишки. Такие деревья редко растут. Кедр растет полосой рядом с елью, а иногда возле лиственниц или в сосняках. На таких деревьях много больших шишек. Соболи любят есть такие шишки, также как и птичка кедровка. Там много соболей бывает, и глухари тоже кушают такие шишки.

Diktak / Диктак	A place with a lot of berries Место, где много ягод

One can find a lot of berries in these places: these can be used to make jam. Such places are situated in wide, flat river basins *(amnunna)*, larch forests, treeless flat lands close to *avlan* rivers, and the tops of small mountains and hills (especially on the sunny sides, where the berries grow well).

Талы тар диктэ балдыдйавки, диктэвэ тэвлэдйэнны, вареньевэ оӈнанны, амнунналду бивки, ирйаӡилду бивки, авларду бивки, ойолду бивки. Или дылачавулын, талы кэтэ балдывки.

Там много ягод растет, там и собирают на варенье. Такие места бывают на ровных бассейнах рек *амнунна*, и в лиственничных лесах, также как на равнинах без леса возле реки *авлан* или на вершинах гор и холмов, особенно на солнечной стороне, где ягоды хорошо растут.

Sehi / Сэhи	A dense forest with a lot of bushes that is difficult to cross Густой не проходимый лес с кустами

Sehi (dense forestation) is mostly found in spruce forests; in such locations, there are a lot of large, dense bushes. Dogs can stop the elk in one place by barking and staying in such locations (such big animals cannot easily escape from this dense forestation), but they are unable to keep pace with the sable which live there. Wild reindeer can hear predators (such as bears) when the latter first start to approach and thus have sufficient time to run away. Lots of lichen, horsetail grass, and long grass grows in these areas, which allows domestic reindeer to graze freely. Wild reindeer come to such places in the summer to escape from the heat, because the atmosphere is fresh. *While documenting Evenki observations of climate and environmental changes, we noticed that most nomads and villagers remarked that the vegetal cover is becoming ever denser (i.e. it is turning into* sehi*). They link these changes to warming, rising humidity,* *and increases in the winter and summer precipitation. Some herders believe that another factor behind the denser vegetal cover is the decline in the reindeer population: these animals are trampling down less of the vegetal cover, allowing it to grow more densely. Let us again note the Evenki understanding of the agency of animals in the process of landscape transformation. (cf. Evenki climatology, Topographic typology: kudu)*

Сэһил бивкил аһиңилдулы, таду тар окталыккур бивкил кэтэкун, һэгдыкур һоктолыккур. Талы тар нинакир коңнокурбэ илэвувкил, андаһийэ этан боконо нинакин. Бэйур талы дыкинчэдйэвкил, бэйӊа эмэллэкин нуӊантыкин долдывувкил талы бэйӊал ӊэнэдйэми, бэйур урувкил амикардук. Оңко таду бивки, сивак бивки, чука бивки. Орор нйан тала оңкодйовкил. Сэһилду бэйур дыкурэкачэвкил окудукин, таду соңутмар бивки.

Густой лес *сэһи* бывает большинство в ельниках, в *сэһи* бывают много кустов и больших кустов. В таких местах собаки ставят лосей, зато никак соболей им там не догнать. Дикие олени там прячутся, ведь сразу услышат, если хищник к ним подойдет; например, дикий олень может успеть убежать от медведей. В таких лесах много оленьего корма, также как и хвощ полевой *сивак* и длинная трава *чука*. Олени туда тоже приходят кормоваться. Дикие олени там прячутся от жары летом, потому что там прохладно.
При нашем исследовании о наблюдениях изменений климата и окружающего мира многие эвенки, живущие в тайге или в поселках, замечают, что тайга все больше становится сэһи *– замечают, что становится все больше кустов и густого и высокого леса. Связывают это изменение растительного покрова с потеплением температуры и с увеличением влажности и (зимних и летних) осадков. Некоторые таежники считают, что фактором роста растительного покрова является также спад оленьего поголовья – ведь меньше стало оленей в тайге, и значит меньше растопчут и сломают растительный покров, которой все гуще становится. Здесь снова можно заметить, как традиционное знание эвенков принимается в счет воздействия животных на трансформацию ландшафта. (см. куду с. 101–102, с. 159–202)*

Oŋko / Оӊко	Reindeer grazing pastures with lichen (among others plants) Олений корм, в том числе ягеля и других растений

Oŋkokun / Оӊкокун	Substantial reindeer grazing pastures containing lichen Много оленьего корма, в том числе ягеля

Cf. diagrams, Sivak grazing, Reindeer grazing versus topography, Part III: diagram analysis of grazing pastures. (см. с. 146, 152–158, 378–399).

Dulgikte / Дулгиктэ
An alder tree
Ольха

The wood and bark of the alder tree are used to smoke meat and fish, thereby adding to the flavour.
The Even of Kamchatka and northern Yakutia, as well as the Evenki of southern Yakutia in the past, use the bark of alder trees gathered in the spring to treat and colour leather.

Дулгиктэдук оллойо нуныпчадйаңнанны и уллэйэвэт, уллэйэ айакикин дйэпча бивки.

Ольху употребляем, чтобы рыбу сушить или мясо; мясо очень вкусное бывает. Сжигают тогда дрова вместе с корой.
Эвены Камчатки и Северной Якутии, и раньше Эвенки в Южной Якутии употребляют кору ольхи, собранную весной, для обработки и окраски кожи.

Lulgikte, liuvukte / лулгиктэ, лйувуктэ
Black or bright green lichen hanging from tree branches
Черный и ярко зеленый ягель висящий на ветвях

Lulgikte lichen grows on larch and spruce: it is eaten by wild and domestic reindeer, *makchika*, and red deer. In dense *sehi* forests, one can find a lot of such lichen because the wind is unable to knock it off the branches. It makes excellent toilet paper and is wonderful for kindling fires within stoves and camp fires, even during the rain. (cf. diagram Reindeer grazing versus topography)

Лулгиктэ ирйактэлду балдывки, аhикталду балдывки, орор тара айамамат дйэпдйэвкил, макчикар дйэпивкил, бэйур нйан дйэпивкил, кумака нйанэ дйэпивки. Талы тар иду сэhитмар талы лулгиктэ кэтэтмэр балдывки, эвки адын анара, эвки тыкивканэ. Окадйэнны айакикин. Тыгдадйэми гулувур айакикин, имакунди дэгдэлэвки.

Ягель типа *лулгиктэ* растет на лиственницах и на елях, олени очень любят такой ягель, также, как и животное *макчика* или дикий олень и благородный олень. Там, где густой лес типа *сэhи*, много такого ягеля потому, что ветер его не ломает и не роняет. Ну и это является также хорошей туалетной бумагой. Дополнительно это употребляется для разжигания печки и костра, даже при дожде, костер быстро разжигается. (см. схемы, с. 152–158)

Liavukte / Лйавуктэ	Lichen growing on the ground Ягель на земле

Liavuktakso / Лйавуктаксо	Place with a lot of lichen Место где много ягеля

This is lichen which grows on the ground: wild and domestic reindeer graze upon it, but elk do not eat it. (cf. diagrams Sivak grazing, Reindeer grazing topography, Part III: diagram analysis of grazing pastures)

Лйавуктэ – оңко. Оңково орор дйэпивкил, бэйур-да. Коңнокул тара эвкил дйэптэ.

Ягель является одним из кормов оленей. Его кушают и домашние и дикие олени. Лоси такое не кушают. (см. схемы, с. 146, 152–158, 378–398)

Abdanna / Абданна	Leaves Листы от всех растительностей, в том числе кустарников

Animals with cloven hooves can graze in this kind of vegetal cover. (cf. oktalyk)

Тара дйэпивкил бэйур, кумакал, орор, коӊнокул, гипчар, макчикал.

Это является едой для парнокопытных животных. (см. окталык, с. 129)

Ireni kachikar, Kachikar / Ирэни качикар, качикар
1 Buds from either pussy willows or other plants; 2 'puppy'.
1 Почки вербы и от любого растения; 2 щенки.

The buds of pussy willows *(ireni kachikar)* and other plants *(kachikar)* mostly grow on plants belonging to willow species. The buds start to flower during the spring, producing very white and fluffy buds which resemble newly-born puppies. Reindeer feast on these buds in the spring. (cf. diagram Vegetal cover varia)

Качикар сйэкталду балдывки, мартын багдамакар йувкил качикарҕачир. Орор нуӊанман дйэпдйэвкил, нйан айавувкил, дйэпчавэр нйэӊнэ. Качикар йувкил нйэӊнэ.

Почки вербы *(ирэни качикар)* и любого растения *(качикар)* растут большинство на растениях из семейства тальников. Почки выходят весной очень белые и пушистые, как маленькие щенки. Олени кушают эти почки весной и от них хорошо насыщаются. (см. схемы, с. 147–151)

Chuturbadieren / Чутурбадйэрэн	When the environment becomes green: many grasses and flowers start to grow
	Когда зеленеет везде, много травы и цветы выходят

Chuturbadieren designates the period when the leaves on trees and bushes start to emerge, when larch needles grow and turn bright green, and when *sivak* grass and flowers begin to appear.
The elders say that nature (and the spirits that govern the environment) is 'decorating itself and re-birthing' during this time of the year: the shamans believed that this was the appropriate period for refreshing themselves and redecorating their costumes and attributes. Women and men used to make metal, fur, and leather decorations which they hung from the shamans' costumes during specific shamanic rituals (Lavrillier 2005). (cf. diagram Vegetal cover varia, Evenki Climatology: seasonal chain)

Чутурбадйаран – абданнал балдыдйара, ирйактэни мучуктэлтын балдыдйара, тара тыка гунэвкил, чукал балдыдйара.

Чутурбадйаран говорят, когда листья деревьев и кустов распускаются, когда хвоя лиственниц вырастает ярко зеленой, когда трава как и цветы растут.
Пожилые люди рассказывали, что в это время шаманы считали своей обязанностью в связи с возрождением окружающего мира в знак уважения и почитания духов природы, которая как считали «изукрашивается», также возобновить свой шаманский костюм и атрибуты новыми украшениями. Мужчины и женщины в это время изготовляли новые металлические части и висюльки из шерсти или ткани, которые добавляли к костюму шамана при специальном камлании. (см. схемы, с. 147–151, 164–172)

Chokchoko iriakte / Чокчоко ирйактэ

A crooked tree on a slope or (more often) an *emker* steep river bank

Кривое дерево стоящее на склоне, часто на крутом берегу реки *эмкэр*

Such crooked trees are found on *emker* steep river banks or on small hills with long flat tops *(kerain)*. They are very useful as landmarks for determining one's path.

Эмкэрилдули чокчоко ирйактэ бивки, кэраилдули чокчоко ирйактэ бивки. Талы тар ичэчэӈнэнны һокто или ӈэнэвки, сачйадйаӈнэнны.

Кривые деревья есть на крутом берегу реки кривые деревья бывают *эмкэр*, также на маленькой горке с плоской вершиной *кэраин*. Это является хорошим ориентиром чтобы узнать свою дорогу.

Nirgakta / Ниргакта
Eriophorum vaginatum, tussock cotton grass
Пушица

Eriophorum vaginatum grows in *kever* tussock fields on the ends of grass stems. Reindeer eat a great deal of it and thus put on weight. This cotton grass appears at the beginning of the spring, just as the snow melts away (cf. Topographic typology: kever, iaŋ). *This grass is so important for reindeer herding that herders establish their spring camps close to the tussock fields where it grows. It allows them to keep reindeer close to the encampment. Indeed, by the end of the winter, the reindeer herd spreads out (in ground of one to five individuals) in all directions in the search for this grass, which is full of protein.*

Ниргактэ балдывки чумникаӡилду йудйэвкил чукакар. Тара тар орор айавувкил. Иманна ундйэрэкин, таргачир йудйэвкил нйэӈнйэ. Орор айалэвкил таду.

Пушица растет на кочкарниках на кончиках трав. Олени насыщаются таким кормом. Как только снег тает весной, пушица вырастает и тогда олени ее кушают и поправляются. (см. кэвэр, йаӊ, с. 60, 90)

Chumnike, chumnikar / Чумникэ, чумникар	Tussocks on kever tussock fields
	Кочки на кочкарнике кэвэр

(cf. nirgakta, Topographic typology: kever)

Buɣar (UN), Kurung (Iengra) / Буӡар (УН), Куруӊ (Иенгра)
Burnt-out forest
Горелый лес

Buɣarykte / Буҕарыктэ	An old burnt-down forest where lichen and various grasses have just started to grow again Старый горелый лес, где ягель только начал расти или другая разная зелень

Buɣar designates a burnt-down forest caused by forest fires. Some of these, called *buɣarykte* can be ten years old: in such instances, lichen and various grass species will have begun to appear again. Both domestic and wild reindeer like to feed on such food sources. In addition, one can take firewood from such places, since the very dry wood burns very well.
Throughout this section on vegetal covers, we have seen that particular landscapes are used by both wild and domestic reindeer for grazing. This demonstrates the equal importance of both kinds of reindeer for the Evenki and that they can use landscapes for hunting and herding simultaneously.

Буҕар тар иду тар которочо дуннэ, ирйактэл олгочол. Ноноптыл тар которол буҕарил нйан бивки. Таду тар элэкэс оӈко балдыдйавки, тара орор айавувкил дйэпчавэр, бэйур нйанэ. Молывкил буҕардук, мол олгокил бивкил, айат дэгдэвкил.

Буҕар это, где лес с землей сгорел, где деревья совсем сгорели, засохли. Есть еще и старые горелые места *(буҕарыктэ)*, например от пожара, произошедшего 10 лет тому назад, на котором только начинает расти ягель. Таким кормом олени хорошо насыщаются, также как и дикие олени. От таких мест мы берем дрова, потому что там дрова сухие и очень хорошо горят.

Duɣan / Дуҕан	The tops of trees or bushes Вершина деревья и куста

The tops of trees, firewood, bushes, or items. The word also refers to the top of the small larch pole that runs horizontally along the roof of a tent and acts as a hanger for clothes and other items *(argi)*: it is located at the centre of the tent and must be oriented towards the back *(malu)*. A strong proscription regulates this rule.

Ирйактэни дуҕаван гундиҕарэ, мова дуҕаван гундиҕарэ.

Вершина (кончик) дерева или куста, или любого другого предмета, сделанного из растительного материала. Вершина маленькой лиственницы, которая служит горизонтальной (привязанной к потолку) вешалкой в центре палатки *(арги)* должна быть направлена в сторону задней стены *(малу)* – место для священных вещей в прошлом и для почетных гостей. Строгий запрет регулирует это правило.

Kapurgaran / Капургаран	Drying out (can refer to any tree or plant) Высыхает (о любом дереве, растении)

Everything dries out eventually: trees, wooden items like sledges, or plants.
Экунда капургавки, ирйактэ-да капургадиҕан, сырга-да капургадиҕан.
Все высыхает когда нибудь, деревья, растительность, даже нарты.

Supikte / Супиктэ
Dog rose, *Rosa canina L.*
Шиповник

Blue-berried honeysuckle *(Lonicera caerulea L.)* grows close to dog roses. Areas where dog-rose bushes grow are very uncomfortable to walk through because of the thorns. This plant is used as a medical treatment: it is dried, scalded, and then drank to improve one's health. Reindeer eat the young leaves of this plant.

Таду тар утумуктэ балдыдйавки тарҕачирду. Тавар супиктэ гиркудйами эруми, гидадйамивки супиктэ. Орор абданналватын дйэпивкил элэкэс йурэкин.

Рядом с шиповником растет жимолость. По местам, где кусты шиповника, очень неудобно ходить потому, что сильно колется. Используют для лечения, его сушат, потом ошпаривают кипятком и пьют для восстановления здоровья. Олени кушают молодые листья.

Devuŋnakte / Дэвуӈнактэ
Mushrooms
Грибы

Reindeer eat mushrooms, as do elk, red deer, and bears. The reindeer gain weight very quickly from eating mushrooms, but can lose it just as quickly. Until quite recently, the nomadic Evenki did not eat mushrooms, considering them to be a food source exclusively for the reindeer.

Орор тара дйэпдйэвкил, коӊнокур, кумакал, амикар дйэпдйэвкил, таду нйан бургувкил таргачиндук, имакунди тиӊнэвкил.

Олени кушают грибы, также как и лоси, благородные олени и медведи. От этого они жирнеют, но потом быстро худеют. До недавних пор таежные эвенки не кушали грибы, считая это оленьим кормом.

Irdia, ardia / Ирдйа, Ардйа

Juniper

Можжевельник

This plant is used to produce smoke, which helps prevent illness among the reindeer. To prevent detrimental spirits from disturbing people with their bad behaviour, the insides of tents are filled with smoke. The same is done when there is really heavy rain. The dogs are covered in juniper smoke when hunting is unsuccessful: this is held to restore good luck.

Ирдйанун улганинэл нунычавкил орор эдан буӡлэрэ, аваһыл эдатын дивэладйэрэ палаткадодун нунычавкил. Манак тыгдадйэрэкин нйан нунычавкил этэдан тыгдадйэми. Нинакирвэ нунычада нада эдатын маӊадйэрэ.

С можжевельником дымокурят, чтобы олени не болели. Чтобы вредные духи *аваһыл* не лезли к людям и чтобы плохо себя не вели, обкуривают внутри палатки. Если будет слишком много дождей, то надо дымить с можжевельником, чтобы дождь перестал, и когда охота плохая, дымокурят собак, чтобы удача вернулась.

Beiŋa diepgan / Бэйӈа дйэпган

Empetrum sibiricum *(empetraceae)* V. N. Vassil.

Вороника, шикша сибирская

This is a medicinal plant which grows on stones. It is used to treat eyes, joints, and bones. It is scalded and attached to the affected area with fabric.

Тар бэӡэ, дйолокилдули балдывки, тары эӡалба, сумулбэ, гарамналва айдйэвки.

Это лекарственный растение, которое растет на камнях, и лечит глаза, суставы костей. Его ошпаривают, заматывают в тряпку и ставят на больное место.

Cheŋkire / Чэӈкирэ

Labrador tea, *Ledum* L., *Ledum groenlandicum, Rhododendron groenlandicum*

Багульник болотный

This plant grows in spruce forests, in larch forests on a soft and dry soil with *ialbuka*, and in open spaces: it is used as a medical treatment for coughs. One must chew it and spit it out. One can also scald it and drink small quantities of the resulting mixture. In its dry form, it is used to ritually purify items, peoples, animals, and places. One must burn it and spread the smoke around. *During traditional weddings, the mother of the bride must burn some of its branches with fat and transmit this household fire to her daughter. These rituals are designated by the verb* ulgani-da. *People also put branches in store rooms containing furs to prevent mice and moths from approaching (Lavrillier 2005).*

Аһиӊилду, ирйакилду, дэбгэлду бивкил таргачир, симкитчэриду айат бэлэвки. Тара сэдйаӊнанны, тумиӊнанны.

Это растет в ельниках, в лиственничном лесу и в мягкой сухой земле с ковром мха *йалбука* и на открытых местах, и употребляется как лекарство, очень полезное против кашля. Надо его жевать и выплевывать. Также делают настойку из него и пьют маленькими дозами от кашля и желудочных инфекций. В сухом виде его сжигают, чтобы провести обряды очищения на людей, на местах и на животных, также как и на оружие и рыболовные сети от вредных духов. *При традиционном браке, сжигают его с жиром для передачи домашнего огня от матери к невестке. Такой обряд называют глаголем* улганида. *Люди также ставят пучки в шкурах, чтобы моль и мыши не подходили.*

Dielge / Дйэлгэ	A gap in the forest vegetation Промежуток в лесу через

Dielge is a gap in the forest vegetation that allows one to see and take aim at distant animals.

Дйэлгэ бидиӊан ирйактэл дйапкалдулитын талы экунавал оддадиӡас, булта-диӡас. Тара тар дйэлгэ гунэвкил.

Дйэлгэ называют промежуток между деревьями, через который можно, что угодно увидеть, стрелять и добыть.

Daldieren / Далдйэрэн	The arrival of the summer Лето настало, растительность выросла, насекомые появились

Daldieren designates the moment in the year when the summer has just arrived: the first horseflies, spiders, butterflies, and various other insects appear. It is an important and long-awaited seasonal switch: several Evenki families will gather in one encampment and move together throughout the summer. During this time, reindeer come to the smoke fires in the camps of their own volition in order to escape the horseflies and mosquitoes (cf. Evenki calendar, Evenki climatology, Lavrillier 2005)

Далдйэрэн – дйуӡани одйаран, чутурбача, иргактакар умукор ичэвуктэдйэрэ, атакил, эл дэӡиктэллэ, лородо.

Далдйэрэн – это когда лето наступает, когда все только, что позеленело, первые оводы летают, пауки начинают появляться, разные букашки начинают летать и бабочки. Это важный и долгожданный поворот между сезонами, когда эвенки собираются несколькими семьями в одной стоянке, чтобы кочевать вместе все лето, когда олени будут сами приходить на стоянку к дымокурам, убегая от оводов и комаров. (см. с. 43–59, 164–165)

Koŋde / Коӊдэ	An empty tree: the heart of the tree has become like dry cotton
	Пустое дерево – это когда внутри дерева сердцевина гниет и становится как вата, высыпающаяся и становится пусто

Koŋde designates a larch which is empty inside. From it, one can manufacture *talki* instruments, which are used to make animal skins softer. Inside such trees, one can find *kuchu* (the heart of the tree has decomposed to the point where it looks like dry cotton). One must take this material and put it within fabric together with duck feathers: the resulting product can be given to babies as bedding, since it will always remain dry. The Even also do this. When cutting down such a tree, one must put a young and small larch inside what remains: otherwise, the children of the one who cut down the tree might lose their intelligence. *Koŋde* can also mean an empty-headed person.

Ирйактэ бивки, долы коӊдэ, нуӊандукин талкийэ можно ода, кучукикин додун бивки. Тара гавкил куӊаркарду чикэнэкин мукамукин олгокин бидан. Коӊдэвэ чукулыдиӊас, таду нада надйэда тала ирйактэкачанэ нйукучукоконэ коӊдэ долан нада мугдэкачанду куӊакар акарыл эдатын бирэ.

Это лиственницы, которые пустые внутри. Из них делают инструмент для обработки кож – *талки*. Внутри такого дерева бывает *кучу* (сердцевина дерева, которое со временем становится как сухая вата). Берут это вещество *кучу*, ставят под уткиными перьями в тряпке и ставят под грудного ребенка, чтобы он оставался сухим. Эвены тоже так делают. Если срубить такое дерево на дрова, обязательно надо ставить внутри оставшегося пенька маленькую лиственницу, чтобы дети не рождались идиотами. *Коӊдэ* говорится также о голове человека, который не думает.

Muchukte / Мучуктэ	1 Young larch 2 Larch needles
	1 Молодая лиственница 2 Хвоя лиственниц

A young larch or larch needle. Larch branches are placed on the floor of the tent all year long to provide protection from the ground; in summer they make a nice smell. Sometimes, one puts them in the smoke fires for reindeer *(samŋin)* in order to produce a good kind of smoke.
The Evenki successfully treated a scurvy epidemic during the Second World War by eating larch needles. In the past, one used the roots of young larches growing in marshes to sew together the pieces of birch bark used to make a Evenki bark mereke. *Young larches were also used during the shamanic ritual for creating a 'road of trees for the spirits'. They are also used to build the tent frame. Many young larches grow on the edges of wide and flat river basins* (amnunna)*: these are destroyed each year by the melting of the river ice during the spring.* (cf. Part III: diagram Landscape transf. due to climate change)

Ирйактэни мучуктэ бивки, ирйактэ гаралду балдыдйавки. Сэктэвэ сэҕиннэнны палаткадодун. Адылдун самңинтыки нодданңнэнны айат дэгдэлдан самңин.

Это молодая лиственница или хвоя лиственниц. Ветви лиственниц стелятся в палатке, и летом от хвои так вкусно пахнет. Иногда его сжигают на дымокурах для оленей *(самңин)* и чтобы хорошо горело.
Эвенки лечили себя, кушая хвою и избегали смерти во время эпидемии цинга после второй мировой войны. В старину употребляли корни молодых лиственниц, растущих на болотах, чтобы сшить вместе пласты бересты, которые составляли лодку мэрэкэ. *Также употребляли молодые лиственницы во время шаманских камланий, чтобы устроить дорожки для духов. Используется для структуры палатки. Молодых лиственниц много по краям широкого ровного бассейна реки* амнунна, *и ледоход каждый год их ломает.* (см. схемы, с. 438–443)

Takin / Такин	A fallen tree Упавшее дерево

If someone is forced to spend a night outside in the winter, he/she can find two fallen trees, set them on fire, and sleep between them: this creates a very warm environment.
Here again (cf. bolgik), we can see the Evenki art of surviving even during harsh colds by using the natural environment.

Туҕэни бирэкин дйур такинэ бакаңнанны, анңачалми тулинду. Тар дйур такинмэ иланнэнны дулиндулин аһинэнны – нйама бивки.

Если приходится зимой ночевать на улице, то можно найти два упавших дерева. Надо их зажечь и ночевать между ними – очень тепло бывает.
Может видеть еще раз (см. болгик, с. 130–131) *искусство эвенков выживать даже при морозе, используя элементы природы.*

Sivak grazing

Sivak (horsetail) grazing and large river basin (*amunna*) / Сивак амнуннаду / Хвощ и широкий бассейн реки (*амнунна*)

Reindeer eat horsetail. They go to large river basins independently / Орор сивакдйэрэ. Мартын амнуннала ҥэнэрэ / Олени кушают хвощ (*сивак*), сами на бассейн реки спускаются.

Horsetail grows on river banks. Reindeer only graze upon fresh horsetail / Сивак бира дйапкадун балдывки. Орор дйэпивкил элэкэс балдычалба сиваккэ / Хвощ растет вдоль рек. Олени кушают только свежий зеленый хвощ.

Reindeer lie on the ice of a large river basin so the mosquitoes do not bite them / Орор тэђэчэрэ дйукэду, дйукэду иҥин – мармактал эвкил киктэ / Олени лежат на льду в широком бассейне реки, потому что комары их там не кусают

This horsetail was eaten last year by wild reindeer, which why there is so little of it/ Тар сиваккэ бэйур гйэванҥан дйэпчал – олыhин нйукучукор / Этот хвощ был съеден в прошлом году дикими оленями, поэтому маленький

Vegetal cover – varia (1)

Reindeer herding and grazing in spring / Оленеводство и пастбище весной

A grandmother checks to see if horseflies larva have caused an infection (a frequent occurrence in the spring) / Эвэ оронми одуладйаран куиктэлбэ / Бабушка оленя проверяет насчет болячек от личинок оводов (весной гниют).

Reindeer always eat their fill of horsetail; in the winter, they gain weight from it / Орон сиваккэ айавувки окинда; туҕэниду сивактук орор бургувки / Олени всегда наедаются хвощем и зимой от этого поправляются.

Horsetail grows on large river basins (*amnunna*) / Амнуннаду сивак / На широких бассейнах рек растет хвощ.

Vegetal cover – varia (2)

Rebirth of the vegetal cover in springtime / Весеннее возрождение флоры

The leaves of the dwarf birches and bushes have just started to appear / Окталыкту абданнал элэкэс йулдйэрэ / На кустах листочки только выходят.

Young pussy willow branches / элэкэс балдырэ ирэл / Свежие кусты вербы.

Pussy willow buds have just started to appear / Ирэни качикар йучал / У вербы сережки появились.

Young needles emerge from larch branches/Ирйакталду мучуктэ элэкэс йудйэрэн / Молодая хвоя лиственниц.

Vegetal cover – varia (3)

Rebirth of the vegetal cover in springtime / Весеннее возрождение флоры

Long grass has just started to grow / Чука элэкэс йуча / Длинная трава только выходит.

The lycoperdon: in the past, the Evenki smoked it / Кукку дамган – нонон эвэнкил дамгатычал тара / кукушкин табак, раньше его курили.

The snowdrop (cuckoo's hat) and the lycoperdon (cuckoo's tobacco) grow close to each other / Кукку авунын, кукку дамган даҕадун бивкил / Подснежник (кукушкина шапка) и дождевик (кукушкин табак) растут всегда рядом.

Reindeer like to eat snowdrops (cuckoo's hat) / Кукку авунын – орор аявувкил / Олени любят кушать подснежник (кукушкину шапку).

Vegetal cover – varia (4)

Rebirth of the vegetal cover in springtime / Весеннее возрождение флоры

Old tree / Мугдэкэ / Старое дерево.

The bog blueberries did not perish during winter, so they will provide berries (thus allowing the normal hunting of sable)/ Диктэ эчэ бэҕирэ балдыӊат / Голубика не заморозилась, значит ягоды будут (и охота на соболя).

Trees become green / Ирйактэл чутурбачал / лес позеленел (украсился).

Dwarf birch (*oktalyk*) leaves appear /Абданнал йучал (*окталык*) / Листья карликовой (*окталык*) березы вышли.

Vegetal cover – varia (5)

Rebirth of the vegetal cover in springtime / Весеннее возрождение флоры

Beiŋa diepgan, Бэйҥа дйэган, Empetrum sibiricum (*empetraceae*) V.N. Vassil. The plants like to grow close to spruce in the shade / Бэйҥа дйэпган / Вороника, шикша сибирская растет, где ель растет, тень любит.

Rhubarb *Rheum rhabarbarum* appear where spruce grow: they like shade and humidity / Тапчакакты дйологду эвкил балдырэ / Ревень, дикий щавель – растут где елки, тень и любят влагу.

Some butterfly larva have eaten through these leaves and then fell to the ground/ Тар абданнакунмэ дйэпчэрэ лородони куликар, тадук саҥарэйэ оҕа дуннэла тыкивки / Кушают эти листья какие то черные черви от бабочек, дырявят насквозь и падают на землю.

Reindeer grazing versus topography (1)

Quality of grazing lands according to: topography, main winds, vegetal cover, snow depth and quality
Корм оленей согласно: топографии, главных ветров, растительного покрова, глубины и качества снега

GRAZING IN DENSE FOREST WITH BUSHES (*SEHI*)
ОҤКО СЭҺИДУ
ПАСТБИЩЕ В ГУСТОМ НЕПРОХОДИМОМ ЛЕСУ (*СЭҺИ*)

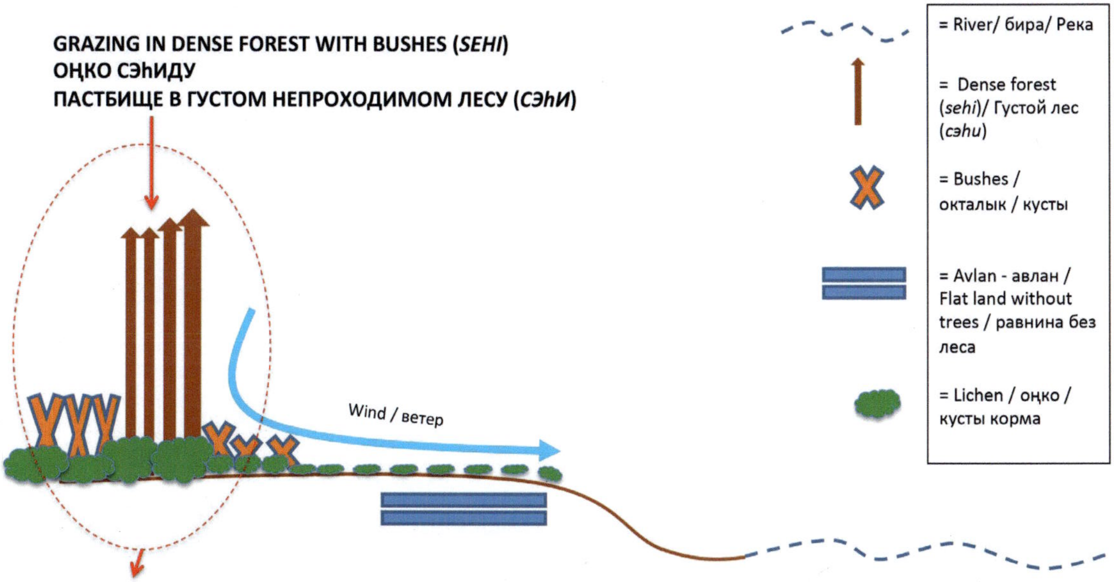

In the dense forest and bushy places (*sehi*), lichen and other grazing vegetation grow in vast quantities, since they are not disturbed by the winds. In such places, grazing pastures are always old and 15-20 cm high. Reindeer do not like old lichen / Сэһиду, окталыкту оҥко һэгдыкун балдывки таду эвки адынэ, сагды оҥко таду 15-20 см. Орор сагды оҥково эвкил айаврэ /
В густом лесу (*сэһи*) или в кустах ягель растет большой, укрытый от ветра, всегда старый, высотой см 15-20. Олени не едят старый ягель.

Reindeer grazing versus topography (2)

Quality of grazing lands according to: topography, main winds, vegetal cover, snow depth and quality
Корм оленей согласно: топографии, главные ветра, растительного покрова, глубины и и качества снега

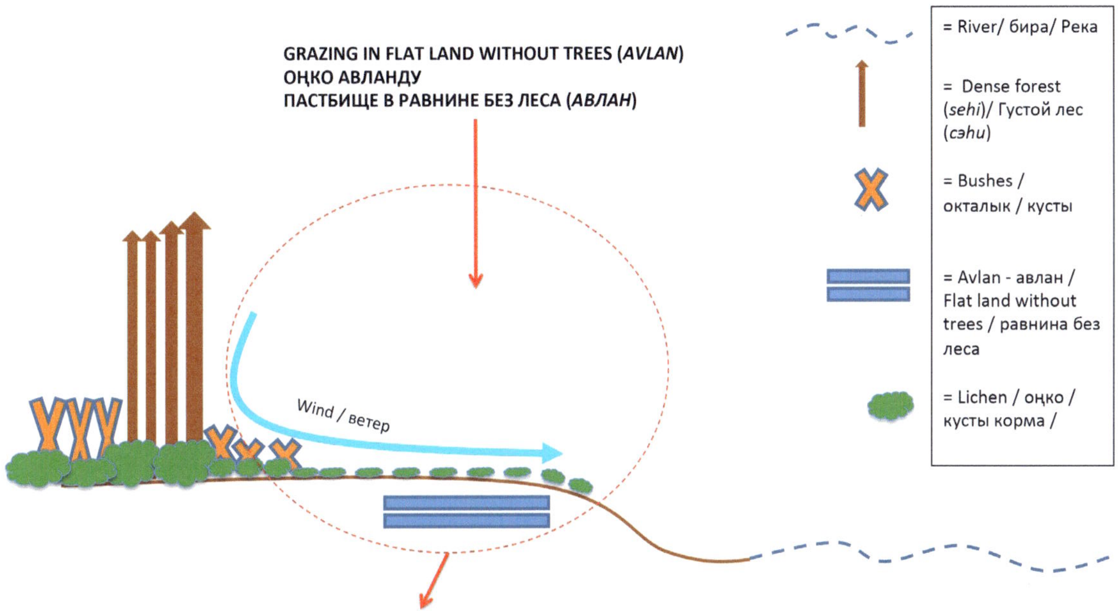

In flat lands without trees (*avlan*), the cold and winds do not allow the pastures and bushes to grow very much, since the wind breaks the ends of smaller branches. There are always fresh pastures, which the reindeer like: they thus gain weight / Авланду, оҥко адынмубдйэки ойолын иҥинидйэнэ, адын укчавки. Оҥко элэкэс балдыдэвки, орор дйэпийочэвкил айавувки/

На равнинах без леса (*авлан*), холод и ветер не дают оленьему корму слишком много расти (ломая кончики), как и кустарникам. В таких местах всегда свежий корм высотой см 5. Олени любят такой корм и наедаются им.

Reindeer grazing versus topography (3)

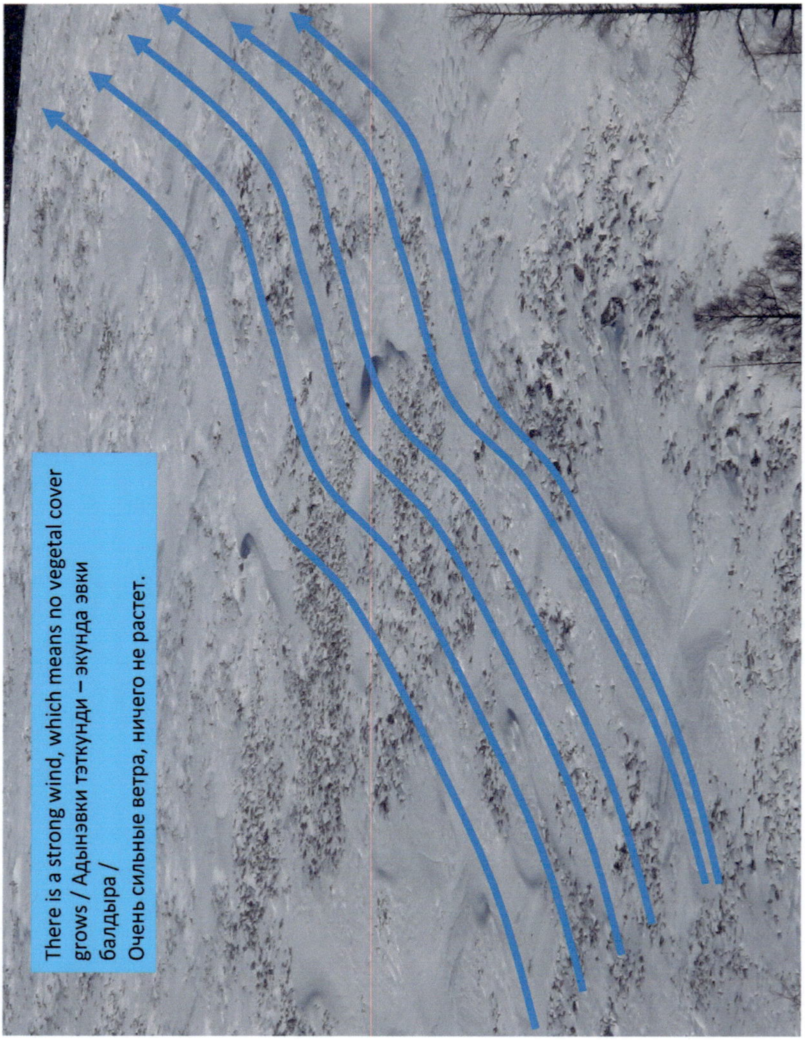

© A. Lavrillier, S. Gabyshev, 2013

Reindeer grazing versus topography (4)

Quality of grazing lands according to: topography, main winds, vegetal cover, snow depth and quality
Корм оленей согласно: топографии, главные ветра, растительного покрова, глубины и качества снега

2 flat lands without trees (*avlan*) – always good grazing: Wind breaks the protruding parts of the vegetal cover, which means they are always growing afresh: the reindeer like this. HOWEVER, one needs to take into account the irregular quality of the snow cover: some zones with hard snow (*chuiur*) do not allow access to the pastures, while other zones with soft snow (*duiukun*), particularly where *pinus pumila* and larch grow, do allow this / Авлакурду – айа оңко /

2 равнины без леса (*авлан*) - Корм всегда хороший: Ветер ломает кончики растительного покрова – всегда вырастает свежий корм и невысокий – олени любят такой корм. НО, снег не везде ровного качества – где-то твердый снег (*чуйур*), где-то мягкий снег (*дуйукун*) особенно, где стланик и лиственница (*болгиктэ, ирйактал*). (cf. diagram Analysis of grazing pastures)

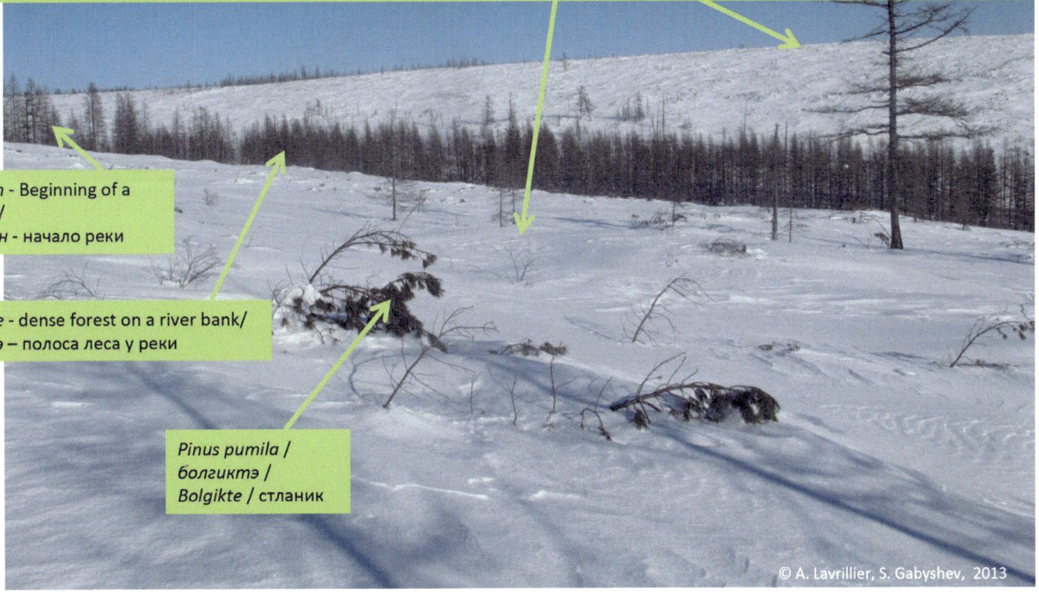

Reindeer grazing versus topography (5)

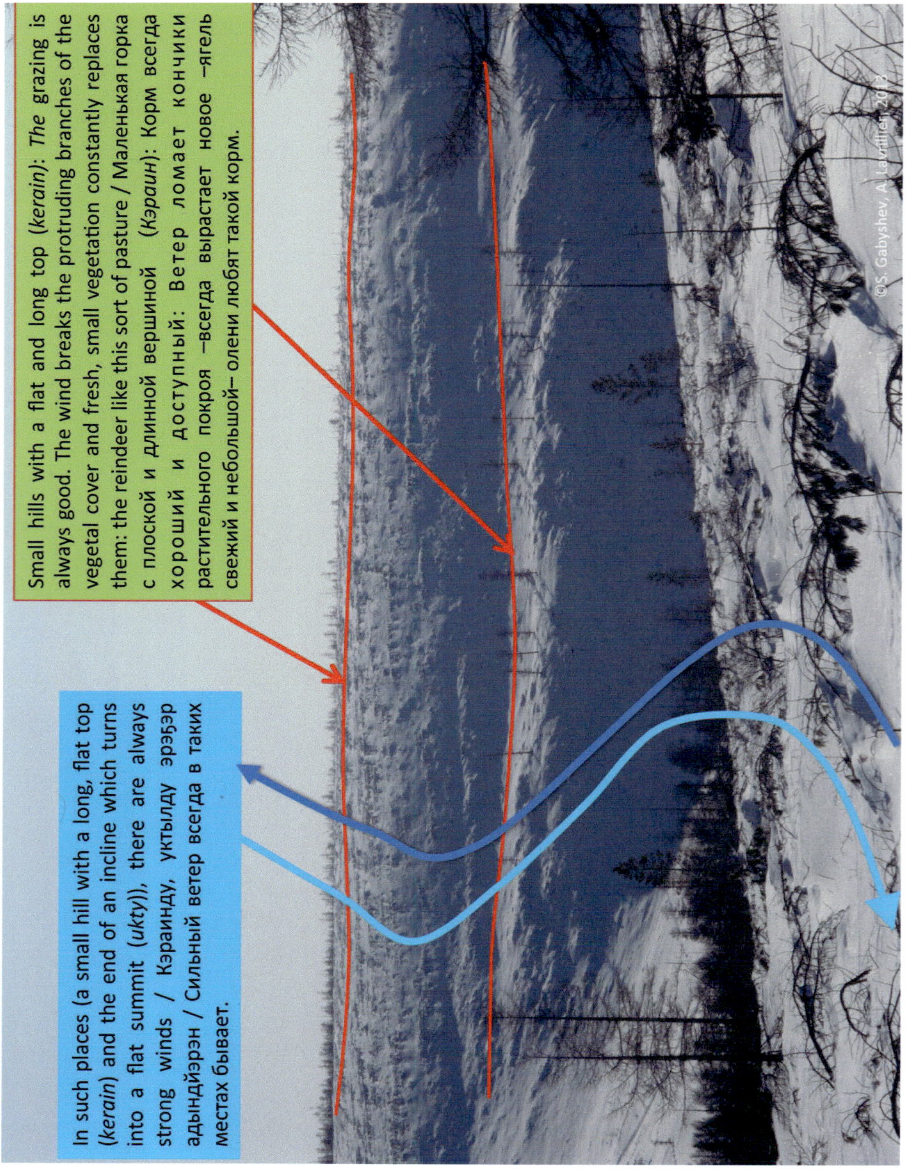

Reindeer grazing versus topography (6)

On the river bank, one finds ground, stones, and horsetail grass (*sivak*). For a long time, the reindeer are able to gain weight from this grass. The quality of the snow varies: on the bank, the snow is soft (*duiukun*), while in the middle of the river, there is a hard layer (*chuiur*). Where the larch are small, there is hard layer of *chuiur* snow, but horsetail grass also grows there /

Бира дйапкадун сивакун балдывки. Тара дйэпми орор горо эвки тиӊнэмэлчэрэ. Иманна бира дйапкадун дуйукун бивки, дулинду чуйуркун бивки. Иду ирйактал нйукучукокор, таду сивак балдывки /

На берегу реки – земля, камни, там растет хвощ – от этого олени хорошо поправляются и не худеют. Качество снега разное: по берегам мягким (*дуйукун*). Где деревья маленькие растут, по середине реки есть наст (*чуйур*), но там хвощ (*сивак*) нету.

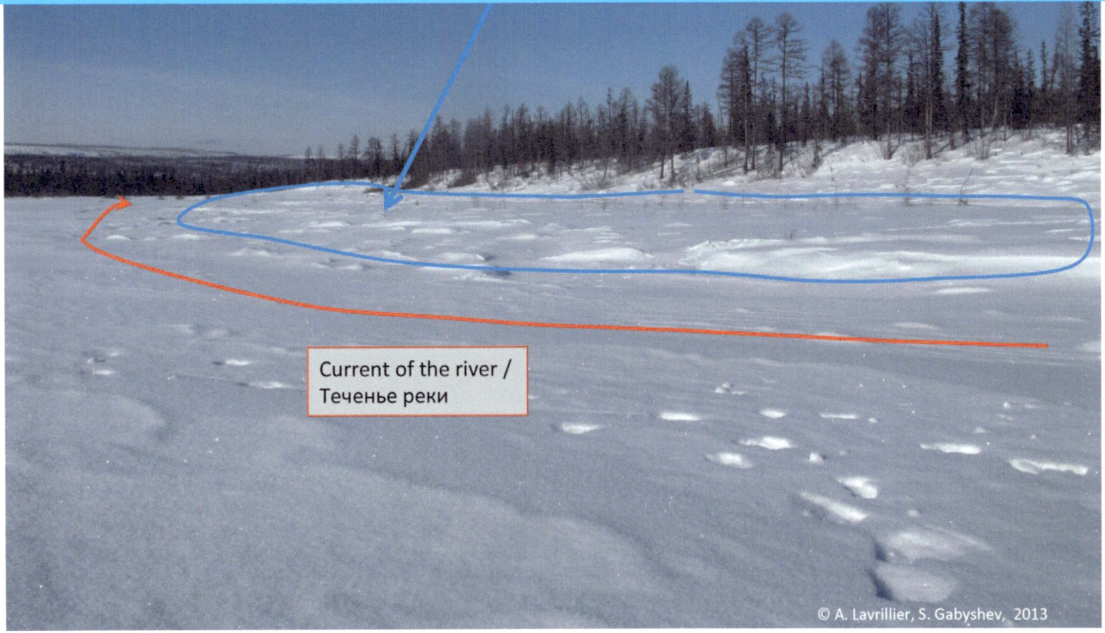

Reindeer grazing versus topography (7)

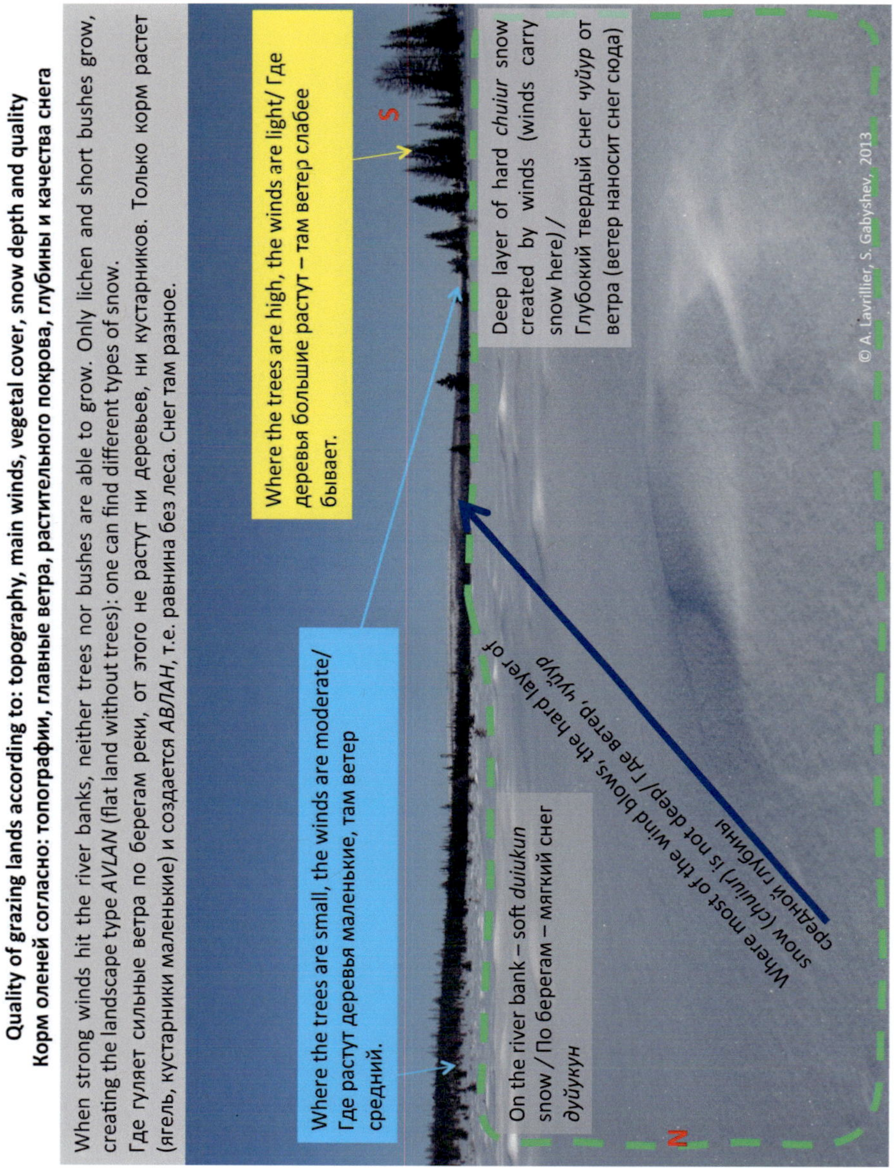

2.3 Indigenous science of climate
by A. Lavrillier and S. Gabyshev

In this chapter, we will first examine the chain of seasonal and inter-seasonal shifts with some additional explanations and concepts. Then we will clarify the Evenki system of temperature measurement and focus on the knowledge nomads have of microclimates and the wind and how they use them. We will then present several systems of weather forecasting. After this, we will detail knowledge about clouds, precipitation, and the circulation of cold and warm air, finishing with a transition to the snow and ice typologies.

2.3.1 Evenki climatology
by A. Lavrillier

One of the specificities of Evenki climatology is that it takes into account not only the weather or climatic events, but also the behaviour of other elements of the natural environment and of human beings.

According to our analysis of the 48 months of daily observation tables from 2013 to 2016, we see that Evenki constantly observe various aspects of their environments in order to predict the weather.

As we hope to show in this book, the Evenki are not only forecasting weather, but also producing hypotheses about changes in the climate several years ahead; these hypotheses could be compared to a kind of 'modelling'. (cf. Part III and conclusion)

It is surprising that the nomads, as with many indigenous peoples in the world, observe the entirety of the environment's behaviour and changes to it in order to forecast weather and climate: from the sky to the ground, from animal behaviour to changes in the vegetal cover, from elements of the topography to the behaviour and feelings of human beings. For forecasting weather, they of course observe weather features (clouds, winds, precipitation, etc.), but also consider birds, fish, reindeer, dogs, the circulation of sounds and air, the behaviour of fire (sounds, smoke, combustion process, coal size), changes in the moon, sun, and stars, and human feelings.

It is not then surprising that the Evenki do not have isolated words for only designating the climate or weather. When they want to ask about the weather, they use the word *buya, буҕа:* 'What is the weather?', 'How is the weather going?' *(Ekudy buya? Экуды буҕа?* On *buya ocha? Он буҕа оча?* – in Evenki). As I will explain later, the same word is used to designate the climate when talking about climate change, like in the expression 'the climate is losing its logic' *(Buya ukchapcha, Буҕа укчапча,* literally 'the natural environment is broken') (cf. Snow and ice typology: snow and extreme event). *Buya, буҕа* has multiple meanings in the Evenki language. It designates, depending on the context of the sentence, the entirety of the biophysical natural environment, the spirits inhabiting it, the main spiritual entity in control of the natural environment, the sky, and one's native land (Cincius 1975, 1977; fieldwork data; Lavrillier 2005). The use of this unique word for weather, climate, sky, and the entire biophysical environment certainly confirms the holistic and systemic perception of the Evenki.

In addition to this, Evenki have a specific understanding of the Russian term *klimat* (basically, the equivalent of the English 'climate'). Their understanding includes many elements. For instance, nomads and villagers see a relationship between climate and industrial development:

> 'The climate entirely changed. There is probably now a lot of radiation because humans "opened" the earth everywhere to extract titanium, gold, coal, and diamonds; they also opened a new spaceport and so on. Radiation came out of the ground. This is why our reindeer became fragile or ill.'

We can see here that while nothing is said about the weather to explain that the climate has changed, the clarifications about the natural environment (and especially about reindeer) are global. The factor consider responsible for environmental changes is extractive industries. They mention local industries more often than industry on a worldwide scale. This is not so far from the IPCC conclusion on the anthropic effects on climate changes (IPCC 2014). However, the Evenki also sometimes worry a great deal about worldwide industrial development and speak of the end of the world (Lavrillier 2013). This can be done with humour: in 2014, Petr Vasilev, evoking a contemporary shamanic legend about the disappearance of the planet (which will burn) and the appearance of a new planet, stated, *'it is said that we will have a new planet. Why? It will be useless, we will probably pierce it in every possible place in order to extract resources and spoil it, just as we have almost destroyed our current planet'* (laughing). Another recurrent focus in the discourses about climate change is the earthquakes, which the Evenki (villagers and nomads) attribute to climate change.[1] Thus, the Evenki understanding of the word *klimat, климат,* (climate) is so holistic that it includes the entire natural environment and human activities (indigenous and non-native, local and global), thereby focusing on the interplay between the elements of the environment (including climate) and human factors. It is also telling that this understanding of *'klimat', 'климат'* truly corresponds to their conception of the universe itself, *buya, буҕа* (i.e., a holistic one focused on the interactions between the elements). It is very important to say that in Siberia in general and among this Evenki group in particular, there has been no mediatisation of popular-scientific notions of climate change, so we can consider these perceptions to be 'purely' Evenki in origin.

Nomads also observe the weather by forecasting other events in the rest of the natural environment, like the appearance of insects, the circulation of sounds, the circulation of warm, cold, humid, and dry airs, the lifecycles of reindeer or fish, etc.

1 This is also the case among the Tuva (Rojo et al. 2016).

In addition to this concrete and empirical knowledge, there is a symbolic idea that humans (shamans and ordinary people), animals, and other elements in the natural environment can change the weather.

According to the nomads, the knowledge and ability to predict weather are both inherited from previous generations and enriched by each individual's experience through the *'permanent observation of the smallest details'* (as the nomads explain it). Among others, Oleg Iakovlev said:

> 'To predict the weather, we need to remember the knowledge transmitted by our ancestors, but also to observe daily everything in the environment, even the smallest details and even those details unrelated to our economy. We also need to study how events and changes go together, and the most important thing is to remember all of these observed small details in order to see afterwards whether they are related or not and how they are related.'

Evenki climatology contains several important and complex concepts that prove that indigenous climatology is highly systemic and analyses not only visible elements (clouds, precipitation, and so on), but also invisible ones like air and sounds. We have documented some of these throughout the several years we spent researching. In this book, we attempt to explain some of them through diagrams, since it is almost impossible to express such knowledge through simple texts.

Take, for instance, the concepts *namuscheren, намусчэрэн / lamuscheren, ламучэрэн* (changing winds that bring snow) (cf. Evenki climatology, Winds and airs typology), *uŋnia, уӈнйа* (that shows the Evenki conceptualise of the cycle of water-snow and ice), and their complex snow typology (which proves the Evenki have established a physics of snow and ice (cf. Snow and ice typology)). The nomads also have very interesting knowledge about the circulation of warm and cold airs that contains the concept *salgyn, салгын,* an understanding of the circulation of smoke from fire, and the concept *idia, идйа,* which is related to the circulation of cold and warm air in relation to topographic specificities and the movement of the sun (Winds and airs typology: diagram Idia). In addition, they have developed very interesting knowledge concerning the circulation of sounds in accordance with weather conditions (cf. Bira doldybdieren soloki; Winds and airs typology: diagram River sounds weather). Other concepts, like *irganen, ирганэн,* focus on the relationships between weather or climate, insects, the reindeer reproduction cycle, and birds. (cf. Evenki climatology here)

The chain of seasonal and interseasonal shifts

Evenki climatology includes a set of concepts about seasonal shifts: these have detailed inter-seasonal steps and an understanding of a chronological chain that admits variability in the expected arrival dates of seasonal or inter-seasonal events.

According to our study of daily observations, the subdivision of the year into months in the traditional Evenki calendar seems to be less significant than its subdivision into large seasonal and inter-seasonal periods. According to various ethnographic sources, the Evenki calendar includes from 11 to 13 months: the months are named according to the main activity or event then taking place (among others Vasilevich 1969: 42–53). This is shown in the table below, which also includes the names of the months for the various Evenki groups of Eastern Siberia.

MONTH	TUNGIR-OLËKMA EVENKI GROUP	ALDAN-UCHUR EVENKI GROUP
1	*Giravun, Гиравун,* 'period of walking'	*Mire, Мирэ,* '(when the snow is falling on) the shoulders'
2	*Mire,* '(when the snow is falling on) the shoulders'	*Conception exists but no name*
3	*Ektenkire, Эктэнкирэ* 'snow on the branches' *Nelkini, Нэлкини,* 'first spring'[1]	*Turan, Туран* 'arrival of ravens'
4	*Ovilahani, Овилахани* 'period of hard snow'	*Conception exists but no name*
5	*Sonkan, сонкан,* 'period of domestic reindeer calving' *Ovilaha, Овилаха,* 'period of hard snow'	*Ovilaha, Овилаха,* 'period of hard snow' *Nengneni, неӈнэ,* '(second) spring'
6	*Irbelehe, ирбэлэхэ* 'spawning period of fish'	*Muchun, мучун,* 'young larch needles'
7	*Milehen, милэхэн* (no translation) Irbeleheni, ирбэлэхэни 'spawning period of fish'	*Ilkun, илкун, ilaga, илага,* 'the berries are ready to be eaten'
8	*Boloni, болони,* 'autumn' *Hunmilehe, хунмилэхэ,* 'period of midges'	*Irkin, иркин,* 'the berries are ready to be eaten'
9	*Sirudian, сирудян, Sirulahan, сирулахан,* 'mating period of domestic reindeer'	*Sirudian, сирудян,* 'mating period of domestic reindeer'
10	*Sigelehe, сигэлэхэ,* 'period of snow and of the freezing of the river'	*Sigelehe, сигэлэхэ,* 'period of snow and of the freezing of the river'
11	*Seteppi, сетепи,* from the Yakut language	*Conception exists but no name*
12	*Conception exists but no name*	*Conception exists but no name*

1 Vasilevich does not give a translation for this term, so the translation is mine. This name is used today by the nomads.

One may note in this table that some months do not have names (12). Other months use the names of the seasons (3, 5, 8). Equally, there is great variability in the names of the months between the calendars of different regional groups. The only regularity between regional groups is the period of hard snow (5), the mating period of domestic reindeer (9), and the installation of the snow (10). In addition, for the name of one and the same month within one and the same regional sub-group, one finds two different names referring to two different events (which is not contradictory). For instance, May (5) is 'the calving period' and 'the period of hard snow' at the same time, while July (7) is both the spawning period for fish and when larch needles appear. (Lavrillier 2005: 188)

Nowadays, the nomads do not use the Evenki names of the months (it is impossible now to even document the Evenki month names when talking to the elders). However, even when they use the Russian names, they pay great attention to the cycle of the moon, as we will see. In contrast, the nomads only use their traditional understanding of the seasons and inter-seasonal periods, as well as the attached Evenki names. According to Lavrillier's research between 1994 and 2003, the Evenki nomads counted six seasons in the past: *nelkini, нэлкини,* (the first spring when the snow is in the process of thawing), *neŋneni, нйэӈнэни,* (the second spring without a snow cover which lasts to the first song of the cuckoo), *diuya, дйуҕа* (the summer), *bolo, боло* (the first autumn without snow), and the second autumn with the installation of the snow cover. (Lavrillier 2005: 187–210)

What is important for the nomads is that climatic events or events linked to the animal and vegetal realms can vary in terms of the dates of their arrival by 15–20 days from one year to the next (i.e. their norm admits such variability). In any event, the nomads are not very concerned by official calendar days: they only pay them heed so they know what is happening in the 'external world' (the settled world of villages and towns) or in order to arrange meetings with other herders.

In addition to the quoted concepts (*idia, идйа, namusteren, намусчэрэн,* etc.) that Lavrillier and Gabyshev documented through co-production, the Evenki lexicon focuses on the process of seasonal changes itself. There are specific terms for each of the seasonal shifts. For instance, ***daldyn*** *(далдын)* means that 'the seasonal shift towards summer has been performed' (i.e., summer has started). ***Tuyerburen*** *(туҕэрбурэн)* means 'it is becoming winter', ***tuyenikun oren*** *(туҕэникун орэн)* that 'the winter has happened', ***neŋnerburen*** *(нйэӈнэрбурэн)* that 'the second spring has happened', and ***bolorburen*** *(болорбурэн)* that 'the autumn is happening'. Each shift period is expected by the nomads, who observe all the details of the environment to determine the passage to the next season or inter-season. These passages often induce adaptations to transportation methods and changes in their main activities, social organisation, modes of hunting, and so on.

In addition to the subdivision into seasons and shift periods quoted above, the Evenki divide each period and shift period into a chain with the expected changes, as we show in the table below. Here, we attempt to show the relationship between the seasons, changes, the shift periods, and the activities of humans.

With this co-documentation by both an ethnographer and a nomad(s), we have come to understand that Evenki knowledge about the calendar of the weather and the natural environment is much more detailed, complex, and rich than a simple list of months and that all the elements of the environment are engaged. (see below)

Seasonal chain with expected changes
by A. Lavrillier and S. Gabyshev

The table starts with the season considered by the Evenki to be the New Year (i.e., the re-birth of the natural environment after the long snow period).

Most of the terms quoted appear either in the chapter on Evenki climatology or in the other typologies developed in this book.

SEASONAL SHIFTS ПЕРИОД СЕЗОННЫХ ИЗМЕНЕНИЙ	EXPECTED INTER-SEASONAL EVENTS МЕЖСЕЗОННЫЕ ИЗМЕНЕНИЯ ПРИРОДЫ	USES / MAIN ACTIVITIES ИСПОЛЬЗОВАНИЯ И ЗАНЯТИЯ
Marmaren – the start of the green spring.	*Marmaren* – Period in the springtime when leaves and mosquitoes appear.	Moving: good because it is possible to cross all the marshes (they are frozen); there is still snow, the river current is not too strong, and the ground is dry. Herding: bad because the predators kill a lot of reindeer and the reindeer are dispersed in groups of 2–5. Hunting: a difficult period because prey is rare and the wild reindeer have not yet come to the *nire* marsh.
Мармарэн – начало озеленения природы весной.	*Мармарэн* – часть весны, когда распускаются листья и комары появляются.	Транспорт: хорошо потому, что можно переходить болота замерзшие; еще снег лежит, течение реки не слишком большое и земля сухая (лужи высохли). Оленеводство: плохо потому, что хищники убивают много оленей; олени пасутся по 2–5 голов. Охота: трудный период из-за редкой удачной охоты и дикие олени еще не спускаются на болота *нирэ*.

Daldieren – the installation of the summer has started. *Далдйэрэн* – начало установления лета.	*Chuturbadieren* – the vegetal cover is turning green, flowers appear. *Чутурбадйэрэн* – Растительный покров зеленеет, цветы появляются. *Atakil ichevulle dyudu* – spiders appear. *Атакил ичэвуллэ дйуду* – пауки появляются. Fish appear, the first mosquitos appear. Рыба появляется; первые комары появляются.	Moving: very little nomadisation, riding and packsaddle. Herding: main attention is on the herd, daytime is spent gathering the herd (the reindeer eat grasses); we wait for the arrival of horseflies. Hunting: no hunting, but fishing every day. Транспорт: редкие кочевки, переходят на езду верхом и на перевозку груза вьюками (нарты оставляют). Оленеводство: главное внимание на оленей, весь день посвящен на то, чтобы стадо собрать (олени кушают траву); ждут появления оводов. Охота: не охотятся, но рыбачат каждый день.
Daldyn, Dallen – the summer is installed. *Далдын, Даллэн* – лето установлено.	*Kuranen* – very hot weather. *Куранэн* – лето очень жаркое. Lots of mosquitos, lady-birds, butterflies. Много комаров, божьих коровок, бабочек. *Irganen* – many different species of horseflies appear after heavy seasonal thunderstorms. *Ирганэн* – много разных видов оводов появляется после сильной грозы.	Moving: riding and packsaddle. We very often nomadise so that the reindeer will be always be on fresh land and will not get sick. Herding: we change camps often so predators will not find the herd. Smoke fires are made (reindeer come to the camp independently as they flee from horseflies), milking. Hunting: from time to time, collective hunting (just enough for personal needs). Gathering: berries. Транспорт: Езда верхом и перевозка груза вьюками. Часто кочуют, чтобы олени всегда были на свежей земле и таким образом не болели. Оленеводство: Часто кочуют, чтобы хищники не заметили оленей. Олени сами приходят на дымокур; дойка. Охота: Иногда коллективная охота, только на собственное употребление Собирательство: Ягоды.

Bolorburen – the autumn is happening. *Болорбурэн* – начало установления осени.	*Agdykun irgaktalbi varen* – heavy thunderstorms kill the horseflies (August). *Агдыкун иргакталби варэн* – сильные грозы убивают оводов (август). *Syngarburen* – the vegetal cover turns yellow (cf. picture below). *Сыӊарбурэн* – растительный покров становится желтым.	Moving: riding and packsaddle. We usually travel to the village to buy food and goods. Herding: many blackflies mean the reindeer come to the smoke fires in the camps; breeding reindeer scratch the skin off their young antlers *(irgilevkil)*; milking (the milk is rich in this season); the reindeer eat mushrooms. Hunting: same hunting activities as in the summer. Транспорт: Езда верхом и перевозка груза вьюками. Обычно ездят в поселок за продуктами и товарами. Оленеводство: Много мух и олени сами приходят на дымокуры к стоянке; производители вычесывают шкуру молодых рогов (иргилэвкил); дойка (богатый удой молока в этом сезоне); олени кушают грибы Охота: та же охота как и летом.
Bolony oren – autumn is installed. *Болони орэн* – Осень установлена.	*Abdannal tykchal* – foliage falls down. *Абданнал тыкчал* – листья и хвоя выпали.	Moving: riding and packsaddle, little nomadisation. Preparation for the winter: making new sledges, preparing and testing guns. Herding: breeding reindeer (bulls) mate and keep the herd gathered. Herders rely on the bulls to keep the herd gathered. Preparation for the winter: making new lassos, sledge reins, and harnesses. Women sew winter hunting shoes, hats, and gloves. Hunting: wild breeding reindeer come to the domestic reindeer: the men kill the wild reindeer to prevent the domestic herd from following them. Транспорт: Езда верхом и перевозка груза вьюками. Немного кочевок. Подготовка к зиме: ремонт и изготовление нарт, подготовка оружия.

	Abdannal tykchal – foliage falls down. Абданнал тыкчал – листья и хвоя выпали.	Оленеводство: У производителей гон и держат стадо собранным. Оленеводам удобно, что олени в это время находятся возле производителей. Подготовка к зиме: делают новые арканы, лямки для нарт и недоуздки. Женщины шьют зимнюю обувь, шапки и рукавицы. Охота: Дикие олени производители приходят к домашним маткам, эвенки за них и охотятся, чтобы они не уводили домашнее стадо.
		Moving: riding and packsaddle, little nomadisation because of the many technical tasks.
Bolony oren – autumn is installed. *Болони орэн* – Осень установлена.	*Lebgaren* – first snow falls and melts. *лебгарэн* – первый снег падает и тает.	Herding: most of the time is spent surveying the herd (to protect them from male wild reindeer and predators). Hunting: from the first snow, hunting to train the dogs takes place. Транспорт: Езда верхом и перевозка груза вьюками; мало кочевок из-за многочисленных технических работ в процессе подготовки к зиме. Оленеводство: Большая часть времени тратится на осмотр и защиту оленей от диких оленей и хищников. Охота: По первому снегу охотятся, чтобы учить молодых оленей.
	Dunne doŋkotoron – the ground freezes. *Дуннэ доӈкоторон* – земля замерзает.	Moving: riding and packsaddles, nomadisation to the place where people will meet the snow. Herding: because of the frozen ground, the reindeer do not leave tracks – it is hard to follow the herd. Hunting: waiting for the snow in order to hunt sable; hunting for wild reindeer. Транспорт: Езда верхом и перевозка груза вьюками; кочевка до стоянки, где люди будут ждать установления постоянного снежного покрова.

	*Dunne doŋkotoro*n – the ground freezes. *Дуннэ доӈкоторон* – земля замерзает.	Оленеводство: Трудно следить за стадом из-за замерзшей земли, в которой олени не оставляют следов. Охота: Ждут снег, чтобы охотиться на соболей; охота на диких оленей.
Bolony oren – autumn is installed. *Болони орэн* – Осень установлена.	*Imannaren* – it is snowing (winter snow). *Иманнарэн* – (зимний) снег идет.	Moving: shift to sledge transport. The nomadisations are organised according to the presence/absence of sable. Herding: each day, the reindeer are led to the camp, where herders choose which reindeer they will ride during hunts. The mating ritual has started or will soon start. Hunting: start of sable hunting with dogs; the wood grouse *(tetrao urogallus)* are fat and are hunted. Транспорт: Переход на транспортировку груза нартами (ездовые и грузовые). Кочевки большей степени организованы в соответствии присутствия или отсутствия соболей. Оленеводство: Оленей гонят каждый день на стоянку, где выбирают на каких оленях поедут на охоту. Гон начался или скоро начнется. Охота: Начало охоты на соболей с собаками; глухари жирные и на них охотятся.
Tuɣerburen – it is becoming winter. *Туҕэрбурэн* – начало установления зимы.	*Iŋinillen* – it is getting cold. *Иӈиниллэн* – холодно становится.	Moving: sledge. Nomadisations are organised according to the presence/absence of sable and towards places where there is less snow (allowing hunting with dogs). Herding: each day, the reindeer are led to the camp in order to go hunting. Hunting: mainly wild reindeer hunting and sable hunting with dogs and traps.

Tuyerburen – it is becoming winter. *Туҕэрбурэн* – начало установления зимы.	*Iŋinillen* – it is getting cold. *Иҥиниллэн* – холодно становится.	Транспорт: Транспорт на нартах. Кочевки в большей степени организованы в соответствии присутствия или отсутствия соболей и к местам, где поменьше снега (что позволяет охотиться с собаками. Оленеводство: Оленей пригоняют каждый день на стоянку, чтобы уехать на охоту. Охота: Большей частью охота на диких оленей и охота на соболей с собаками и капканами.
	Tuyeni oren – see also the many snow types and complex snow physics in snow and ice typology. *Туҕэни орэн* – см. тоже эвенкийские различные типы снега и сложную физику снега.	Hunting: if it is too cold, there is no wild reindeer hunting: because of the very noisy snow, the game will run away long before the hunter can approach. Fishing: net under ice (rare). Охота: если слишком холодно – не охотятся на диких оленей потому что снег очень шумный (дичь будет убегать от охотника). Рыбалка: ставят сетки подо льдом.
Tuyenikun oren – the winter happened. *Туҕэникун орэн* – зима установлена.	*Salgyn* – 1st warm and dry air. *Салгын* – первый теплый и сухой воздух.	Moving: sledge. Nomadisations organised according to the requirements of hunting. People organise transport to the village to purchase food and goods that will last to the end of the summer. Herding: each day, the reindeer are led to the camp in order to go hunting. The reindeer do not go too far. The females are not used for hunting anymore because they are pregnant. Hunting: easier to hunt wild reindeer (warmer and less noisy). No more sable hunting, the furs are no longer suitable for sale. Less wild reindeer hunting because the meat will dry out and turn black from the *salgyn* air: it is not good enough to sell. Fishing: a lot for food and selling. Mastering: the skins dry well and become soft.

Tuүenikun oren – the winter happened. *Туӈэникун орэн* – зима установлена.	*Salgyn* – 1st warm and dry air. *Салгын* – первый теплый и сухой воздух.	Транспорт: Нарты. Кочевки в большей степени организованы в соответствии правилам охоты. Люди собираются ехать в поселок, чтобы набрать продуктов и товары до конца лета. Оленеводство: Оленей пригоняют каждый день на стоянку, чтобы уехать на охоту. Олени далеко не уходят. Маток больше не используют в транспорте, потому что они беременные. Охота: Легче стало охотиться на диких оленей, потому что теплее и не шумный снег. Перестают охотиться на соболей, потому что шерсть уже не выходная (из-за потепления). Меньше охотятся на диких оленей на продажу, потому что мясо сохнет и чернеет от *салгын* воздуха и не годится для продажи. Рыбалка: много ловят для еды и для продажи. Мастерство: Шкуры хорошо сохнут и становятся мягкими.
Nelkini oren – the first spring is being installed. *Нйэлкини орэн* – Первая весна устанавливается.	*Dullen, imanna unillen* – the snow melts from the spring sun. *Дуллэн, иманна униллэн* – от весеннего солнца снег начинает таять.	Moving: sledge. Transport of food/goods to the storage houses for next year's nomadisation cycle. Herding: the reindeer do not move too far, they are sleepy and lazy. Males (castrated or otherwise) are separated from the females. Calves are born. Hunting: wild reindeer hunting stops for the mating season. People only hunt the wood grouse. Fishing: under the ice with nets. Транспорт: Тяжело – снег начинает таять, становится липким, мокрым, тяжелым, а ночью замерзает и создает корку. Тяжело ехать по полустаявшиему снегу и по неровной земле. Лужи от таяния снега везде появляются. Олени лениво идут.

Nelkini oren – the first spring is being installed.	*Dullen, imanna unillen* – the snow melts from the spring sun.	Оленеводство: Олени далеко не ходят, они ленивые и полуспящие. Производители или быки живут отдельно от маток). Период отела начался.
Нйэлкини орэн – Первая весна устанавливается.	*Дуллэн, иманна униллэн* – от весеннего солнца снег начинает таять.	Охота: На диких оленей уже не охотятся, только охотятся на глухарей. Рыбалка: рыбалка подо льдом.
Nieŋnerburen – 2nd spring is beginning.	*Imanna unen* – the snow is melting away.	Moving: nomadisation to the 2nd spring camp. Change to riding and packsaddle transport. Prepare enough firewood for until the advent of summer *(daldieren)*. Herding: the last calves are born. All human attention will be on the herd. Hunting: wood grouse *(tetrao urogallus)* hunting (main food of the season), very few wild reindeer hunted (prefer to hunt males, avoid females) for personal consumption (because of the breeding season). The female wild reindeer have migrated away.
Нйэӈнэрбурэн – Вторая весна начинается.	*Иманна унэн* – снег совсем растаял.	Транспорт: Кочевка на стоянку второй весны.[1] Переход на езду верхом. Готовят дрова в необходимом количестве к началу лета *(далдйэрэн)*. Оленеводство: Последние телята рождаются. Все внимание людей сосредоточено на оленях. Охота: на глухарей (главная еда сезона с рыбой), и иногда на диких оленей, но редко, из-за отела (предпочитают самцов убить и все равно самки мигрируют к Северу раньше самцов).

1 Согласно эвенкийскому календарю есть две весны:
1) нэлки – весна со снегом,
2) нйэӈнэ – весна без снега.

Nieŋne oren – the spring is installed. *Нйэӈнэ орэн* – весна установлена.	*Kukku kukurren, dunne unen, biral mudal* – The cuckoo sings, the ground melts, and the rivers become wide. *Кукку куккурэн, дуннэ унэн, бирал мудал* – Кукушка кукует, земля тает, реки большие.	Moving: the rivers are too wide (flooding), no nomadisation. Herding: the calves grow stronger. The reindeer walk a lot and spread out in different directions. Herders spend their time gathering the herd and leading reindeer back to the camp in the evening. Predators begin to attack the herd. Hunting: only wood grouse hunting. No other hunting (the wild reindeer are breeding and leaving for their summer lands; there is no time because herding takes up all attention). Транспорт: Реки слишком большие (наводнения) и не кочуют. Оленеводство: Телята становятся сильнее. Олени далеко уходят по 2–3 особи в поисках свежей зелени. Оленеводы тратят свое время, собирая оленей, чтобы пригнать стадо на стоянку. Хищники начинают атаковать. Охота: только на глухаря. Другой охоты нет (у диких оленей отел и они мигрируют на Север к их летним землям и все внимание людей обращено на домашних оленей).

And the cycle begins again…

Syŋarburen / Сыӈарбурэн (© photo Vasilii Gabyshev, 2016)

Some seasonal shift concepts require more explanation, like *marmaren, марамарэн, irganen, ирганэн*, and the deductions made from the evolution of larch needles. One can see below how traditional Evenki weather knowledge is based on interactions between several elements of the natural environment (vegetal cover, insects, fish, and so on).

Marmaren / Мармарэн	Period of the springtime when leaves and mosquitoes appear Период весны когда появляются листья и комары

Marmaren, marmandieren is said when larch needles, long grass, and bush leaves start to appear: the upper layer of the ground thaws and the spring puddles dry out. Mosquitoes start to fly around. Young people have forgotten this term.

Мармарэн, Мармандйэрэн гунэвкил мучуктэ, чука, абданна элэкэс йудйэрэкин, дуннэ ойгу унэкин, таир олгороктын. Мармакатал дэҕиктэлэвкил. Эдырил тар турэнмэ омҥочол.

Мармарэн, Мармандйэрэн называют когда хвоя, длинная трава и листья кустарников только начинают выходить, также когда верхний слой земли оттаял, и весенние лужи высохли. Тогда отмечается появление комаров. Молодые люди забыли это слово.

Irganen, Irgaŋdieren / Ирганэн, Иргаҥдйэрэн	A strong wind is a sign of the appearance of horseflies (in June) Ветер сильный – знак появления оводов (в июне)

At the beginning of the summer, when leaves and needles appear on the bushes and trees at the end of June *(daldyn)*, large clouds come and bring thunder and small typhoons. After this thunder, the Evenki say *'irganen'*, meaning that the horseflies will appear soon and sting the reindeer. The reindeer will come to the encampment to be close to the smoke fires: the herders will no longer have to search for reindeer all day long in the forest. In addition to the horseflies, ants and other insects come from the sky during this period. After this, it hails and thunders.

One may note here the systemic character of indigenous knowledge, with this perception of interactions between clouds, winds, the growth of vegetation, and the appearance of insects after the long winter. Evenki knowledge is centred on the interaction between climatic, vegetal, and animal elements.

Элэкэс далдйэрэкин мучуктэл чукал йучалйатын агдыкучивки сэҕиндйэнэкун тар амардукин ирганэвки. Иргактал дэҕиктэлэвкил. «Иргандэрэн» тара гунэвкил. Туксукун эмэдйэрэкин сэҕикучивки; таду тар иргактал, ириктэл буҕадук тыкивкил. Тадук бонакучивки һэгдыкур тыкивкил агдыдйанакун.

В начале лета когда листья на деревьях распустились и трава выросла в конце июня *(далдын)*, приходит большая туча с громом и ураганом. После такого грома, эвенки говорят *«ирганэн»* – это означает, что скоро появятся оводы, которые будут кусать оленей. Те будут приходить на стоянку к дымокурам, и пастухи могут больше не ходить за оленями в лес. Кроме оводов, муравьи и другие насекомые падают с неба. После идет град с громом.

Muchuktel ulargara, burullu – ollol iiergille – daγaly imannaldiŋan / Мучуктэл уларгара, буруллу – оллол ийэргиллэ – даҕалы иманналдиҥан	When the larch needles turn yellow and fall to the ground, it means that the fish will go downstream: this means that it will snow soon. Когда хвоя желтеет, потом падает, рыба спускается по течению, значит, что скоро снег будет.

For as long as larch needles fall down to the ground, the fish will go downstream. Just as the needles start to fall, the Evenki set up nets to catch the fish. This lasts for three to four days.

Адыллэйэ мучуктэл тыктэл, тарбанма оллол ийэргивкил. Тадук эвэнкил мучуктэ тыкчэрэкин адылчавкил оллолво дйавадйэнэ. Тара илалэйэ, диҕилэйэ бивки.

Сколько дней хвоя падает и столько дней рыба спускается по течению вниз. Как только хвоя начинает падать, эвенки ставят сетки, чтобы ловить рыбу. Это происходит в течение 3–4 дней.

Indigenous temperature measurements

by A. Lavrillier and S. Gabyshev

Traditionally, the Evenki did not measure temperatures with numbers or a numerical system. Even recently, the nomads did not own thermometers and did not pay attention to the number of C degrees; they have other ways of measuring the temperature and the cold.

It seems that this measurement system is based on signs visible in the surroundings and on the human body. For instance, they observe the extent to which the branches of the pinus pumila are raised or lowered, frost on plants and trees, the behaviour of birds, the various states of snows and ices, good visibility or different kinds of fogs, the circulation of sounds, and their own skin or breathing (cf. Indigenous Science of climate).

The measurement of 'cold', 'warm' and 'hot' in this region (where temperatures reach +35°C) on a scale between 0°C and -50°C/-60°C without a thermometer is expressed in the following ways:

- *Soŋun, soŋukikin / соӈун, соӈукикин* – (all seasons) – cool, very cool – прохладно.
- *Eltandieren / элтандйэрэн* – (all seasons) – weather in the morning and evening is colder than during the rest of the day – холодает утром и вечером.
- *Iŋinie, iŋiniekun / иӈинйэ, иӈинйэкун* – (in winter) – cold, very cold (in the 1990s, from -40°C; in the 2000s, from -30°C) – морозно, очень холодно (в 1990-ые годы: от -40°C; в 2000-ые годы: от -30°C).
- *Ikterebda / иктэрэбда* – (all seasons) extreme cold and freezing: people freeze and their faces and clothes are covered by hoar frost (in the 1990s, from -45°C; in the 2000s, from -35°C) – замерзать (о живом существе), сильно холодает, много иней на лице и одежде (в 1990-ые годы: от -45°C; в 2000-ые годы: от -35°C).
- *Dullen, dulcha / дуллэн, Дулча* – springtime warmth from the sun in clear weather that makes the snow melt (around +10°C) – весной днем, когда солнце греет сильнее, из-за тепла начинает таять снег.
- *Oku- (okullen, okulkuchivki) / Оку- (окуллэн, окулкучивки)* – it is hot, very hot in summer (between +25°C and +30°C) – жарко, очень жарко летом (от +25°C +30°C).
- *Kuraŋ, Kuranen / Кураӈ, куранэн* – heat wave in the summer (from +35°C) – пик жары летом (от +35°C) (for further details, see section winds and airs).

According to Lavrillier's fieldwork from 1994 to 2017, the various notions or feelings of 'cold' and 'freeze' have shifted together with the development of climate change. Two things may have caused this change: because of general warming, people are no longer used to harsh frost (which they face more and more rarely because of climate change), and/or the cold is strongly augmented by humidity in the air. Interviews given by the herders seem to support both explanations. (cf. Winds and airs typology)

Knowing and 'using' the microclimates

The Evenki organisation of physical space allows for the discrete occupation of the area and the sustainable use of natural resources, be they hunted or allotted to domestic life (domestic reindeer, encampments). This management of natural resources is optimised by perfect knowledge of the many local microclimates and the principle that the Evenki express when they say '*extend the best part of each season*' or '*travel in time*', which is based on the use of microclimates. They use this principle in two ways: in the context of a normal year, they benefit for longer from the better parts of the seasons, while, in the context of an abnormal year, they adapt by using microclimates. The positions of seasonal camps in particular are chosen according to this knowledge.

For example, a camp of the second spring *(nieŋneni, нйэӊнэни)* is situated at the midstream of a river. In several weeks, the foliage develops there: the forest becomes green, flowers grow, and temperatures rise. Since the reindeer need fresh foliage (i.e., foliage which has just flowered) and do not like hot temperatures, the Evenki then nomadise upstream towards cooler zones, but only by several dozens of kilometres. In this camp, the snow still covers the ground: it is still the first spring *(nelkini, нэлкини)*. Nomads and reindeer can once again benefit from spring processes and grazing soils, especially the young protein-rich tussock cotton-grass *(nirgakte, ниргактэ)* and the young leaves of dwarf birch *(oktalyk, окталык)*. (cf. Vegetal cover typology)

Some months later, in the summer camp, the season comes to an end: the forest turns yellow and brown and the ground starts to freeze. The nomads decide to re-live some of the summer and go downstream to a camp where it is still summertime.

So, the taiga is not a 'savage' place, occupied and travelled over in an aleatory fashion: it is definitely an organised, managed, and sometimes even maintained space.

Let us now present in more detail the knowledge about microclimates (including snow conditions). Our work of co-productive documentation has given us a more detailed explanation of such knowledge, as is shown in the diagrams. One can see in these diagrams (cf. diagram Microclimates) that the topography of the landscape plays an important role in the existence of such microclimates (temperatures and snow depth and qualities) and that this is conceptualised by the Evenki. As the Evenki explain it, in the **winter** a camp situated at a higher altitude (upstream) is (relatively) warmer and there is more snow than downstream. In the winter, the reindeer will lose weight if there is too little snow and it is cold because they will walk around a lot to stop themselves from freezing. In contrast, in order to maintain their weight the reindeer need warm weather (they will not have to walk a lot to get warmer) and a lot of snow (to keep them grazing in one place). In addition, the Evenki explain that a thick snow cover provides insulation from the frozen ground and offers warmer air. So, during the winter the Evenki often like to choose places at a higher altitude or upstream. The other advantage of such places is that the thick snow cover protects the reindeer against wolves, which cannot walk in deep snow (cf. Topographic typology). In spring, at an altitude and upstream, the snow melts later and it is colder than downstream, while downstream the springtime process of the appearance of the foliage starts earlier. The nomads will tend to move to such places at the beginning of the spring. Then, as mentioned earlier, to extend the best part of the springtime and offer new pastures to the reindeer, they will move upstream. In the **summer**, it is much fresher at a higher altitude, so the Evenki will tend to stay there (cf. Topographic typology). After this, they can stay move a short distance downstream to benefit a little more from the end of the summer end. In the **autumn**, at a higher altitude or upstream it is much colder and the snow falls earlier then downstream, where it is warmer and the snow is installed later. The nomads will stay at a higher altitude or upstream to start the snow season early. (cf. diagram Microclimates)

Knowing and observing the winds

As the reader can see throughout the chapter on Evenki climatology (including snow) and also in the typologies of topography and the vegetal cover, the nomads know the winds and how they change. The study presented below was made by S. Gabyshev and is a combination of the traditional knowledge he received from elders, his own observations as a herder-hunter before his contribution to the BRISK project, and his daily observations in the framework of BRISK (cf. Winds and airs typology: diagram Compass rose). As he has stated, his participation in BRISK monitoring has increased the intensity of his observational abilities and, consequently, enriched his knowledge (cf. Introduction). Gabyshev and Lavrillier co-realised the presented diagrams. Here, we can definitely refer to the co-production of knowledge, with indigenous knowledge subdivided into several layers or kinds: 1) knowledge inherited from elders, 2) one herder-hunter's traditional observation and experimentation according to ancient nomadic ways of doing things, 3) hybrid observation and experimentation that combine a traditional approach with new multidisciplinary scientific methods developed in BRISK, and finally 4) an individual conclusion made by a herder involved as co-researcher in an international project (cf. Introduction). The results (shown in the diagram Compass rose) demonstrate that in the winter and snowy spring, there are regular and ordered changes in the winds. This contrasts with the situation during the summer, when there is no specific regular order (that was observed, at least). In the winter and snowy spring, one finds the phenomenon of *namuscheren, намусчэрэн* (a change in the wind that brings warmth), which is well known by the Evenki. Thus, the observations of Gabyshev and the hypothesis provoked by daily observation in the context of BRISK monitoring show the following specific order of winds. First, a wind comes from the south *(namuscheren, намусчэрэн)* and brings warmth; second, the wind comes from the east and brings clouds and light wind; third, a northern wind brings snow. Then the wind changes, comes from the west, and brings clear weather, and so on. In the summer, with the exception of *namuscheren, намусчэрэн,* which brings warmth and rain, the same winds bring other events: the wind coming from the north brings enduring rain, while wind from the west brings, most of the time, thunderclouds (cf. Winds and airs typology: diagram Compass rose, Evenki climatology: namuscheren, Clouds typology). More detailed documentation of winds and warm and cold airs is provided in the section 'Winds and airs typology'.

'Knowing that' and 'knowing how'?

Regarding the theoretical discussion on the features of cognitive systems or systems of knowledge, let us note here that, as in other parts of indigenous or scientific knowledge, the rules for wind changes belong to 'knowing that' and not to 'knowing how', because no explanation is given. In this book, we will describe other aspects of knowledge that belong to 'knowing how' (for instance, regarding the circulation of air, cf. Winds and airs typology: diagram Idia) or the circulation of sound in relation to future weather (cf. Evenki climatology: River sounds weather). In the diagram on rivers, sounds, and weather, one can see the complexity of Evenki 'know-how', with a sophisticated explanation of the interactions between future weather, the present humidity of the air and its sources (sky, river), air pressure, human blood pressure and bone pain, and the circulation of sound from up- and downstream.

Means of forecasting the weather and climate: the notion of acting upon the weather

by S. Gabyshev and A. Lavrillier

Weather forecasting from observing the wind and air

Diuɣadieren / Дйуҕадйэрэн	When the air in the summer becomes as light as smoke/fog after rain. This is a sign of good, bright weather.
	Когда летом воздух становится, как дым, мутным, после дождей. Это признак к наступлению хорошей ясной погоды.

Only old people know this word nowadays.
Только пожилые люди знают такое слово сегодня.

Lamuscheren, namuscheren / Ламусчэрэн, намусчэрэн	Abrupt changes in the winds leading to warming (with rainfall or snowfall)
	Резкое изменение ветра к потеплению (с дождем или снегом)

A warm wind blows from the south towards the north, and the sky is covered with semi-transparent clouds. The sun can be observed with difficulty through the clouds. Such an event is called *lasmuscheren* and brings rain in the summer and snow in the winter. (cf. Winds and airs typology: diagram Compass rose)
Югтук адын нйамакикин адынэлэвки севертыкаки туҕурбувки. Дылача тэпиривур туксувэ ичэвувки нэһилэ – тара гунэвкил ламусчэрэн тыгдалдиҥан-у, иманналдиҥан-у.

С южной стороны теплый ветер дует на север и небо покрывается просвечивающими тучами. Солнце сквозь тучи еле видно – эвенки это называют *ламусчэрэн* – это к дождю или к снегу. (см. схемы, с. 223–225)

Lamuscheren iŋinikatchane / Ламусчэрэн иӈиникатчанэ	Northern wind with snow, a sign of future cold Северный холодный ветер со снегом – к холоду

This expression designates the moment when light snow *(keyaradieren)* falls and there is a relatively moderate northern cold wind. It is a sign that the harsh cold will start very soon. While waiting for the harsh cold, the Evenki hunt wild reindeer.

Иманнадйэнэ кэраҕадйэрэн адын севердук иӈинидйэнэ адыктэллэн – иӈинилкучиӈан. Эдйэлин иӈиниллэ экуравэр бултада бэйуктэмнэк.

Это выражение указывает на момент, когда идет очень мелкий снег *(кэраҕадйэрэн)* и ветер северный холодный несильно дует – это знак, что холод наступит. Пока холода не настали, эвенки охотятся на диких оленей.

Uŋniarikteren / Уӈнйариктэдйэрэн	Changeable strong winds that bring brief snowfalls, rain showers, or hail Часто меняющиеся сильные ветра, ведущие к кратковременным осадкам (снегу, дождю, граду)

Ungniariktadieren means: 1) When there are strong changeable winds that bring clouds, short snowfalls or showers will follow; 2) Brief snowfall (fluffy/bushy snow), rain showers, or hail; 3) Muddy weather during snowfalls in springtime or rain showers in summer.
If there is *uŋniariktadieren* in the evening, one knows there will be clear weather all night and for the next three or four days. (cf. Snow and ice typology, Precipitations typology)

«Уӈнйарикдйэрэн» тара гунэвкил адынэвки, тадук иманнавки эмиски нйэгдэлэвки тар инэӈнун бивки. Сачйадйаӈнаны кэҕа нйэгдэлэдиӈан долбонывэ. Тадук илалэйэ, диҕилэйэ уӈнйаридиӈан. Чаҕидадун нйэгдэлучэвки.
«Уӈнйариктэдэрэн» это: 1. Когда сперва сильный ветер дует, принося тучи, потом перестает, а потом идет кратковременный снегопад или ливень; 2. Кратковременные снегопады (пышный густой снег) или дожди (ливень или мелкий

град); 3. 'Мутная погода' во время весенних снегопадов или летних ливней. Тогда уже знают, что к вечеру будет ясная погода на всю ночь и продолжится на трое или четверо суток. (см. с. 243–367, 213–216)

Weather forecasting by observing the sun

Dylacha diulacha – iŋineldiŋan / Дылача дйулача – иӈинэлдиӈан	Literally 'the sun has built a house for himself'. If there is a ring (rainbow-like) around the sun, it means that it will be very cold (around -35/-40°C).
	Буквальный перевод: «Солнце себе домик сделал»; Смысл: если кольцо радуга (гало) сформировалось вокруг солнца, значит очень холодно будет (примерно -35/-40 °С).

This term designates that a rainbow surrounding the sun is a sign of future harsh cold for several days. When it is very cold, the Evenki repair sledges and make various items at home.

Тар сэкалаврэн дылачаду мурули кэҕа тэҕэлтэнэ иӈинилдиӈан адылладудэ. Сыргалва иӈиниллэкин авйэстанарум. Дйуду авалыктадйанарум.

В зимнее время когда вокруг солнца образуется радуга, значит холод наступит на несколько дней. Эвенки в это время ремонтируют нарты и дома работают.

Ulaly dylacha iendieri – adynnildiŋan; Dylacha ulaly iendieri – tygdaldiŋan / Улалы дылача иэндйэри – адынилдиӈан; Дылача улалы йудйэрэкин – тыгдалдиӈан	A red sunset means that the wind will blow; a red sunrise means that it will rain
	Красный закат солнца – к ветру; Красный восход солнца – к дождю или к снегу (летом и зимой)

Weather forecasting from observing the human body and feelings

Beiel amadiere imannadiŋan badaɣa / Бэйэл амадйэрэ иманнадиӈан бадаҕа	When people feel drowsy, it means that there is good chance that it will snow
	Когда у людей вялость, значит снег будет, наверно

When people want to sleep and feel drowsy, in particular during the springtime, it is a sign that it will probably snow.

Окин бэйэл амадйэрэ, нэһилэ горидйэрэ, иманналдиӊан-у, тыгдалдиӊан-у бадаҕа?
Когда у людей вялость, значит снег будет, наверно; физическая слабость – к снегу, особенно весной.

Algarbi okudiere imannaldiŋan /
Алгарби окудйэрэ иманналдиӊан

If one's legs feel warm unexpectedly, it means that it will snow soon

Если ногам становится жарко без причин, значит, что скоро снег будет

If in the evening a hunter feels his/her legs burning (i.e., warming up), it means that there will be snowfall and he/she will suffer a headache.
In several of the quoted means of prediction, we note that the Evenki (as with other indigenous peoples of the north) use their body and feelings as barometers for predicting the weather. In addition, they use (as do many indigenous peoples in the world) the behaviour of animals, be they domestic or wild.

Кэҕа уклаһинми алгары окудйэрэ – иманналкучиӊан, дылис онулдиӊан.

Если охотник вечером ложится и ноги «горят» – это к снегу и голова болеть будет.

Weather forecasting from observing the moon

Bieγa diulacha – iŋineldiŋan /
Бйэҕа дйулача – иӊинилдиӊан

Around the moon, there is a rainbow-like ring – it will be very cold. Literally, 'the moon made a house'.

Вокруг луны образовалась радуга (гало) – это к холоду; буквально «луна себе дом сделала».

If the hunter comes back home in the evening and there is a rainbow-like ring around the moon (i.e., if the 'moon built a house'), it means that it will be very cold and that one needs to prepare more firewood.

Долбо мучудйами бйэга дылача иӊинэлдиӊан молыда дйуду надо.

Вечером когда охотник возвращается с охоты, если луна «дом себе сделала», т.е. если вокруг луны есть радуга кольцо (гало), значит холода наступят и надо дров готовить домой побольше.

Bieɣa oŋkaptyki baldycha – iŋineldiŋan /
Бйэҕа оӊкаптыки балдыча – иӊинилдиӊан

If the moon appears in the sky upside down (bottom down, slightly to the side), it means that there will be bright weather and harsh cold for a long time.

Бйэҕа оӊкаптыки балдыча – горокунмэ нйэӊдэлучэдиӊан энэ иманнара иӊиниктэдиӊан.

Если луна родилась лежа спиной и слегка набок, это к долгой ясной погоде и похолоданию (зимой).

Bieɣa ilgimaptyki baldycha – niamaldiŋan /
Бйэҕа илгимаптыки балдыча – нйамалдиӊан

If the moon appears in a standing position, it means that temperatures will rise, which is good weather for hunting wild reindeer.

Бйэҕа илгимиптыки балдыча, нйамалдиӊан, бэйуктэдйэми айа.

Луна родилась стоймя, это знак что потеплеет и что будет хорошо охотиться на диких оленей.

Bieγa uklaptyki baldycha /
Бйэҕа уклаптыки балдыча

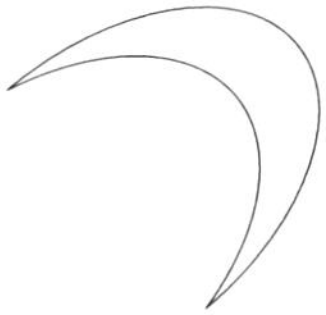

If the moon is lying down (bottom up and slightly to the side), it means that it will either rain or snow. If there is snow, the Evenki hunt wild reindeer and sable. (cf. Precipitation typology: tygda)

Бйэҕа уклаптыки балдырэкин иманналдиӈан-у, тыгдалдиӈан-у. Экурвэл удйарадиҕарэ бултанавкил бэйурэ, андаһилэ.

Если луна лежа родилась и слегка набок – это к снегу или дождю. Всякий зверь оставит след на свежем снегу и эвенки едут на охоту на дикого оленя или на соболя. (см. тыгда, с. 213)

Bieγa umkulbudiene baldycha diugha –
tygdaldiŋan ereγer tar beγadu;
Bieγa tygdarevachechan /
Бйэҕа умкулбудйэнэ балдыча дйуҕа
– тыгдалдиӈан эрэҕэр тар беҕаду;
Бйэҕа тыгдарэвачэчан

During the summer, if the moon
appears upside down (bottom up)
at the beginning of the month,
it will be rainy throughout the month;
the moon caused the rain (it will rain).

Если летом луна в начале месяца
родилась «выливая» (т.е. дно вверх,
то все лето будут идти дожди;
Луна дождь сотворила (дождь будет).

In the summer, if the moon at the beginning of the month looks like it is turned upside down (as if it has been 'poured out'), it means that the entire month will be rainy. For the Evenki, this is a good thing, since the reindeer will be able to graze and gain fat. Indeed when it rains, there are much fewer mosquitoes and horseflies: the reindeer can graze rather than spend all day at the centre of the encampment close to the smoke fires.
In several expressions about weather forecasting, it is said that the elements of the natural environment or animals are willingly changing the weather through their behaviour. Here it is said that the moon provokes the rain.

Бйэӡа умкулбуптыки балдыча дйуӡа тар бйэӡаду тыгдамар бидиӊан, орор бургудиӡарэ тар айа.

Если летом луна в начале месяца родилась сверху вниз, т.е. как будто «выливая», то весь месяц будут дожди. Для эвенков это хорошо, значит, что олени поправятся и жирными будут. Ведь при дожде намного меньше комаров и оводов летают, и олени могут ходить кормоваться, и не стоять у дымокуров в центре лагеря и кормиться не только вечером и утром, а весь день.

Bieγa ilgimadiene baldycha diuγani – aia bidiŋan: dychakan, tygdakan tar bieγadu /
Бйэӡа илгимадйэнэ балдыча дйуӡани – айа бидиӊан: дылачакан, тыгдакан тар бйэӡаду

In the summer, if the moon appears in a standing position at the beginning of the month, it means that throughout the month there will be good weather (a good balance of sun and rain). This is good weather for the reindeer.

Бйэӡа илгимаптыки балдыча дйуӡа – айа бидиӊан: нйэӊдэлучйэнэда тыгданэда – ороpду айа.

Если летом луна появилась в начале месяца стоймя, то будет хорошая погода, т.е. достаточно солнца и достаточно дождей. Оленям хорошо будет.

Bieγa ukladiene baldycha diuγani – okukun bidiŋan tar bieγadu /
Бйэӡа укладйэнэ балдыча дйуӡани – окукун бидиӊан тар бйэӡаду

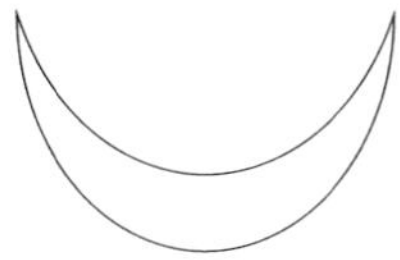

In the summer, if the moon appears to be lying down (bottom down) at the beginning of the month, it means that there will be a heat wave throughout the month.

Бйэӡа укладйэнэ балдыча дйуӡа тар бйэӡаду окукакун бидиӊан.

Летом если луна появилась лежа (дно вниз), значит в этом месяце жара будет.

Bieγa duγalin emeril manapdierekin – iŋinekicheren, taduk etan imannara /

Бйэҕа дуҕалин эмэрил манапдйэрэкин – иӊинэкичэрэн, тадук этан иманнара

If the moon shows its sharp tips at the end of the month, it will be very cold and will not snow.

Бйэҕа дуҕалин эмэркур манапдйэрэкин иӊинилдиӊан этан иманнара.

В конце лунного месяца, если луна показывает острые концы – холодно будет и снега не будет.

Bieγa manavullen imannaldiŋan-u, tygdaldiŋan-u /
Бйэҕа манавуллэн иманналдиӊан-у, тыгдалдиӊан-у

When the moon dwindles, the Evenki always expect several days of snow or rain.

Когда луна кончается, эвенки всегда ожидают снег или дождь на несколько дней.

Changing weather: Human-environment interactions

Nakata amikanni gibda imannadan /
Наката амиканни гибда иманнадан

One shakes a bear skin to provoke snowfall and wind in the winter and rain and wind in the summer

Трясут медвежью шкуру – чтобы вызвать снег и ветер зимой или дождь и ветер летом

If one shakes a bear skin, it will provoke strong snowfall and wind in order to hide the bear's tracks. If a bear leaves a den or if it is frightened by someone, the power to provoke strong snowfall and winds is attributed to the animal: the weather will thus hide the bear's tracks from hunters.
In the nomadic world, people (and, more rarely, elders in villages) still use this method when they really want to change the weather, for instance, when there are forest fires close to the village (as in 2015).

Амикан накатван гибдами иманналкутчиҥан адындйэкутчиҥан, адыкун удйалва унңэдиҥан. Амикан дйудукви йуми-у, илбэвми-у буӡавэ иманналивканэвки адын микачэвки удйави дйайукачэвки.

Если трясти медвежью шкуру, будет сильный снег и сильные ветра, которые заметут его следы. Если медведь с берлоги вышел, или кто-то его спугнул, медведь может вызвать сильные снегопады и ветра, чтобы свои следы замести от охотника с помощью снегопада и пурги.

Kikarida adyneldan / Кикарида адынэлдан	One whistles to cause wind Свистят, чтобы вызвать ветер

One whistles in order to make the wind blow. If there is no wind, the Evenki whistle: it is held to attract the wind, which will help the smoke fires produce good smoke and thereby protect the reindeer from horseflies. Sometimes, one whistles when it is too hot so that one can be cooled by the wind.

Кикарими адыниливки. Эрэкин адынэ кикаривкил самңил айат нуныдатын. Адылдун кикарими олус оку бирэкин айа овки.

Свистят, чтобы вызвать ветер. Если нет ветра, эвенки свистят и таким образом вызывают ветер, чтобы дымокуры задымили хорошо и чтобы оводы оленей не кусали. Иногда свистят, когда слишком жарко бывает, чтоб прохладно стало.

Weather prediction according to the behaviour of birds

Chipkachar deɣiktediere ergili – tygdakichare / Чипкачар дэӡиктэдйэрэ эргили – тыгдакичарэ	Birds fly low, a sign of rain or snow Птички летают низко – к дождю или снегу

When small birds fly low, it is a sign that it will rain. All items likely to get wet must be gathered into the storage houses.

Чипкачар эргили дэӡиктэдйэрэ тыгдарикитчэрэ. Улапңатылэ тэткэлэ сайбаду навкил.

Когда птички низко летают – это к дождю. Все, что может намочиться, собирают в амбары.

Chipkachar deγipchediere uγil eγedienel – nieŋdelekichara /
Чипчачар дэҕипчэдйэрэ уҕил эҕэдйэнэл – нйэӊдэлэкичара

If small birds fly high in the sky, it is a sign that there will be clear weather (a good time to dry clothes).

Чипкачар уҕили дэҕиктэдйэрэ нйэгдэлэдиӊан, тэткэлвэ олгида.

Когда птички летают высоко – это к хорошей погоде, тогда эвенки в это время сушат вещи.

Olyl uγili deγiktediere, moŋnoldionol – imannaldiŋan-u, tygdaldiŋan-u /
Олыл уҕили дэҕиктэдйэрэ, моӊнолдодйонол – иманналдиӊан-у, тыгдалдиӊан-у

If the ravens fly high in the sky whilst playing and cawing loudly, it is a sign of rain or snow.

Олыл уҕийали дэҕиктэдйэрэ моӊнолдодйонол – тыгдалдиӊан-у, иманналдиӊан-у.

Если вороны летают высоко, балуясь, и кричат – это к дождю или снегу.

Kukku olgokin modu teγetcheren – nieŋdelediŋan / **Кукку олгокин моду тэҕэтчэрэн – нйэӊдэлэдиӊан**	Cuckoos sit on dry trees, a sign of clear weather Кукушка сидит на сухом дереве – к ясной погоде

In addition, when the cuckoo stops cuckooing, the berries are ready to be gathered. *The cuckoo is considered to be the shaman-bird, or the spiritual 'double' of the shaman, and is able to predict the weather and important life events. During the springtime, the Evenki await the first song of the cuckoo with eagerness: it is held to start the long-awaited summer re-birth of the natural environment. Many proscriptions, prescriptions, and habits are attached to the shaman. In addition, many plants are thought to be attributes of the shaman: a lasso (the larch ivy), a hat (the snowdrop), tobacco* (Lycoperdon, *see picture), etc. (Cincius 1975: 426; Lavrillier 2005).* (cf. Vegetal cover typology)

Кукку этэрэкин куккурэ – диктэл ирчал.

Когда кукушка перестает куковать, это значит, что ягоды поспели и что можно их собирать.

Кукушка считается птицей-шаманом, или двойником шаманов, которая может погоду предсказать, также как и другие важные события. Весной таежники ждут с нетерпением ее первое пение, которое считается стартом долгожданного летнего возрождения природы и многие приметы связаны с кукушкой. Кроме того много элементов природы считаются атрибутами кукушки, как например его лассо (плющ), его шапка (подснежники), его табак (дождевик/табачный гриб). (см. с. 149)

Kukku ulapkun modu teɣetcheren – tygdaldiŋan / Кукку улапкун моду тэҕэтчэрэн – тыгдалдиӈан.	If a cuckoo is sitting on a fresh and wet tree, it is a sign of rain Если кукушка сидит на живом мокром дереве – это к дождю

Weather prediction according to sounds from the river

Bira doldybdieren soloki – tygdarikitcheren / Бира долдыбдйэрэн солоки – тыгдарикитчэрэ	The river is making noise upstream, a sign of rain Река шумит в верху течения – это к дождю

If one hears the noise of the river and it sounds as if it is coming from upstream (even though one is standing by the middle of the river), it is a sign of rain. (cf. diagram River sounds weather)

Уригдан дулиндун-у, бираду-у тэҕэтчэми бираду солоки эйан долдыбдйэрэн тыгдалдиӈан.

Если посередине плеса реки или посередине реки стоять и услышать шум течения, исходящий как будто снизу реки, то это знак к заморозкам и ясной погоде. (см. схемы, с. 201–202)

Bira doldybdieren edili – eltandiŋan / Бира долдыбдйэрэн эдили – элтандиӈан	The river makes noise downstream, a sign of clear weather Река шумит внизу – это к ясной погоде

If one hears the noise of the river and it sounds as if it is coming from downstream (even though one is standing by the middle of the river), it is a sign of clear weather and frozen conditions. (cf. diagram River sounds weather)

Бира эдили долдыбдйэрэкин элтандэҕан нэгдэлючэдэҕан.

Если посередине плеса реки или посередине реки стоять и услышать шум течения, исходящий как будто снизу реки, то это знак к заморозками и ясной погоде. (см. схемы, с. 201–202)

Weather forecasting from observing fire (shapes of the coal and sounds)

Chatukur / Чатукур

When the burning coals are big, a harsh cold will arrive.

Печка чатурбурэн (һэгдыкур чатукур ора) – иӊинэлдиӊан.

Когда в печке горячие угли большие – это знак что сильно холодно будет.

Otu aiat degdedieren – niamakitcheren /
Оту айат дэгдэдйэрэн – нйамакитчэрэн

During the winter, if the wood burning stove is burning well and makes sounds like 'pfu!, pfu!, pfu!', it is a sign of warming.

Оту айат дэгдэдйэрикин долдыбдйэнэ «пфу пфу пфу» – нйамакитчэрэн.

Если зимой печка хорошо горит, производя звуки похожие на «Пфу! Пфу! Пфу!», это к теплу.

Otu adykitcheren eɣediene /
Оту адыкитчэрэн эҕэдйэнэ

The wood burning stove is 'provoking the wind' by whistling and blowing

Печка «поет», ветер делает, это к ветру

If the wood burning stove is blowing and whistling as it burns, it is a sign of winds and storms.

The Evenki think that in each fire there is a spirit which people need to feed (by throwing a piece of human food into the fire) before lunch. The power to predict the weather and important life events is attributed to fire (so long as one makes the offering properly and only if one knows how to understand the signs given by the fire's spirits). Through the fire, one can communicate with the other spirits of the natural environment; if fed properly, they offer good luck in hunting and reindeer herding. In other words, the fire feeds the living and the dead, since it feeds living humans by offering good hunting and one can feed the spirits of the ancestors through it. The fire is considered to be a channel of communication between the three worlds of the universe. For instance, by burning the clothes of a dead person, the Evenki believe they send these clothes to the world of the dead.
This expression literally means that the fire itself provokes storms and strong winds. In most expressions relating to weather forecasting, the literal translations indicate that the Evenki believe that dogs, birds, fire, the moon, and the sun cause the weather.

Оту дэгдэдйэрэн кикариһиндйэнэ, эӡэдйэнэ – адынилкучиӊан.

Если печка горит со свистом и пением (т.е. производя звуки как будто свистит, дуя), это к ветру и сильной пурге.
Эвенки считают что в каждом огне есть дух, которого нужно кормить (бросая немного пищи в огонь) при каждом приеме пищи. Этот дух предсказывает погоду, также как и важные события жизни тому, кто его кормит и умеет его понимать. Через него общаются с другими духами природы, которые при угощении приносят удачу на охоте и к благополучию в оленеводстве. Огонь кормит и живых и мертвых, поскольку живым добычу приносит и через него кормят мертвые души предков. Также огонь считается каналом общения между 3мя мирами вселенной. Например, сжигают вещи покойников, чтобы их передать в мир мертвых.

Weather prediction according to smoke circulation

Nuny uɣiskaki nunydieren – nieŋdelekitcheren / Нуны уӡискаки нуныдйэрэн – нйэӊдэлэкитчэрэн	The wood burning stove is 'provoking the wind' by whistling and blowing
	Печка «поет», ветер делает, это к ветру

If smoke goes from the chimney stack straight into the sky, it is a sign of clear weather.

Нуны турбадук уӡискаки нуныдйэрэн – нэӊдэлэкитчэрэн.

Если дым из трубы идет вверх в небо – это к ясной погоде.

Weather prediction according to the behaviour of domestic animals

Ninakir sogdonovi imannadu oɣidieren, moŋnoldodionol – imannakitcheren adynmykatchene / Нинакир согдонови иманнаду оҕидйэрэн, моңнолдодйонол – иманнакитчэрэн адынмукатчэнэ	When dogs scratch their backs on the snow, it is a sign of snowfall Собака трет спину об снега, это к пурге и к снегу

During the winter, if dogs lie on their backs and scratch them on the snow whilst also playing, it is a sign of snow and wind. If they do the same during the summer whilst also pointing their open mouths at the sky, it is a sign of rain (if their mouths are closed, this is a sign of fire).

Нинакир согдоннотпи иманнаду оҕидйэрэ моңнолдйоно иманнакичэрэн адынмукачэнэ. Нинакин согдонотпи укладэнэ амңави ничэдэнэ уҕискаки дйуҕаниду тыгдалдиңан, амңави саммичэдйэнэ окукулчиңан.

Если собака на спине валяется, чистясь, играя на снегу, зимой это к снегу и ветру. Если в летнее время собака валяется на спине в земле, играя с открытым ртом вверх, это к дождю, а если она это делает с закрытым ртом это к жаре.

Eŋnekar evidiere uktyktedienel – imannakitcheren adynmuketchene / Эңнэкар эвидйэрэ уктыктэдйэнэл – иманнакитчэрэн адынмукатчэнэ	Reindeer jump and play madly, running quickly in all directions on the snow or the ice: this is a sign of snow Олени балуются (бегают, прыгают) на льду или на снегу – это к снегу

Around 20 years ago, the elders told that when reindeer suddenly and madly jumped about and rab quickly in all directions in the springtime, it was a sign that the calves were being born. Similarly, the elders said that calves were born when it snowed during the springtime.

Лет 20 тому назад, старики говорили что когда весной олени балуются (резко прыгают и бегают в разные стороны), это знак к снегу и рождению оленят. Также когда снег шел, говорили что телята рождаются (Лаврилье 1995).

Agdy irgaktalbe vakitchere / Агды иргакталбэ вакитчэрэн	Thunder kills the horseflies, a sign that the summer will soon end
	Гром убивает оводов – конец лета скоро будет

In August, when the lightning is more frequent, the wings of the horseflies crack and dry, causing them to die.

Августу, агдымар бирэкин, иргакталду лэпурэлтын йолдорговкил, катавкил тадук бутэвкил.

В августе, когда грозы чаще бывают, у оводов крылья трескаются, высыхают и они умирают.

Weather prediction according to the behaviour of the vegetation

Bolgikte uɣirburen – niamakitcheren / Болгиктэ уҕирбурэн – нйамакитчэрэн
The branches of the *pinus pumila* rise, a sign that it will warm up
У стланика ветви поднимаются вверх – тепло будет

First of all, let us note that the peculiarity of the *pinus pumila* is that during the summer its branches rise towards the sky, while during the winter its branches gradually lower until they become completely embedded within the snow cover. If the branches of the *pinus pumila* rise up from the snow within which they were encased during the cold winter, it means that the weather will warm up. If these branches are humid, it is a sign of snow (the atmospheric pressure is low). However, if it will be cold, the branches of *pinus pumila* go lower. By observing the extent to which the branches of this small tree are lowered, one can predict how cold it will be. *Pinus pumila* is a good indicator of atmospheric pressure. (cf. Vegetal cover typology: bolgikte)
We note here that, as in indigenous knowledge elsewhere in the world, Evenki informants sometimes link their traditional knowledge to popular scientific knowledge, as we see here with the mention of atmospheric pressure. (cf. also diagram River sounds weather)

Болгиктэ уҕирбурэкин – нйамалдиңан. Болгиктэ дэрбаврэн – иманналдиңан. Болгиктэ тэҕэрэкин – иңинилдиңан.

Если ветки стланика поднимаются вверх от снега – это к теплу. Если ветки стланика влажные – это к снегу (давление низкое). А к холоду стланик оседает. Осмотрев насколько стланик оседает, можно узнать насколько холодно будет. Стланик хорошо определяет давление. (см. болгик, с. 130)

Weather prediction according to the state and behaviour of fish

Ollol tygdarivacheren – doγu sulukta avulevki /
Оллол тыгдаривачэрэн – доҕу сулукта авулэвки

When you catch a fish and its intestines are inflated, it is a sign of rain.
Оллойо дйавами сулукта доҕун авулча – тыгдалдиңан.
Когда у пойманной рыбы надутая внутренняя кишка – то это к дождю.

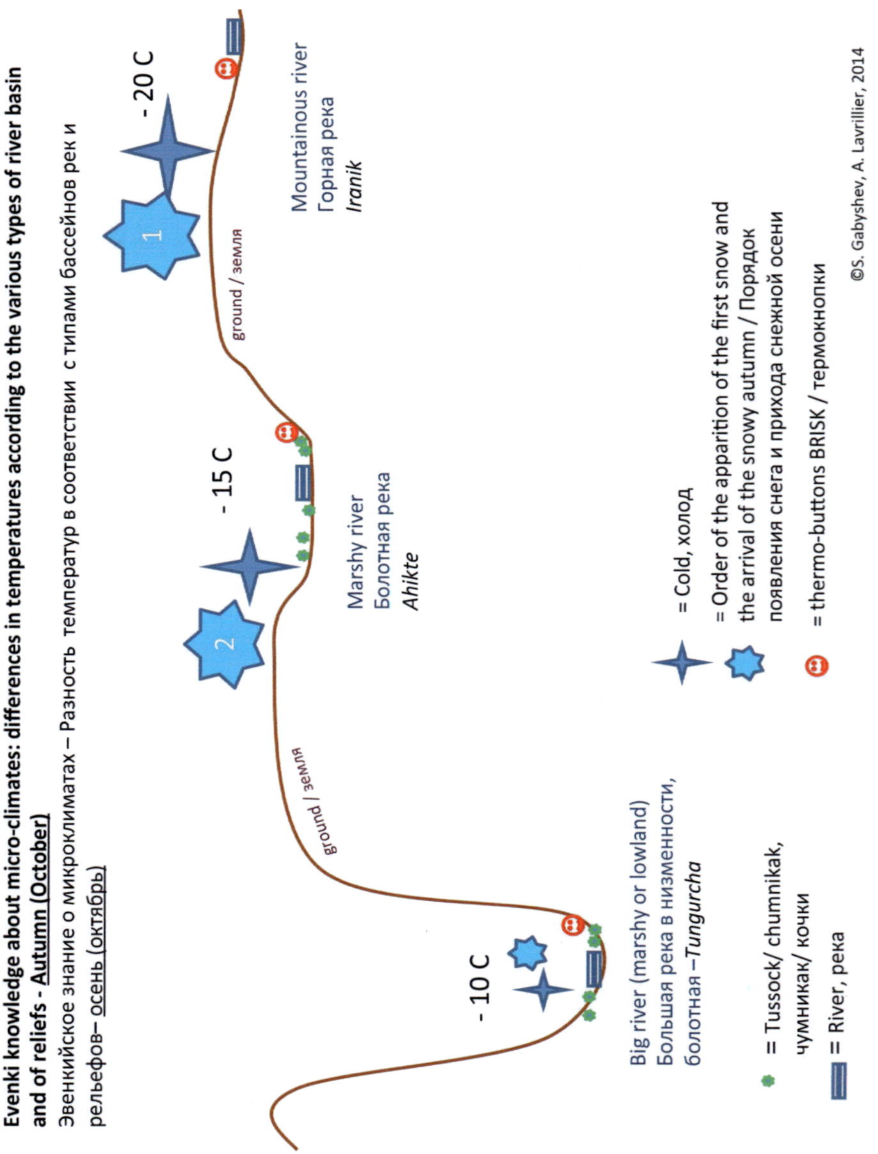
Evenki knowledge about micro-climates: differences in temperatures according to the various types of river basin and of reliefs - Autumn (October)
Эвенкийское знание о микроклиматах – Разность температур в соответствии с типами бассейнов рек и рельефов– осень (октябрь)
- 10 C
- 15 C
- 20 C
2
1
ground / земля
ground / земля
Big river (marshy or lowland)
Большая река в низменности, болотная –Tungurcha
Marshy river
Болотная река
Ahikte
Mountainous river
Горная река
Iranik
= Tussock/ chumnikak, чумникак/ кочки
= River, река
= Cold, холод
= Order of the apparition of the first snow and the arrival of the snowy autumn / Порядок появления снега и прихода снежной осени
= thermo-buttons BRISK / термокнопки
©S. Gabyshev, A. Lavrillier, 2014

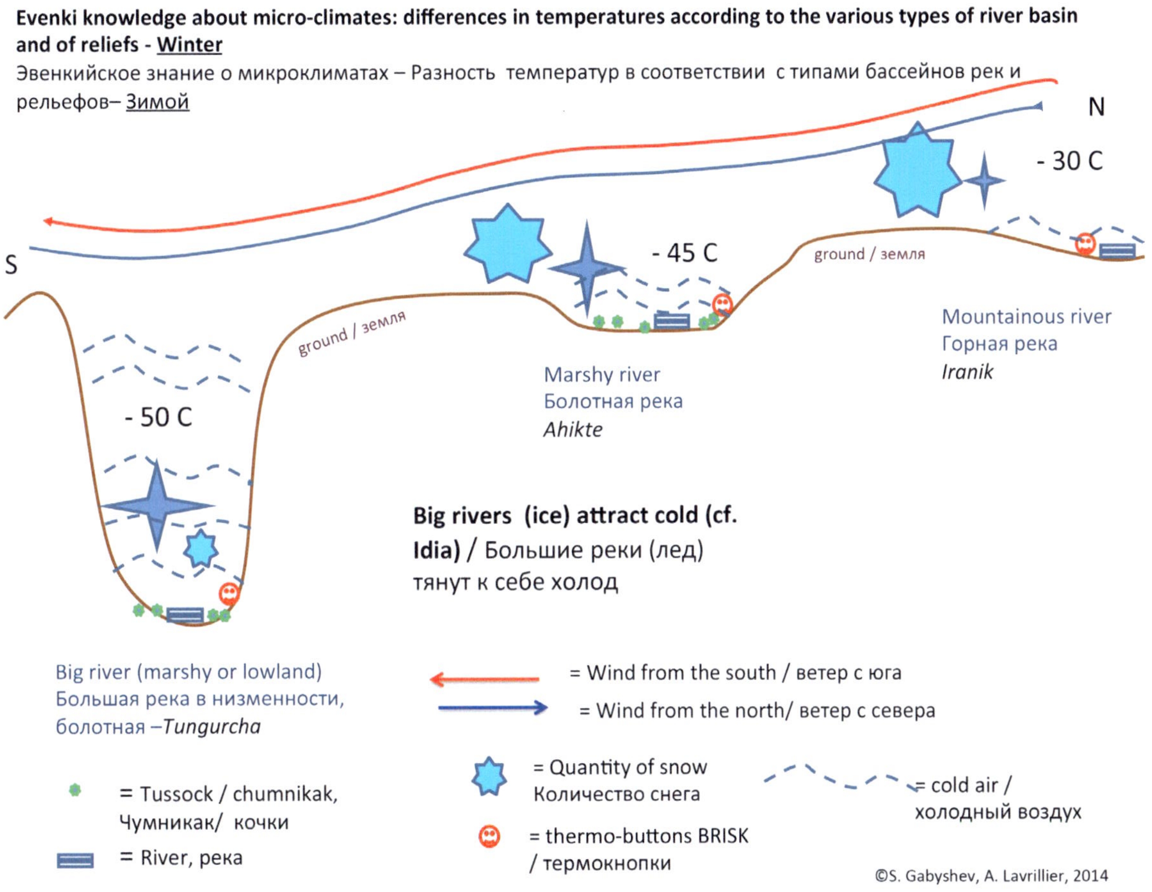
Evenki knowledge about micro-climates: differences in temperatures according to the various types of river basin and of reliefs - Winter
Эвенкийское знание о микроклиматах – Разность температур в соответствии с типами бассейнов рек и рельефов– Зимой
S
N
- 30 C
- 45 C
- 50 C
ground / земля
ground / земля
Marshy river
Болотная река
Ahikte
Mountainous river
Горная река
Iranik
Big rivers (ice) attract cold (cf. Idia) / Большие реки (лед) тянут к себе холод
Big river (marshy or lowland)
Большая река в низменности, болотная –Tungurcha
= Wind from the south / ветер с юга
= Wind from the north/ ветер с севера
= Quantity of snow
Количество снега
= cold air /
холодный воздух
= Tussock / chumnikak, Чумникак/ кочки
= River, река
= thermo-buttons BRISK / термокнопки
©S. Gabyshev, A. Lavrillier, 2014

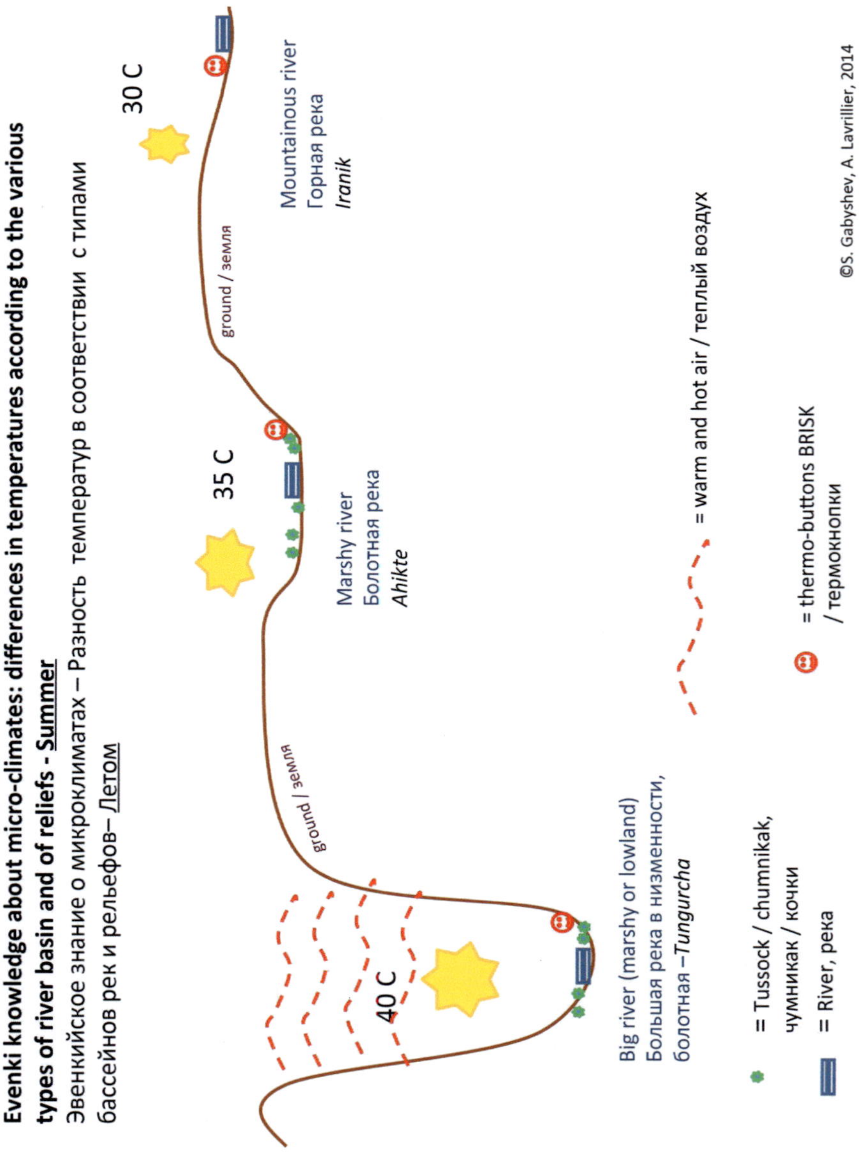
Evenki knowledge about micro-climates: differences in temperatures according to the various types of river basin and of reliefs - Summer
Эвенкийское знание о микроклиматах – Разность температур в соответствии с типами бассейнов рек и рельефов– Летом
30 C
35 C
40 C
ground / земля
ground / земля
Mountainous river
Горная река
Iranik
Marshy river
Болотная река
Ahikte
Big river (marshy or lowland)
Большая река в низменности,
болотная –Tungurcha
= Tussock / chumnikak,
чумникак / кочки
= River, река
= warm and hot air / теплый воздух
= thermo-buttons BRISK
/ термокнопки
©S. Gabyshev, A. Lavrillier, 2014

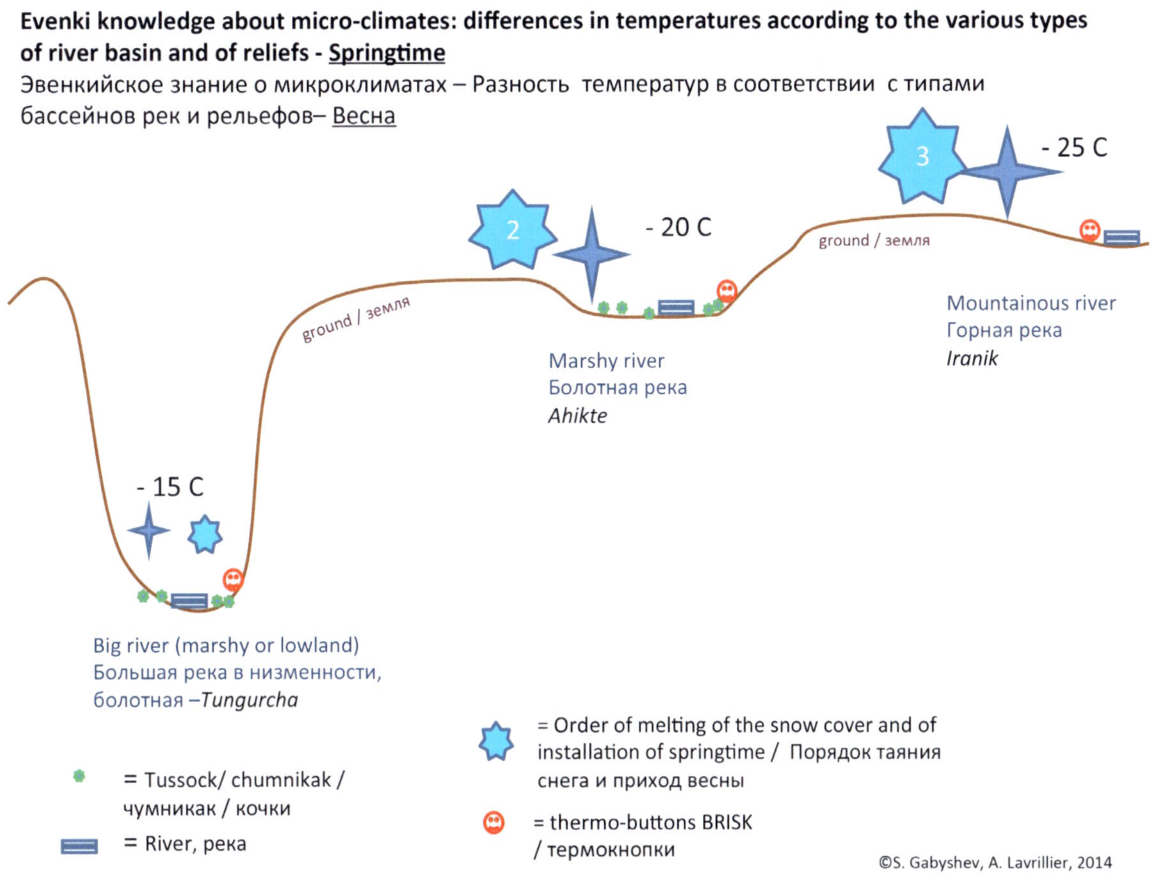
Evenki knowledge about micro-climates: differences in temperatures according to the various types of river basin and of reliefs - Springtime
Эвенкийское знание о микроклиматах – Разность температур в соответствии с типами бассейнов рек и рельефов– Весна
- 15 C
- 20 C
- 25 C
2
3
ground / земля
ground / земля
Big river (marshy or lowland)
Большая река в низменности,
болотная –Tungurcha
Marshy river
Болотная река
Ahikte
Mountainous river
Горная река
Iranik
= Tussock/ chumnikak / чумникак / кочки
= River, река
= Order of melting of the snow cover and of installation of springtime / Порядок таяния снега и приход весны
= thermo-buttons BRISK / термокнопки
©S. Gabyshev, A. Lavrillier, 2014

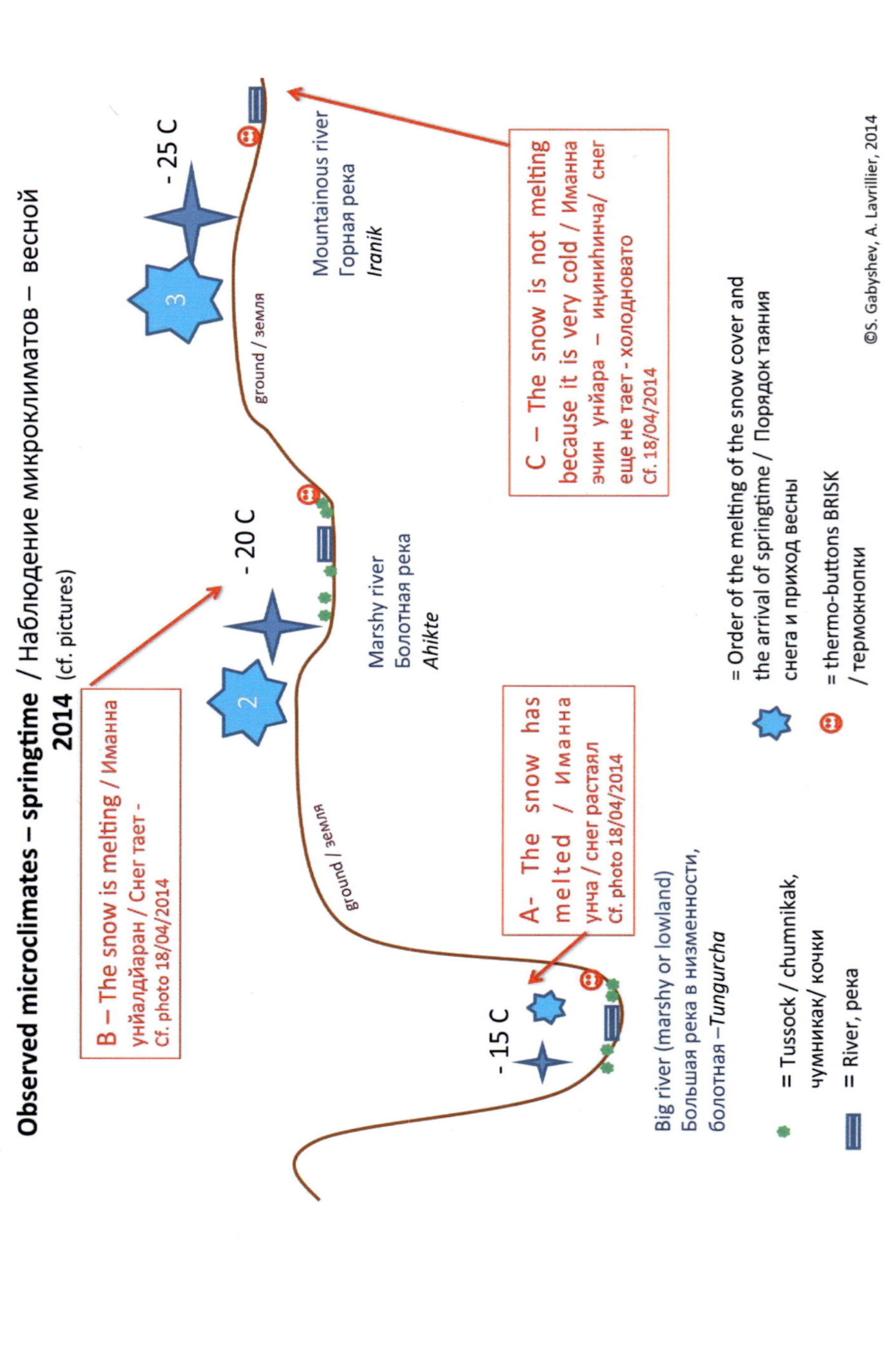
Observed microclimates – springtime / Наблюдение микроклиматов – весной 2014 (cf. pictures)
B – The snow is melting / Иманна унйалдйаран / Снег тает - Cf. photo 18/04/2014
- 15 C
- 20 C
- 25 C
2
3
ground / земля
ground / земля
A- The snow has melted / Иманна унча / снег растаял Cf. photo 18/04/2014
C – The snow is not melting because it is very cold / Иманна эчин унйара – иӊиниӊинча/ снег еще не тает - холодновато Cf. 18/04/2014
Big river (marshy or lowland) Большая река в низменности, болотная –Tungurcha
Marshy river Болотная река Ahikte
Mountainous river Горная река Iranik
= Tussock / chumnikak, чумникак/ кочки
= River, река
= Order of the melting of the snow cover and the arrival of springtime / Порядок таяния снега и приход весны
= thermo-buttons BRISK / термокнопки
©S. Gabyshev, A. Lavrillier, 2014

Observed microclimates – springtime / Наблюдение микроклиматов – весной **2014** (cf. diagram)

B

C

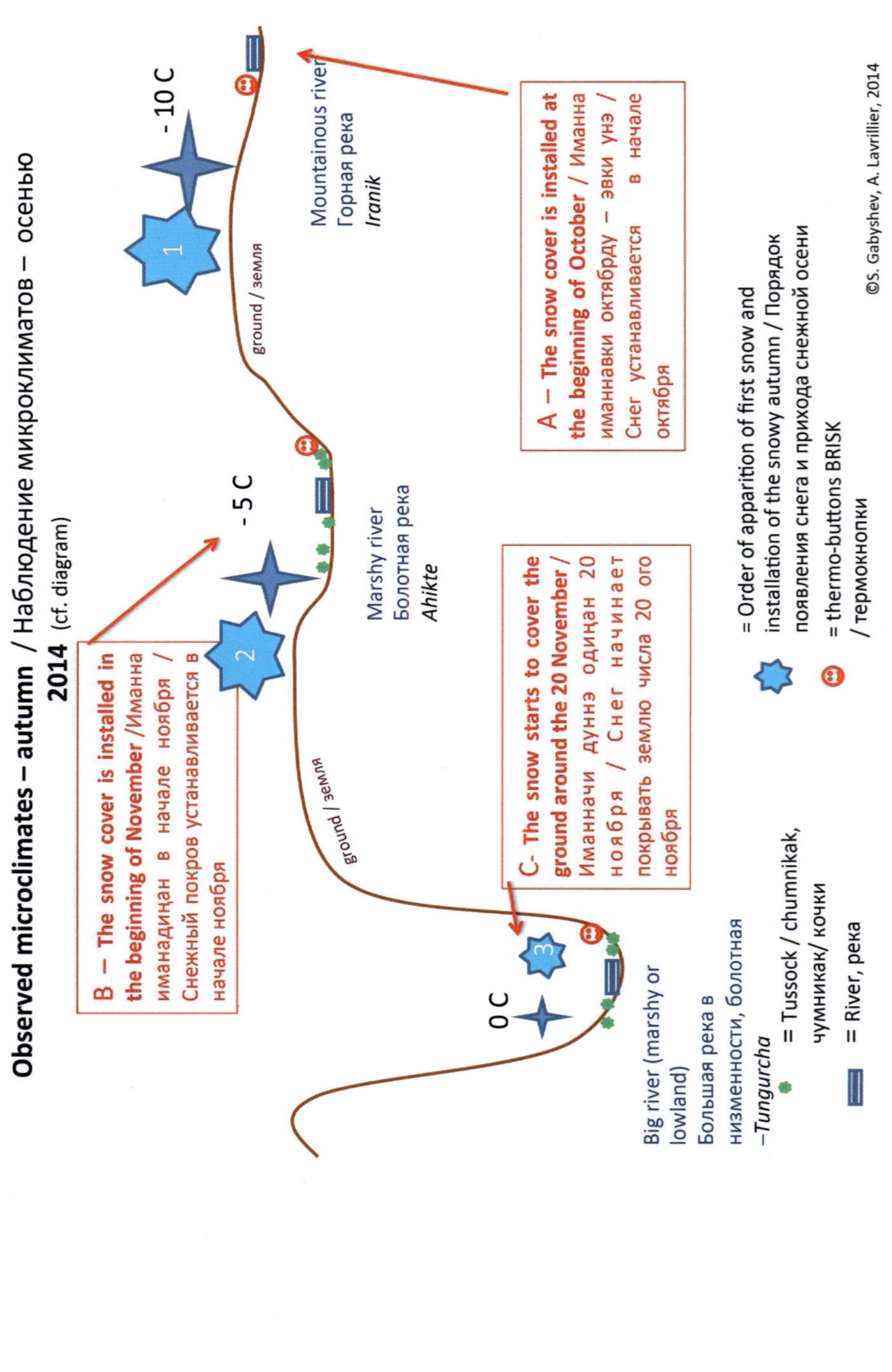
Observed microclimates – autumn / Наблюдение микроклиматов – осенью
2014 (cf. diagram)
B – The snow cover is installed in the beginning of November /Иманна иманадиңан в начале ноября / Снежный покров устанавливается в начале ноября
C- The snow starts to cover the ground around the 20 November / Иманначи дуннэ одиңан 20 ноября / Снег начинает покрывать землю числа 20 ого ноября
A – The snow cover is installed at the beginning of October / Иманна иманнавки октябрду – эвки унэ / Снег устанавливается в начале октября
0 C
-5 C
-10 C
1
2
3
ground / земля
ground / земля
Big river (marshy or lowland)
Большая река в низменности, болотная
–Tungurcha
Marshy river
Болотная река
Ahikte
Mountainous river
Горная река
Iranik
= Tussock / chumnikak, чумникак/ кочки
= River, река
= Order of apparition of first snow and installation of the snowy autumn / Порядок появления снега и прихода снежной осени
= thermo-buttons BRISK / термокнопки
©S. Gabyshev, A. Lavrillier, 2014

Sound river weather (1)

Circulation of sound in cases of approaching rain and wind / Циркуляция звуков в случае скорого дождя с ветром

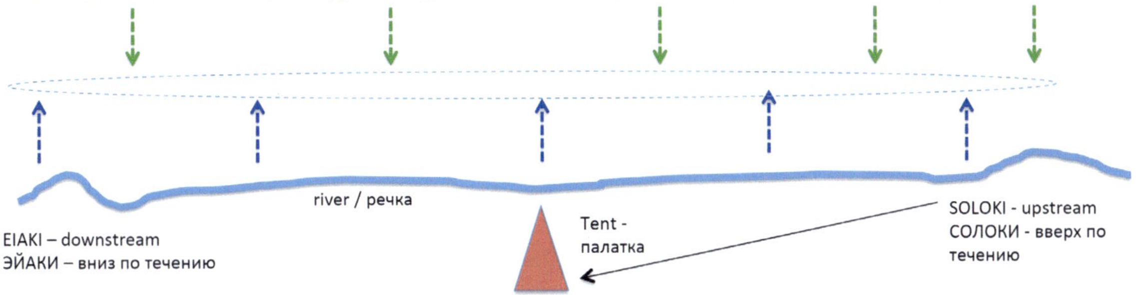

= compression of the air creates pressure / раздавленный воздух, плотный воздух создает давление
= audibility / слышимость
= humidity from the river / влажность с речки
= humidity from the air / влажность воздуха

***Bira doldybdieren soloki – tygdarikitcheren* = When the river makes noise upstream, it is a sign of rain in the near future** (cf. Evenki climatology)
When rain approaches in the morning and evening, the humidity from the surface of the river and the sky becomes dense. One can hear the noise of the river from upstream (the current bears the sound down stream). The dense humidity is a sign of a future change in the weather and of precipitation. The humidity from the sky and the river compress the air, creating low atmospheric pressure. This causes pain to people who suffer from high blood pressure or from bone fractures. In such conditions, one can expect several days of rain and wind.

Когда дождь намечается, вечером или утром, из-за давления влажности и с неба и с речки происходит уплотнение общей влажности. Звук идет от верха реки, потому что речка идет вниз (течение несет звук вниз, а шум, который внизу реки, не слышно –его уносит вниз с течением).
Именно это двойная влажность является признаком будущего изменения погоды к осадкам. Влажности с неба и с реки создают низкое атмосферное давление, тогда у кого тахикардия– поднимается давление, у того у кого суставы больные и бывшие переломы –появляются боли – это знак к дождю.
Следует ожидать изменения погоды на несколько дней (дождь, ветер).

Sound river weather (2)

Circulation of sounds in case of future good weather / Циркуляция звуков в случае скорой хорошей погоды

Nieŋdelekikin – Clear weather in the second spring and summer / Нйэӈдэлэкикин – ясная погода во второй половине весны без снега, летом

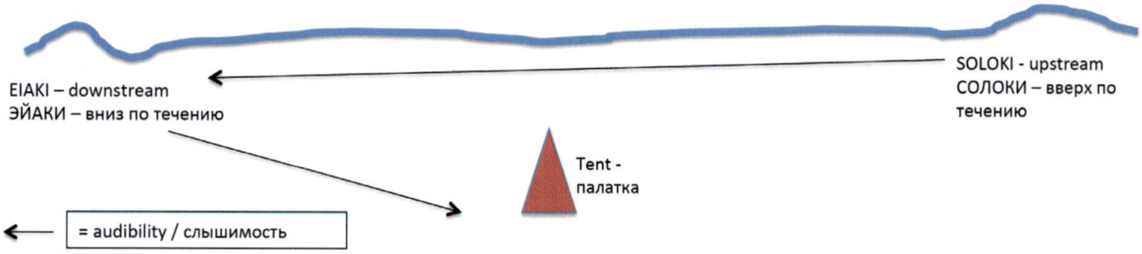

When clear weather approaches (*nieŋdelekicheren*), there is no humidity: all sound moves downstream. The air and sound move freely because they are not compressed by the humidity and kept in one place. This means that sound comes from downstream, which leads one to expect clear weather for several days

Когда ясная погода намечается (*нйэӈделекитчэрэн*), влажности совсем нету, только сухой воздух и поэтому весь шум уходит вниз по реке. Воздух (и звуки) гуляет свободно, потому что влажность его не давит и не держит, значит весь шум слышится с низу. Тогда можно ожидать ясную погоду на несколько дней.

2.3.2 Clouds typology

by S. Gabyshev and A. Lavrillier

Tuksu, Tuγu / Туксу (УН), Туӻу	Cloud; when there are small clouds everywhere in the sky that produce light rain or snow. Туча; когда везде маленькие тучи, вызывающие дождь или снег.

When there are such clouds, the entire sky is covered. If it is not snowing or raining, peoples can hunt wild reindeer or sable. During the winter, when there are such clouds, temperatures are not very low, making it very comfortable to move about. During such weather and temperature conditions, the wild reindeer will not be able to hear hunters. *The Evenki know about the link between air temperature and sound speed: when temperatures are very low (around <-45°C), even a quiet crack can be heard throughout the entire taiga, which endangers hunting. During cloudy weather, and the warmer temperatures associated with it (>35°C), audibility is much less important. In summer, if the clouds do not bring too much rain, reindeer can graze well and gain weight. There are far fewer mosquitoes and horse flies during rainy weather than during dry weather. This means the reindeer are not obliged to stay close to the smoke fire in the encampment and can graze in the surroundings.* In addition, clouds and rain lead to the growth of mushrooms, which are a good source of sustenance for the reindeer. In the past, the Evenki did not eat mushrooms; however, Russian influence has turned them into one of their sources of food.

Туксу бирэкин буӻа упкат туӻурбувки. Эрэкин тыгдарэ бэйэл бултактэдйэвкил, бэйуктэнавкил. Туксуду эвки иңиникуттэ туӻэ, айакикинди ңэнэктэдйэңнэнны, эвкил долдырэ синэвэ бэйур. Дйуӻаниду элэкин тыгдарэ, орорду айакикин оңкодйодо, айат бургудйэвкил. Тыгдадйэрэкин нйан айа, дэвуңнактэ балдыдйавкил, дэвуңнактэвкил: орор-да, бэйур-да. Тыгдадйэрэкин дйуду авалдйаңнанны.

Когда такие тучи бывают, все небо ими покрывается. Если дождь/снег не идет, то люди охотятся на диких оленей или на соболей. Зимой при пасмурной погоде холодно не бывает и очень удобно ехать. Итак, при такой погоде и температуре дикие олени не слышат охотника. *Эвенки знают о существовании интересного феномена, связывающего температуру воздуха и слышимость звуков: когда сильно холодно любой мелкий треск слышится по всей тайге. Наоборот, когда пасмурная погода и тепло – слышимость намного уменьшается.* Если тучи приносят не слишком много дождей в одно лето, то олени хорошо кормятся и поправляются за лето. Когда дождь идет долгое время, комаров намного меньше, чем при сухой погоде, значит оленям не обязательно проводить весь день на стоянке возле дымокуров (где они не могут пастись). (см. нйэӊдэлэ, с. 210)

Tuɣupdieren, tuksupdieren / Туӻубдйэрэн, туксупдйэрэн	The sky is becoming cloudy (if clouds arrive rapidly)
	Становится облачно (если тучи быстро идут)

High winds can make the skies cloudy very quickly: sometimes, this is like a storm. People do not travel in such weather. If this happens in summer, it means that there will be heavy rain; if it happens in the winter, it signifies heavy snow fall.
When it is snowing heavily, the Evenki usually do not move around except when the snow types offer specific services. (cf. Snow and ice typology: keraɣadieren)

Туӻубдйэрэкин адындйавки, адынмар, илйакат бэйэ этан уруру, тыгдалдиӊан бэйэ, гунчэдиӊан дйуӻа.Туӻэ иманналдиӊан гунчэӊнэнны, илйакат эӊнэнны ӊэнэктэрэ.

При становлении облачной погоды, ветер очень сильно дует, как пурга, и люди никуда не ходят / не ездят. Летом это значит, что скорее всего сильный дождь пойдет, а зимой – что будет сильный снег, поэтому никуда не ездят при такой погоде. (см. кэраӻадйэрэн, с. 260–261)

Tuɣupcho, tuksupcho / Туӻупчо, туксупчо	The entire sky is covered in large clouds without rain or snow
	Все небо полностью покрывается большими тучами – без дождя / снега

When the sky is covered in large clouds and it is starting to rain a lot, the Evenki do not go anywhere. When similar conditions occur in the winter and produce heavy snow falls, they also remain in the encampment. This term is used for designating large clouds.
There are very few types of snowfall that can keep a herder or a hunter at home; usually, only heavy snowfall during strong winds or wet snow can persuade them to remain at the camp.

Туӻуврэкин тыгдаӊат бирэкин, эӊнэнны илйа-да нйан уруру. Туӻэ иманнаӊат бирэкин, нйан илйа-да эӊнэнны уруру. Туӻупдйорокин тар туксукун оча.

Как только все небо покрывается большими тучами и должен пойти дождь, мы никуда не едем. Зимой то же самое, если сильный снег идет, никуда не едем. Этот термин мы употребляем, когда очень большие тучи покрывают все небо.

Tuksukon / Туксукон

Small cloud

Маленькая тучка

A small cloud which brings brief spells of rain or snow. In such weather conditions, one can hunt wild reindeer and, in the summer, fish. (cf. Snow and ice typology)

Нйукучукокон туксу уңнйариктадиңан. Таргачинду бэйуңнэнны, олломодйуңос дйуӈа.

Небольшая туча, из которой идет кратковременный дождь или снег. При такой погоде можно охотиться на диких оленей и летом рыбачить. (см. с. 243–367)

Tuksukun / Туксукун

Big cloud

Большая толстая тучка

Just like large clouds in the summer, these clouds signify the arrival of thunder and lightning. It is necessary to stay at home in such conditions and to put on a hat to protect one's head against the thunder, especially if one is going to leave the tent. In such weather conditions, one should not hold metallic items. In the past, the elders sat awake throughout the night to implore the thunder spirit: 'Go away! Go away! Do not hit close to us! Do it very far from us!' In some cases, they sat under their blankets in their tents, afraid of getting burnt by the thunder. Today, as in the past, the Evenki remove pieces of bark from all four sides of those trees close to the tent (the sides represent the four corners of the world): this is so the thunder does not hit these trees and does not endanger either households or people. In addition, guns are buried into the earth. During such weather, one is forbidden from doing anything: one cannot cook, sew, or even read a book. One can only sit within the tent.

Let us note that thunder and lightning in the taiga are very impressive and dangerous, as proven by terrible accidents and, occasionally, deaths. In addition, in the Evenki shamanism of this region, the strongest and more powerful spirit is the spirit of thunder and lightning, erki / agdy *(Mazin 1984 and authors fieldwork) (cf. Erki). It is believed that one can convince these spirits to change their path away from the encampment. There is a similar ritual regarding typhoons (cf. Winds and air typology: seyin).*

Туксукун эмэллэкин агдыкутчиӊан дйуҕа, дйуду тэҕэтчэӊнэнны, авуна тэтэһинчэдйэӊнэнны. Тулиски йуми авунма тэтэӊнэнны эдан дыл онудйэрэ. Сэлэйэ эдэ дйавалаттэ. Нонон атыркачар тэҕэтчэпкир долбонывa бичал, эӊкитын укладйэрэ турачинӊкитын «Горолы-горолы, дарыла ӊэнэдйэкэл!, даҕалы энэ иктэрэ!, горокундулы ӊэнэнэ!» или марбэр уллалди тэҕэтчэнэл каиӊкитын палаткаду тэҕэтчэнэл. Ирйактэлбэ алдыҕнанны илкэһиндйэнэнны эдан агды иктэрэ. Пэктэрйавурбэ дуннаду наӊнэнны эдан уҕиски бирэ. Экунэда эӊнэнны ора, эӊнэнны ирирэ, эӊнэнны кэнигэвэ таӊна, тэҕэтчэксода нада.

Когда большие тучи приходят летом, значит будет гроза, тогда дома надо сидеть и сразу надеть на себя шапку. Выходить на улицу надо в шапке, а то голова будет болеть (от грозы). При такой погоде не надо держать никакое железо. В прошлом старики сидели всю ночь, не спали и говорили грозе «Подальше, подальше уходи, в сторону! Не бей рядом, делай это очень далеко от нас!», или сидели, прятались под одеялом в палатке (боялись сгореть от грозы). Сегодня также, как раньше делали, у деревьев, которые стоят рядом с палаткой, убирают часть коры во все четыре стороны света, чтобы гроза не ударила в палатку и нас. И оружие прячем в землю, чтобы ствол в земле был. Ничего нельзя делать, нельзя варить, нельзя даже книгу читать, надо просто сидеть. *Надо сказать, что гроза в тайге очень впечатляет и очень опасная и бывают страшные несчастные случаи. Помимо этого, в эвенкийском шаманизме данного района самый сильный дух считается* эрки / агды *дух грозы (Мазин 1984), (см. также арги/эрки, с. 215). Считается, что можно этих духов убедить мимо стоянок пройти. Похожий обряд проводится с вихрем (см.* Сэҕин, с. 219*).*

Tamnaksa / Тамнакса

Fog

Туман

During summer fog, visibility is limited: this can help a hunter approach wild reindeer, which will not be able to see him. The reindeer graze well in such conditions because there are no horse flies. In the winter, fog means that it is very cold and it is better not to travel. When the fog is very dense, visibility is extremely limited and sounds travel much further because of the humidity of the air: as such, hunting is impossible. It is so very important in such conditions to chop firewood into small logs so that the stove will burn well and heat the tent properly in such harsh cold. (cf. Winds and airs typology: tamna)

Тамнакса эрут ичэвдйэвки, нйурмадйами бэйуртыки айакикин дйуҕа. Орор айат оңкодйовкил, иргактал ачэр. Туҕэниду тамналлэкин иңикун овки, илйакат эңнэнны ңэнэктэрэ. Эрумит экунда ичэбдиңан; Эруми бултадйэми, мова кэтэкуна колтоҕоңнэнны.

Летом при тумане плохо видно и это очень удобно, чтобы прицелиться в дикого оленя – он не увидит охотника. Олени хорошо кормятся – оводов нету. Зимой при тумане очень холодно бывает, никуда лучше не ехать. Ничего не видно, ни на кого не поохотишься из-за плохой видимости и хорошей слышимости (из-за влажности). И еще надо дрова мелко расколоть, чтобы печка хорошо горела – ведь холодно бывает. (см. тамна, с. 231)

Tamnakso odiaran / Тамнаксо одйаран	The fog is rising Туман встает

The fog rises on summer mornings from lakes and rivers and reaches high altitudes. In winter, fog also appears, but only on very cold mornings and evenings.

Тамнаксо одйавки ңарилдйэрэкин уҕирипдйэвки. Туҕэниду, кэҕа нйан элтанэкин, ңарилдйэрэкин тамнаксо уҕиривулэвки.

Туман появляется, как только рассветает, от рек и озер летом и поднимается вверх. Зимой, вечером и утром, когда наступает стужа, то бывает туман и поднимается вверх.

Eltan-da / Элтан-да	Bright weather (with a cloudless, clear sky) and cold in the morning or evening, in all seasons Утренняя и вечерняя ясная холодная погода в течение всех сезонов

In the winter, the trees and bushes are covered by hoar-frost during such weather conditions. In the summer, these are good conditions for preparing wood for the smoke fires. When the sun has only just begun to rise, one must go to see whether the wild reindeer have come from the mountains down to the *nire* swamp. In the winter, one must leave early during this morning weather in order to get back home earlier (the days are very short). During the summer, the berries ripen very quickly: by the autumn, they will have begun to rot. When such weather conditions occur during the snow period, one can treat skins with smoke (cf. also Precipitations typology: silaksa) so long as the reindeer have left the encampment (the smoke has a bad effect on their lungs). (cf. Winds and airs: eltandieren)
This type of weather (in particular in the summer or autumn) implies seasonal change. The nomads note this in their daily observations, saying: 'Eltandieren – diuγa manapdieren' *(the* eltan *weather is becoming more frequent and stronger: this means that the end of the summer has arrived).*

Элтанэкин соӈукикин бивки. Самӈинду мойа бэлэкэӈнэнны. Ӈариллэкин нирэлбэ ичэнаһидйэӈнэнны бэйур эврэ эчэл-у? Туӻэниду элтанду урудо нада, имакунди ӈэнэктэммэн. Дйуӻа элтанэкин диктэвэ ирипканэвки. Боло тэпаргавкандиӈан диктэвэ. Элтанэкин нутиӈнэнны айаксот орор урурyктын.

Когда бывает ясная утренняя или вечерняя погода зимой деревья и кусты покрываются инеем. Летом удобно при такой погоде готовить дрова для дымокуров. Как только солнце встает, также надо идти осматривать болота типа *нирэ*, чтобы увидеть, спустились ли дикие олени с гор кормиться. Зимой надо уехать в это время при такой утренней погоде, чтобы пораньше приехать из-за того, что дни очень короткие. Летом при такой погоде ягоды очень быстро поспевают, а осенью быстро портятся. При такой погоде (в течение периода без снега) удобно шкуры обрабатывать дымом (см. силакса, элтандйэрэн, с. 214, 228), как только олени ушли со стоянки (ведь дым для них вреден).
Этот тип погоды, особенно летом и осенью, является знаком перемены сезонов, как видно в таблицах наблюдений, которые ведут оленеводы: Элтандйэрэн – дйуӻа манапдйэрэн, *т.е. 'погода* элтан *все чаще и сильнее – значит лето кончается'.*

Nuny / Нуны	Smoke from forest fires (this term is used for all smoke, including smoke from the stove/camp fire). There is a different term, *imty*, which designates the smoke from the fat/oil used for ritual offerings.
	Дым от пожара (этот термин употребляется для любого дыма, в том числе от печки/костра), в отличии от дыма с жиром или маслом, который употребляют для обряда угощения *(имты)*.

Sometimes, when it is very smoky, this can mean that there is a forest fire somewhere: this usually happens during summer heat waves. For the reindeer, such weather is very bad because the heat and smoke causes them to lose weight. *Nuny* smoke also comes from the smoke fires made to protect the reindeer against mosquitoes. In the summer, one always makes smoke fires.

Идувэл дэгдэдйэрэкин кураӊивки гунмачиӊнэрэм, окулкучивки, пожарил дэгдэдйэвкил. Ороpду эрукун, орор тиӊнэвкил окукунду. Самӊиндук нуныпчадйаӊнанны ороpбо эдатын мармактал кикчарэ. Самӊилва илатчавки окинда.

Иногда сильно дымно, если где-то горит при летней жаре. Для оленей это очень плохо, они худеют от сильной жары и дыма пожара. Дым *нуны* также называют то, что идет от дымокуров, которые делают для оленей, чтобы защитить их от комаров. Летом дымокурят постоянно.

Nunykun / Нуныкун	Very smoky because of forest fires Очень дымно от пожара, сильный дым от пожара

Sometimes forest fires create very dense smoke which severely limits visibility and makes it difficult to breathe; people stay at home. One must endure this since the smoke is almost impossible to escape from. In addition, such smoke causes the clouds to evaporate (i.e. there will not be any rain), which means that many lands will burn. Some nomads have provided narratives about groups which tried to escape from the smoke, but, because of the poor visibility, went directly towards the fire: they somehow managed to change direction at the last minute and got away from the fire.

Нуныкунду экунда эруми ичэвувки, экунэда этанны ичэрэ, эруми эрида, дйуду тэҕэчэннэнны. Нуныкундук онда этанны туксанэ. Нуныкундук туксук чумдйэвкил, горокунмэ нйэӊдэлэчэвки, кэтэ дуннэ котоpолдиӊон. Умнакан бэйэл нуныкундук туксаһинчал уӊту дэгдэдйэрилэ даҕамачал, нэһилэ туксаһичал.

От больших пожаров бывает очень густой дым и плохая видимость, тогда и дышать тяжело, и люди дома сидят. Приходится терпеть, от дыма почти невозможно убежать. Когда бывает такой дым, тучи испаряются и дождей нету, значит еще много земли будет гореть. Есть рассказы таежников о том, что убегая от густого дыма, они наоборот шли на пожар и кое-как успевали развернуться и убежать от пожара.

Sekalan / Секалан	Rainbow Радуга

If a rainbow appears, this indicates scattered showers. Such scattered showers occur because the rainbows draw water up from lakes, rivers, etc.
When rainbows appeared in the past, women would hang coloured ribbons on the trees close the tent as an offering to the natural environment (Lavrillier 2005).

Сэкалан оракин тыгда амардукин овки. Сэкаланду уӈнйариктадйэвки, сэкалан мувэ танэвки, нйан тыгдалдиӈан.

Если радуга появляется, значит кратковременные дожди будут. Внутри радуги кратковременные дожди бывают – радуга притягивает воду к себе от рек, озер, и т.д.
В старину, при появлении радуги, женщины должны были вешать на дерево возле палатки разноцветные ленточки-подношения для угощение природы (Lavrillier 2005).

Nieŋdele / Нйэӈдэлэ	Bright weather (during all seasons, day or night). Ясная погода (во время всех сезонов, в течении дня или ночи).

'Nieŋdele' refers to bright weather, especially during the winter. It means that it is very cold. During such harsh cold, there is no point in hunting: sound travels very quickly, allowing the wildlife to hear the hunter before he/she approaches and run away (cf. also tuksu). In such weather, it is only possible to check traps.
These weather conditions allow the Evenki to predict the weather and temperature for the next few days; bright weather announces several days of harsh cold.

Нйэӈдэлэрэкин гунэвкил туӡэ нйэӈдэлэ бирэкин иӈинидйэвки, эӈнэнны бултанарэ экунавал, долдывкил таргачинду, горойодук долдыдиӈан, урудйуӈон синдук. Таргачирду ичэччэӈнэнны капкарби.

Говорят «нйэӈдэлэ», когда зимой стоит ясная погода и, следовательно, очень холодно бывает. Из-за холода, любой звук очень хорошо распространяется и даже не стоит ехать охотиться – зверь сразу услышит охотника и убежит (см. туксу с. 203). При такой погоде можно только свои капканы проверять.

Nieŋdelekikin / Нйэӈдэлэкикин	Very bright weather; relatively warm in winter and not too hot in summer Очень ясная погода зимой или летом – сравнительно теплая зимой и не слишком жаркая летом

In very bright weather, the sky is completely cloudless. During winter, such weather allows one to hunt easily because it is relatively warm: even the birds sing. The reindeer do not become tired when travelling or hunting. This is also good weather for fishing in the summer because of the mild temperatures and the presence of fewer mosquitoes.

Нйэӈдэлэкикинду умукон туксу мат ачэн бивки. Нйэӈдэлэкикин гунэвкил бултанами айакикин бидиӈан, нйамаһинча бидиӈан, чипкачар эƃэдйэвкил таргачинду. Орор эвкил дэрурэ булатанами. Олломоӈнэнны таргачинду дйуƃаниду. Дйуƃаниду олус эвки окукуттэ.

При очень ясной погоде даже ни одной тучи нету в небе. Если такая погода стоит, хорошо ехать охотиться, потому что сравнительно теплая погода, птицы поют при такой погоде. Олени не устают, когда едешь охотиться при такой погоде. Удобно также и рыбачить, комаров мало. Летом при такой погоде не слишком жарко бывает.

Dylacharieren / Дылачадйэрэн	It is sunny; the sun is shining Солнце светит

cf. Nieŋdelekikin, nieŋdele

Dylacha teɣedieren / Дылача тэƃэдйэрэн	Sunset Закат солнца

The Evenki often observe the sunset in order to predict the weather. For instance, if the sunset is red during the summer, it is a sign of rain; but if the sunset is bright, there will not be rain, etc. (cf. Predictions typology). In addition, the specific cold weather type idia occurs in the winter during the sunset. (cf. Winds and airs typology: diagram Idia)

Эвенки часто наблюдают закат солнца, чтобы предсказать погоду. Например, если закат летом красным бывает, значит дождь будет, а если яркий закат, то дождя не будет и т.д.. Зимой при закате всегда холодный воздух типа идя бывает. (см. идйа, с. 232–242)

Identifying temperature by the lightness of the snow

Определить температуру по оттенку цвета снега

If the snow looks bright, it is warmer /
Если снег светлее выглядит, то теплее

If the snow looks dark, it is cold /
Если снег темнее выглядит, то холодно

2.3.3 Precipitations typology
by S. Gabyshev and A. Lavrillier

Tygda (Dygda) / Тыгда (Дыгда)	Rain Дождь

When it rains for a long time, people stay in the encampment and use their time to make reindeer skins, taken from the legs of the animal, softer with the *talki* instrument. It is also good weather for sewing the skins from the torsos of the reindeer into a cover for the packsaddles. This must be done during wet weather because the skin (sewed with the fur inside) must be sewed while wet so that they will cover the entire wooden frame of the saddle. When the skin dries, it will cover the entirety of the saddle frame. It is also ideal weather for treating elk skin with scraping instruments *(chuchun* and *kederia)* in order to make harnesses for the reindeer pulling the sledges. One must decorate these harnesses.

Тыгдадйэрэкин илйакат эӈнэнны ӈэнэктэрэ, оӡалба маннидйэӈнэнны, эмэӡэлбэ уллидйаӈнэнны, айакикиндит улливувкил. Одйаӈнэнны чучундит, одйаӈнэнны найаксолбэ, кэдэрйэӈнэнны лйампилвэ одйаӈнэнны, онйодйоӈнонны. Бэйэл дйуду тэӡэтчэнэл упкатвэ одйавкил дйуду.

Когда начинается обложной дождь, никуда невозможно ехать, только можно шкуры лап оленей размягчать руками или с инструментом *талки*, обшивать вьючные седла (стельной шкурой оленей, перевернутой шерстью внутрь). Шьется в мокром виде, чтобы высыхая, шкура хорошо натянулась на деревянном остове седла. Со скребками *чучун, кэдэрйа* обрабатывают лосиные шкуры, чтобы делать лямки для оленьей упряжки и узорами украшаешь. Люди остаются на стоянке и делают все свои домашние работы дома.

The size of the rain

The size of raindrops can be expressed with suffixes belonging to normal Evenki language.

VERB	ENGLISH	RUSSIAN
Tygdariktedieren Тыгдариктадйэрэн	Light rain is falling	Мелкий дождь идет
Tygdadieren Тыгдадйэрэн	Moderate rain is falling	Средний дождь идет
Tygdakutten Тыгдакуттэн	Heavy rain is falling	Сильный дождь идет (ливень)

Silaksa (silaγa, silakte) / Силакса (силаҕа, силактэ)	Dew
	Роса

Dew appears when the weather becomes colder. Dew creates fog, which then produces clouds: these lead to rain or snow. These weather conditions offer ideal circumstances for making lassos because one can use the dew to moisten and soften the skin, which is then plaited into bands. If this is done, the lasso will stretch well when it dries. Some people prefer to smoke the skins during such weather with *ilte* (rotten larch): this will make the skins stronger and waterproof. This is due to the fact that the smoke travels upwards in the little hut within which the skins are smoked, entirely covering them. This makes them very soft. (cf. Clouds typology: eltan-da)
Note indigenous knowledge about the circulation of water in its different states or positions between the land and the sky (from dew to fog, from fog to clouds and rain). The Evenki possess even more detailed knowledge about air circulation. (cf. Winds and airs typology: namuscheren, diagram Idia)

Силакса бивки тыгда амардукин, элтанэкин-ҕу силакса овки, силаксадук тамнакса овки, тамнаксадук туксу овки, туксу оракин тыгдалдиӊан. Таргачинду мавутэла одйаӊнанны, танэӊнанны олгокинмэ. Тадук, адыл нутивкил таргачирду, нуны уҕискаки урудйувки айакикинди нутивувки, нэлбэрикикин овки.

Роса бывает после дождя, как только похолодает. От росы появляется туман и от тумана появляются тучи, которые приносят дождь или снег. При такой погоде хорошо делать лассо – валяют сырую кожу для лассо в росе, чтобы мягче стало, чтобы заплетать ленты для лассо. Если так делать, лассо хорошо натягивается при высыхании. Некоторые люди шкуры обрабатывают дымом *илтэ* (гнилой лиственницы)для прочности и непроницаемости, потому что при такой погоде дым уходит вверх и хорошо окутывает шкуры дымом. Таким образом шкуры мягкими становятся. (см. элтан-да, намусчэрэн, 207-208, 178; см. схемы, с. 232–242)

Bona / Бона	Hail
	Град

Hail appears during a short rain shower *(uŋnia)*. Occasionally, hail arrives together with thunder. It is sometimes a good idea to hunt during a small hail storm because animals leave tracks in the fallen hail on the ground. However, during big hail storms, one cannot go anywhere because the hail is really painful.
Please note the interesting parallel between a short rain shower and hail (cf. uŋnia). See also the chapter on snow and ice, where the term uŋnia *is also used.*

Бона уӈнйаридйэрэкин овки. Адылдун агдыдйарэкин бона бонайачивки. Адылдун нйукучукокондит бонадйаракин бултанаӈнэнны удйарадиӡарэ. һэгдыкун бонакун эмэдйэрэкин илйакат эӈнэнны уруру – онукун бивки бона!

Град появляется от кратковременного дождя *(уӈнйа)*. Иногда град идет одновременно с грозой. Иногда при маленьком граде хорошо охотиться, потому что тогда следы животных появляются в талом граде, упавшем на землю. Но при большом граде никуда невозможно ехать – град очень больно делает. (см. уӈнйа, с. 179, 216, 268).

Agdy / Агды	Thunder Гроза

cf. Clouds typology: tuksukun.

Erki / Эрки	Lightning Молния

Erkapcha designates larch struck by lightning. Thunder creates lightning. From items touched by the lightning, one can take a sliver and treat a painful tooth by pricking it with this sliver (it is better to use larch). People used to take larch struck by lightning and craft it into the form of an arch: female reindeer encountering problems with giving birth were passed through this arch before calving season in a shamanic ritual. A human passing through the same arch helps guarantee themselves a good life (cf. Clouds typology: Tuksukun).

Эркипча ирйактэ бивки таргачинду, эрки гунэвкил. Агдыдук эрки овки. Эркавчадук тадуккар ниду иктэ онудйэрэкин супиктэвэн ирйактэдук гаӈнэнны, гидалаӈнэнны, этэвки онудйэми. Ирйактэвэ мартын бэйэл йолдолывкил, саманидйарил, тадук талы тар нйамива ӈэнэпканэвкил дулиндулин балдывулдан эӡивэ балдыврэ, бэйэ манын ӈэнэвки-у, айа бидан тар дулиндулин колтолэвкил.

Эркипча называют лиственницу, которую молния ударила. Это гром делает молнию. От объектов, которых молния ударила, берут щепку, желательно от лиственницы, и лечат людей, у которых зубы болят, укалывая больное место этой щепкой. Люди раскалывали лиственницы, тронутые молнией, и при шаманском камлании перед отелом проводили по середине расколотого, как арка дерева, оленью самку, которая не рожает, или человека, чтобы хорошо жил. (см. туксукун, с. 205)

EVENKI	ENGLISH	RUSSIAN	USES / PROBLEMS
Uŋnia / Уңнйа	Short spells of light rain or snow	Кратковременные, небольшие дожди или снег.	This is good for the vegetal cover, giving it the required level of humidity. However, in winter it is bad for hunting because tracks are lost.
Uŋniakun / Уңнйакун	Short spells of heavy rain or a snow storm (15–30 cm).	Кратковременные дожди или снег — ливень или снег сильный (15–30 см).	This is good for the vegetal cover, giving it the required level of humidity. However, in winter it is bad for hunting because tracks are lost.
Uŋniakan / Уңнйакан	Short spells of very light rain or snow	Кратковременные, очень небольшие дожди или снег.	This is good for the treatment of skins: in the winter, one can still read tracks during such rain.

Interestingly, the Evenki have a generic term for snow and rain precipitation: this unique term designates short spells of light rain or snow precipitation. The reader will find more detailed information about short spells of light snow in the section about snow and ice.

Since there is one term for two different kinds of precipitation, it seems that the interesting information for the Evenki is not whether the water is frozen or thawed (snowflakes or rain drops), but the fact that the precipitation is brief and repeated. This cannot be explained by the different activities that the precipitation allows or the different problems it causes. Here again the uses of the suffixes allows them to express the size and intensity of this kind of precipitation.

Uŋnia / Уңнйа © Vasilii Gabyshev (left), Semen Gabyshev (right)

2.3.4 Winds and airs typology (cold and warm, dry and humid)
by A. Lavrillier and S. Gabyshev

The Evenki use their knowledge about the interactions between the specificities of the topography and the existence or absence of wind. In the winter, they will choose places largely without winds (like a river bend or a piece of wood alongside a *teŋke, тэӈкэ* river bank) because the wind reinforces the feeling of cold. More rarely, when there are bad snow conditions they look for grazing pastures with constant winds like *ukty, укты,* (a slope which abruptly turns into a flat summit), since these winds make the lichen accessible (cf. Topographic typology: diagram Ukty). In contrast, in the summer they will look for windy topographic types to protect humans and reindeer from an extreme number of mosquitoes and horseflies (cf. Topographic typology).

Some additional terms can be found in this section about weather forecasting, like *diuyidieren / дйуҕидйэрэн*. The reader will appreciate the detailed and rich knowledge about the regular order of the appearance of winds and its relation to weather changes in the compass rose diagram (cf. Evenki climatology: weather forecasting, pp. 178–193. It is also very interesting that the Evenki distinguish between regular winds that blow all year long and seasonal winds like *salgyn, салгын* (wind in the first spring and first autumn). One may note that the winds which blow throughout the year either do not have specific names (except the indications of their geographical origin) or have names that have been forgotten.

Winds

Adyn Адын	Wind (generic term) Ветер

This is the generic term for wind. It is often adapted by adding suffixes, which allow one to express its intensity and regularity (see below). Sometimes, winds are appreciated, as is the case in summer (against mosquitoes) or during hunting (so the game cannot smell the hunter). In contrast, in the winter the Evenki try to avoid windy places most of the time. Winds harden the snow and create the *chuiur* snow type (cf. Snow and ice typology); they also erase the snow roads and striate the snow: strias are used in the tundra as an axis of orientation. On a spiritual level, the Evenki believe that strong winds are caused by the main spirit of the natural environment *buya* after the death of a human in order to erase his or her tracks from the world of the living. They also believe that one can summon the wind in a variety of different ways. (cf. Evenki climatology: weather forecasting, pp. 178–193)

Адындйэрэкин айа дйуӡа мармактэлвэ анавки ororво эвкил киктэ. Бэйуктэдйэми нйанэ айа сачйадйанны удйанэнны или эвки уӊнэрэ. Адын туӡэниду иманнавэ анавки уктэ овки. Орон таду айат оӊковки. Адын чуйурбэ овки – бэйэ ойолын ӊэнэктэдйэвки. Бэйэ ачэн орэкин адындиоттэн удйаван урувувки.
Общий термин для ветра. Считается полезным эвенками иногда летом против комаров, или во время охоты, когда охотник знает по ветру как идти, чтобы животное не учуяло охотника. Еще зимой ветер убирает снег от *укты* и освобождает корм (см. 105, 114, 152–158). Еще ветра делают снег твердым и таким образом позволяют людям и собакам по верху ходить. Но зимой большинство эвенков избегают ветренные места. Есть поверье у эвенков, что главный дух природы *буӡа* вызывает ветер после смерти человека, чтобы убрать его следы с мира живых. Также верят, что можно вызывать ветер обрядами. (см. с. 178–193)

Adyndiekutten Адындйэкуттэн	Gale, very strong wind Пурга, сильный ветер

When there is a strong storm, people stay at home and work. More courageous people hunt wild reindeer because animals cannot see or smell anything in this weather.

Адындйэкуттэкин, дйуду авалыктадйанарум. Соткур бэйувкил, бэйур эвкил ичэрэ, эвкил амтара.

Когда сильная пурга бывает, люди дома работают. Самые заядлые люди охотятся на диких оленей, ведь животные ничего не видят и не чуют.

Adyndiekutten imannadiene Адындйэкуттэн иманнадйэнэ	Storm with snow Пурга со снегом

People stay at home: there is no visibility because of the snow.
Дйуду тэӡэчэвкил бэйэл, экунда эчэ ичэврэ иманнадйаракин.
Люди дома сидят, ничего не видно из-за снега.

Adyndiekutten ulapkun immananyn **Адындйэкуттэн улапкун иманнанун**	Storm with wet snow Пурга с мокрым снегом

People stay at home and work on various tasks. In wet snow conditions, clothes get wet and freeze by the end of the day.

Бэйэл дйуду тэҕэчэвкил авалдйэнэл. Улапкун иманнадйэрэкин илйа да бэйэ эвки уруру, кэҕа тэткэл доӈкотовкил.

Люди дома сидят, работая. При мокром снеге вечером вещи мерзнут.

Adyndiekattan **Адындйэкаттан**	Light wind Слабый ветер

People hunt wild reindeer; they know from which direction the wind is coming and stay in a position that will prevent the animal from smelling the hunter.

Бэйэл бэйувкил, садйаӈнынны илйа адындйаривэ бэйун уӈнэдиӈан этан-у, нйамаһинча бивки таргачинду.

Охотятся на диких оленей, зная в какую сторону ветер дует, он встает так, чтобы дикий олень не учуял его, т.е. охотника.

Seγin **Сэҕин**	Thyphon, hurricane Смерч, вихрь, ураган

A hurricane occurs when the summer arrives (cf. Evenki climatology: daldyn), bringing with it large clouds and horseflies. The wind blows very hard, uprooting trees, lifting the tents, and bringing horseflies and large winged ants. In addition, however, some hurricanes kill horseflies by breaking their wings at the end of the summer. When a hurricane approaches, some people take a sharpened axe and fix it with a rope to a pole or a tree, with the sharpened part pointing towards the hurricane: this is so it will stay some distance from the camp and not blow too strongly. (cf. Evenki climatology: irganen, irgandieren).
There are many strange stories in the nomadic world about hurricanes; one of them talks about a hurricane that transported a family's traditional kumalan *fur carpet from one of their camps to another. In the past, when a hurricane approached, one of the elders would take an axe, go to the edge of the camp site, fix the axe to a tree, and talk with the spirits of the hurricane in order to persuade them not to lead the hurricane into the encampment. This was noted by Lavrillier in the 1990s, and one can see such a ritual in Semenov's film about the Evenki shaman Matriona Kulbertinova (Semenov 1998)* (cf. *Irgandieren).*

Сэӡин бивки далдйэрэкин, иргандйэрэкин туксукунмэ эмупдйэнэ. Тэтэкунди адынэвки, ирйакталбэ тыкипдйэнэ, палаткалва тэкэӡэвки, иргакталбэ, ириктэлбэ эмувувки. Таргачин нйан бивки иргакта вадйэнэ, дйуӡани манапдйэрэкин. Адыл эмэркун сукэвэ тэлбэду, ирйактэлду уйивкил, дйэйэдин иргит сэӡин эмэвки, дарылы ңэнэдан, манак эдан адыннэ.

Ураган бывает, когда лето наступает *(далдын)*, и приносит большие тучи. Ураган очень сильно дует, вырывая деревья, рвет палатки, принося оводов и больших крылатых муравьев. Также, в конце лета, некоторые ураганы убивают оводов, ломая им крылья. Некоторые люди берут острый топор, привязывают к дерево или к столбу, направляя острый конец топора в сторону урагана, чтобы мимо прошел или не дул сильно (см. иргаңдйэрэн, с. 173, 161, 165).
Многие странные истории ходят по таежному миру об урагане, в том числе об эвенкийском коврике кумалан, *который был транспортирован с одного лагеря в другой. В старину, при подходе* сэӡин *пожилые люди шли в направлении вихря/ смерча с топором в руках, привязывали его веревкой к дереву в направлении вихря/смерча и произносили слова, адресованные духам вихря, прося пройти мимо стоянки. Такой обряд Лаврилье записала в 1990х в истоках Алдана, и можно увидеть в фильме Семенова об эвенкийской шаманке Матрене Кульбертиновой 1996. (*см. tuksukun, с. 205)

Burga, burgadiekutten	Snowy storm
Бурга, бургатйэкуттэн	Вьюга

This term comes from the Russian 'purga' and has the same meaning. Today, it is used in everyday life.

Этот термин от русского «пурга» с таким же значением употребляется сегодня часто.

Airs: warm, cold, dry, humid

This part of Evenki knowledge is crucial for understanding their traditional climatology in general and their knowledge about microclimates in particular (cf. Evenki climatology: microclimates and related diagrams, like Idia). It is particularly interesting, since this knowledge pertains to invisible (or almost invisible) aspects of the environment and takes into account the interactions between warm and cold airs, humidity and dryness, topographic specificities, the sun's rays, ices and snows, and finally the circulation of sounds (cf. diagrams of this chapter). In other words, we could call it an indigenous physics of airs and the cryosphere.

As highly systemic knowledge, it needs to be read together with the following diagrams: River sounds weather; Idia; and part III: Ice anomalies. What is impressive in this *idia* concept is the knowledge about that which is almost invisible (the movements of large quantities of air during very dark moments of the day). How did the Evenki gain this knowledge? By travelling at different moments of the day and night and feeling the diverse temperatures of the air? This knowledge is shrewd enough to distinguish clearly between the laws of large air flows from the sky to the ground and along rivers and their interaction with winds.

A curious aspect is that we could not find a local Evenki term for designating 'air', despite this detailed knowledge. Even in dictionaries, the translation of 'air' (*vozdukh, воздух*, in Russian) is either *salgyn, салгын, halgyn, һалгын* (as some of the elders we asked said, but in reality this designates the spring air and wind (Winds and airs typology) (see here below), the Russian term *vozdukh, воздух,* or a Mongol term *kei, кэй* (Vasilevich 2005: 57; Boldyrev 1994: 48). In dictionaries, *kei, кэй* is attached to 'air' and *erikse, эриксэ,* which means 'respiration', 'air', and, in some dialects, 'one of the parts of the soul'. The root of this term gives other words the meanings 'shout(ing)' and 'air instruments' (Vasilevich 1958: 227, 565). The elders from the Evenki communities we worked with (in South Yakutia and the northern Amur region) do not know the word *kei, кэй,* while *erikse, эриксэ,* and *erinda, эринда* mean 'breath', 'breathing'. However, when they talk about 'air' in their discourse about the circulation of cold, warm, humid, and dry airs, there is a clear difference from breath. In Evenki, the circulation of warm and cold airs are expressed directly with the words 'warm' and 'cold', expect for *idia, идйа,* a specific cold air. (cf. diagram Idia)

According to our analysis, the Evenki pay great attention to the interrelation between cold and warm airs, some types of precipitation, and the rules of the circulation of sounds. This knowledge helps them to make the best decision in terms of their economic productivity (hunting, herding, or travelling). For instance, they know that under certain temperatures sound travels very well, making the forest very noisy (every crack and other small sound is audible from very far away): this reduces the hunter's ability to approach and kill an animal. They also know that relatively warm temperatures in the winter and cloudy weather mean that sound travels very poorly; thus, they can approach game much more easily (cf. Indigenous science of climate: weather forecasting, clouds typology, precipitations typology, snow and ice typology). Evenki knowledge also takes into account the interactions between the dryness and humidity of airs and the rules for the circulation of sound, and they use these features to forecast the weather (cf. Evenki climatology: diagram River sounds weather).

Salgyn Салгын	Warm air in springtime and autumn, or light wind
	Весенний и осенний, теплый и сухой воздух или легкий ветер

Salgyn air occurs in February. Thanks to this air, the Evenki are able to dry animal skins (from torsos and legs), since they become softer. Hands, faces, and lips dry, and faces bronze. The meat dries out and turns brown, so the Evenki do not hunt for the market anymore, just for their own consumption.
In the 1990s, this air appeared from March and in late autumn, while from the mid 2000s it has frequently appeared from January-the beginning of February. This was the case in 2015: it transformed the upper layer of the snow into *tepama imanna* (wet and sticky snow) and made transportation difficult (cf. Snow and ice typology) at the beginning of February. The Evenki say that *salgyn* air/wind 'eats the snow', meaning that it causes the reduction or disappearance of the snow without any melting. This was the case in autumn 2016, when *salgyn* 'ate' a very deep snow layer (40–50cm) at the beginning of October.
One can find this term in several Siberian languages with the same meaning (in Tuva or Yakut, for instance). (Evenki climatology: seasonal chain)

Салгын овки февралду. Таргачинду нарналва, оӡалва тэлэдйэвкил. Салгын иманнавэ дйэпивки (иманна арбавки сокунди), ӊалыл катадйэвкил, дэрэ салгынвувки. Умур катавкил. Уллэ коӊнорговки, катавки, эӊнэнны манак бэйуктэрэ, максодуви. Дйур дйар анӊаны чаӡуду салгын бинкин мартаду, эткан февральду, тадук адылдун январду.

Салгын бывает в феврале. При нем тельные шкуры и лапы натягивают – *салгын* их хорошо сушит, мягче бывают. Руки и губы сохнут, лица загорают. Мясо сохнет и темнеет и поэтому охотятся только для себя, а не на продажу. В 1990-ых годах такой воздух появлялся с марта месяца, с 2000-ых появляется часто в январе- начале февраля, как в 2015ом г.. Это трансформирует снег в *тэпама иманна* (мокрый липкий снег) и создает проблемы для транспорта (см. с. 243–367, 372–377). Эвенки говорят, что *салгын* «кушает» снег, имея в виду, что при *салгын* снег исчезает (хотя не тает). Например, осенью 2016, когда *салгын* «съел» толстый слой снега, который выпал в начале октября.
*Этот термин существует на разных сибирских языках с тем же значением, как например на тувинском или якутском (*см. с. 164–172)

Lamuscheren, namuscheren Ламусчэрэн, Намусчэрэн	Abrupt changes in the wind leading to warming (with rainfall or snowfall)
	Резкое изменение ветра к потеплению (с дождем или снегом)

Compass rose (1)

Regular changes in the winds and its consequences for precipitation and the weather – Winter and snowy spring

Постоянные изменения ветра и последствия на изменение погоды - **Зимой и снежной весной**

1, 2, 3, 4 = the order in which the winds change

→ = Coloured arrows – winds / разноцветные стрелы = ветра

+ also depends on the moon: As the moon dwindles, the weather changes (in the winter and a snowy spring – wet snow; in summer – rain): this does not depend on the direction of the win.

+ От луны тоже зависит – при убывающей луне (бйэӻа балдылэвки и манавулавки) погода меняется (зимой, весной - мокрый снег, летом - дождь) – несмотря на направление ветра.

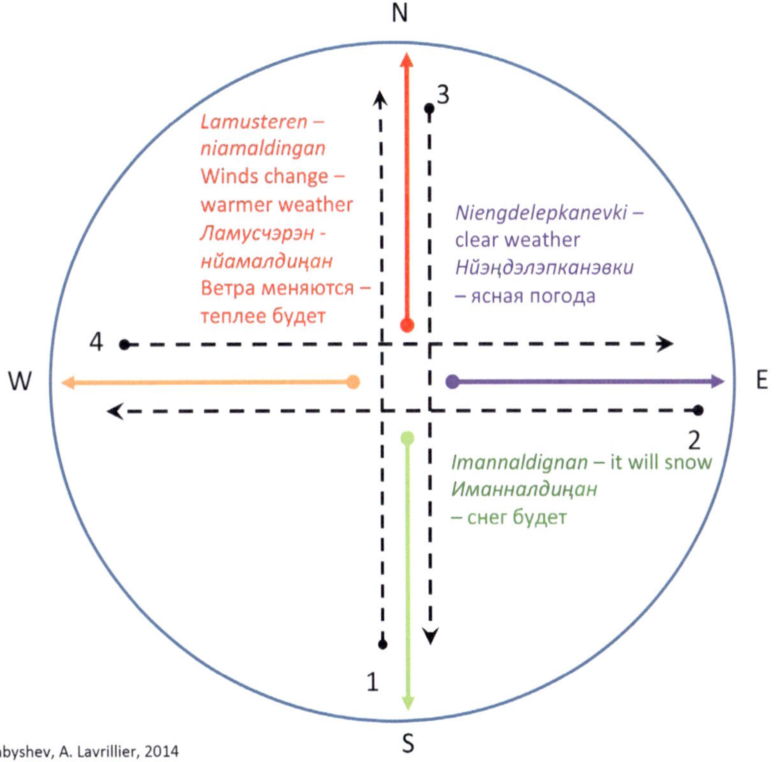

Compass rose (2)

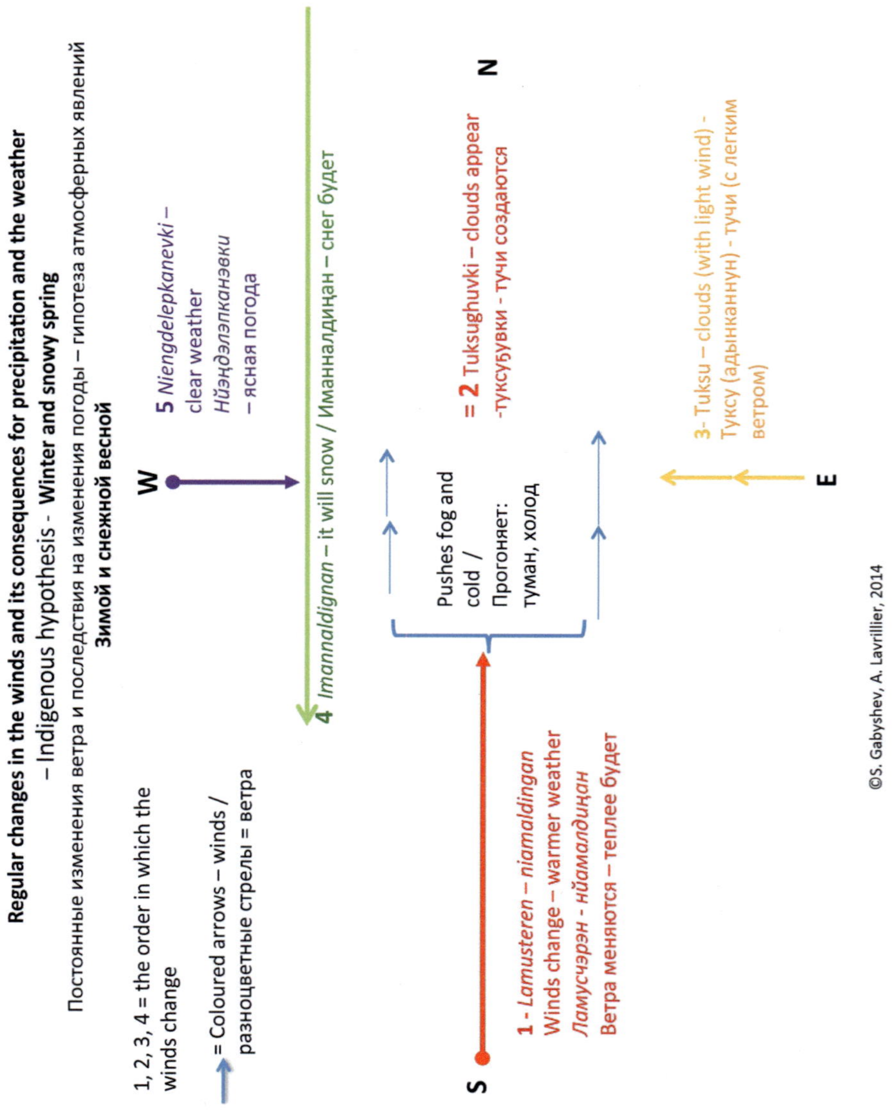

Compass rose (3)

Regular changes in the wind and its consequences for precipitation and the weather – Summer

Постоянные изменения ветра и последствия на изменения погоды - **летом**

There is no order in the appearance of the winds like in winter.
Нет порядка появления ветра как зимой.

+ also depends <u>on the moon</u>: As the moon dwindles, the weather changes (in winter and snowy spring – wet snow; in summer – rain): this does not depend on the direction of the win.

+ <u>От луны</u> тоже зависит – при убывающей луне (бйэђа балдылэвки и манавулавки) погода меняется (зимой, весной - мокрый снег, летом - дождь) – несмотря на направление ветра.

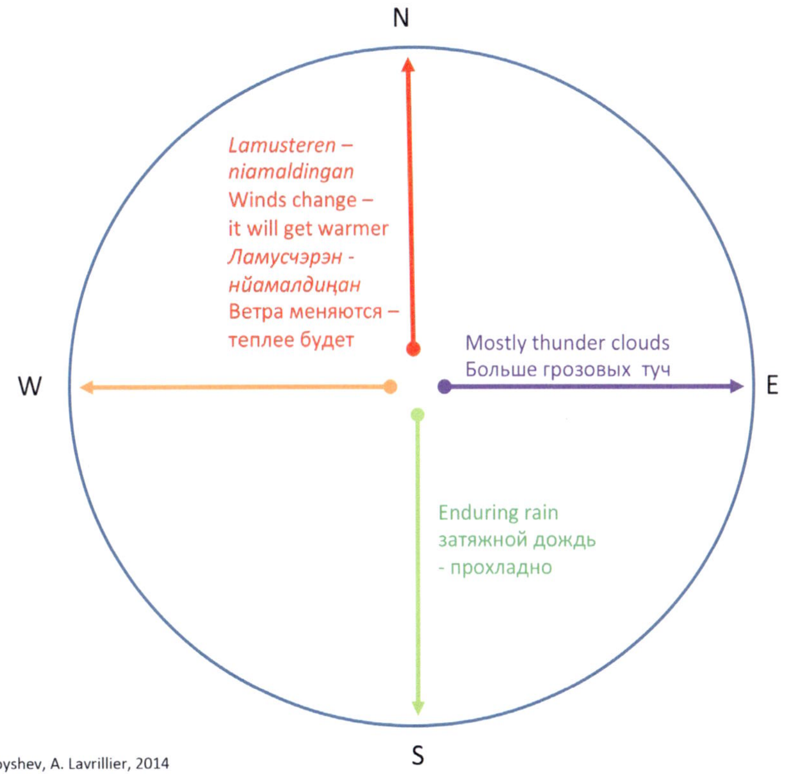

Olgokin Олгокин	Dryness Сухость

Everything can dry: wood, the ground (if there is very hot weather), air, ice, and wind. Even the clouds can dry up, as we see on those occasions when rain does not pour from them. The rivers dry up when the water level decreases and the sky is also dry when there is no rain for a long time. Humans and animals also dry out as they age. When many butterflies fly around, the elders say that the berries are getting ready to be gathered. As the white butterflies dry out, they sit on the *leva* marsh and die there. At this time, the cuckoo stops singing and the Evenki know that it is time to gather berries.

Упкат олговкил: мо, окудйэрэкин – дуннэ, воздух, иманна, дйукэ нйан олгокин, адын нйан олгокин, туксэ нйан олгокин – эвки тыгдарэ, бирал олговкил арбаран, буҕа нйан олгокин бивки горокунмэ нэӈдэлучивки. Бэйэл нйан олговки сагданми. Лородо багдарыр, кэтэкур дэҕиктэлэвкил, гунчэнэнны – диктэвэ ирикитчэрэн. Багдарыр лородол олговкил лэвалду довкил таду дат бутэвкил, диктэвэ тэврилэвкил, кукку этэвки куккудйэми.

Все может сохнуть: дрова, земля, когда жара бывает, воздух, снег, лед, ветер; даже тучи, когда не провоцируют дождь по эвенкийским поверьям; реки сохнут и вода убывает; небо тоже сухое, когда долго дождя не бывает. Люди и животные тоже сохнут, старея. Когда белые бабочки летают много, старики говорят, значит, ягода поспевает. Когда белые бабочки сохнут, они садятся на болотах *лэва*, там же погибают. В это время кукушка перестает куковать и эвенки знают, что пора собирать ягоды.

Derbakun (UN) / Siɣilcha (UN) Дэрбакун (УН) / Сиҕилча (УН)	Humidity Влажно

It is said that the air makes clothes and items humid. This air makes everything humid: wood, the ground, the sky, and so on.

Дэрбакун гунэвкил оҕалба, нарналва дэрбавканэвки. Дэрбаканэвки упкатпа: молбэ, дуннэвэ, буҕава.

Говорится на воздух, которые делает вещи влажными. Этот воздух делает лапы, шкуры, дрова, землю, небо влажными и так далее.

Iŋikun, Iŋinie!, iŋiniekun Иӊикун Иӊинйэ!, иӊинйэкун	Freezing cold, it is cold! It is very cold! Сильный холод, морозно! Очень холодно!

When the sun rises and one can see one's breath upon leaving the tent, we say that it is *iŋinie* cold. People prepare a lot of wood and chop it into small logs for the stove in order to create a lot of warmth. This term is also used in the summer when one can see one's own breath. It was used in the 1990s for temperatures lower than -40°C, while in the 2000s it has been used for temperatures lower than -30°C. (cf. Evenki climatology: Indigenous temperature measurements)

Ӊариллэкин дйудук йуми эриhинми амӊадук тамнаhинэн иӊинилча гунчэннэнны. Бэйэл молывкил мова нэмкуканди колтоҕоӊнэнны айат дэгдэдан оту. Дйуҕани нйан «иӊинйэ!» гунэвки окин амӊадук тамнаhинэн.

Когда рассветает, если при выходе из палатки изо рта пар идет, сразу понимаешь, что холодно. Люди готовят дрова, дрова колят тонкими поленьями, чтобы хорошо горела печка. Этот термин употребляется также летом, когда пар идет изо рта. Говорилось в 1990-ые годы для температуры от -40°C, а в 2000-ые годы говорится для температуры от -30°C. (см. с. 174–175)

Soŋun, soŋukikin, soŋutmar Соӊун, соӊукикин, соӊутмар	Cool, very cool (all seasons) Прохладно, средний холод

For the Evenki, this is the ideal cold temperature: it is neither too warm nor too cold. This term is used for the best winter temperature, during which people can hunt or guide the reindeer to the camp without breaking a sweat. It is also used in the summer when it is fresh and the mosquitoes and horseflies bite less. *Soŋut* is very good for everyone, including animals.
Sometimes, the Evenki use the topography to find such temperatures. (cf. Evenki climatology: Indigenous temperature measurements)

Соӊудйэрэкин айакикин бивки мунду олос эчэ окурэ, эвки иӊинирэ олос эчэ нйама. Таргачинду бэйуктэдйэми айа эӊнэнны нйаҕинэ. Ороврво онодйоми айакикин. Дйуҕа соӊундун мармактал, иргактал эвкил бирэ эмивкил кикчарэ. Соӊун упкатылду со айа бэйӊалду нйанэ.

Самая лучшая холодная температура для эвенков, не слишком холодно, не слишком тепло. Термин употребляется для идеальной температуры зимой, когда человек может охотиться или за оленями гоняться не потея, и летом, когда прохладно, видно потому, что меньше оводов и комаров, меньше кусаются. *Соӊун* очень хороший для всех, в том числе для животных.

Видно в типологии топографии, что некоторые виды ландшафта гарантируют прохладную температуру. (см. с. 174–175)

Emur, emure Эмур, эмурэ	Cold, very fresh
	Холод, холодный, очень холодный, холод

The cold produces *emur,* which contributes to *idia*. Everything can become *emur:* ground, trees, rivers, air, and metal. *Emur* is not visible except in some fogs: mostly, one can only feel it. (cf. diagram Idia)

Эмур иӈйиӈдук овки, эмурдук идйа овки. Упкат эмурэлэвки: дуннэ, ирйактэл, бирал, воздух, сэлэ. Эмурба эвки ичэврэ, адылду тамнаксоду, мэдучидйэӈнэнны.

Холод делает *эмур,* от *эмур идйа* становится. Все может быть холодным *эмур:* земля, деревья, реки, воздух, железа. *Эмур* не видно, его только чувствуют. Говорится в том числе, когда мокнут в ледяной речной воде, также говорится о воздухе холодном. (см. схемы, с. 232–242)

Eltandieren элтандйэрэн	Weather in the morning and evening that is colder than the rest of the day (all seasons)
	Холодно стает утром и вечером

This type of cold / frost is very important in Evenki ecological knowledge. It is thought to play a crucial role in seasonal shifts and the process of snow and ice physics (i.e., in the transformation of one snow type into another). (cf. Evenki climatology: Indigenous temperature measurements, Clouds typology: eltan; Snow and ice typology: diagrams Chuiur, Cheγa, Buldo)

Элтан гунэвкил кэӡарэкин иӈиниливки ӈариллэкин нйанэ элтакучивки дылача йудйэрэкин содын ӈэнэвки. Элтандук иманна булдо овки иманнавэ арбапканэвки.

Элтан говорится на утренний и вечерний холод на рассвете и закате. *Элтан* воздействует на физическую трансформацию снега, например, разные типы снега могут стать *булдо*, и уровень снежного покрова падает. (см. схемы, с. 174–175, 207, 228, 305–314, 319–323, 315–318)

Ikterebda **иктэрэбда**	Extremely cold, freezing: people freeze and their faces and clothes are covered in hoar frost Замерзать (о живом существе), сильно холодно становится, много инея на лице и одежде

Ikterebda means that a person is freezing in the harsh frost. When one walks outside, he/she is covered in hoar frost. This term is used in the winter only. One must make kennels for the dogs so that they will not freeze.
In the 1990s, this term was used when temperatures fell below -45°C; in the 2000s, it has been used for temperatures below -35°C. (cf. Evenki climatology: Indigenous temperature measurements. (pp. 174–175)

Иктэрэбда бэйэ иктэрэвувки. Ӈэнэдйэми саӊивувки иӊикун бивки. Нинакирду нйанэ дйуйатын оӊнанны эдатын иктэрэврэ. Туҕэн таргачин бивки.

Иктэрэбда значит, что человек мерзнет при сильном морозе. Когда идешь, весь инеем покрываешься. Собакам тоже домик делаешь, чтобы не мерзли. Этот термин употребляется только для зимы. Говорилось в 1990-ые годы при температуры от -45°C, а с 2000-ых годах говорится от -35°C. (см. с. 174–175)

Niama, niamakikin **Нйама, нйамакикин**	Warm, very warm Теплый, очень теплый

This term is used very often during the snow period (second autumn, winter, and first spring) in order to say that it has become warmer. It is also used to describe the warmth of a house, especially after a trip in the harsh cold. If the weather becomes *niama* in the winter, the Evenki hunt wild reindeer. Very often, *niama* warmth corresponds to snowfall. This is particularly the case during the spring *niama,* when calves are born. The *niama* warmth can also come from some mountain slopes. In such places, the upper layer of snow is the scab-like *cheya* type, while there is slightly wet snow underneath. Sables like to live in such places during the winter. It corresponds to various temperatures, from -15°C to approximately +5°C +10°C. This has changed over the years thanks to climate change, i.e. with the general warming, temperatures considered warm-*niama* have become warmer. (cf. diagram Idia)

Туҕэ, дйуҕа тара бивки. Нйама оракин туҕэниду тары айа бэйэл бэйуктэливкил. Иктэрэвми дйула ими нйан нйамакикинтыки иннэнны. Дуннэдук нйанэ нйама йувки эллэӊэду, таду ойолын чэҕа. Чэҕа эрэлын унйадйавки иманна. Таргачирду со андаhил туҕэвкил. Нйэӊнэ нйамалэкин эӊнэкар балдылэвкил.

Термин употребляется очень часто во время снежного периода (вторая осень, зима и первая весна), чтобы сказать, что тепло стало или о тепле дома. Если зимой становится тепло, эвенки начинают охотиться на диких оленей. Когда замерз в дороге, в палатку заходишь к теплу. От земли тоже тепло идет от некоторых склонов гор (эллэӈэ) и там снег сверху *чэƃа*, а внутри мокроватый снег всю зиму. Весной, когда тепло, телята рождаются. Совпадает с разными температурами от примерно -25°C до примерно +5/-10°C и это тоже менялось с годами и с изменением климата. (см. схемы, с. 232–242)

Oku- (okullen, okulkuchivki) Оку- (окуллэн, окулкучивки)	It is hot, very hot in the summer (between +25°C and +30°C)
	Жарко, жарко, очень жарко летом (от +25°/+30°C)

When it is hot, the Evenki light smoke fires for the reindeer so that the mosquitoes will not bite them. People make small roofs (hanged fabric) above tables outside, where it is fresh. Women treat skins and sew various items. Sometimes, the Evenki make small roofs on wooden poles *(kaltama)* so that the reindeer can be in the fresh air. (cf. Indigenous temperature measurements, pp. 174–175)

Экудйэрэкин бэйэл самӈирвэ илатчавкил. Нэлбэйэ овкил таду соӈун бивки аƃал оƃала манидйавкил уллидйэвкил. Адылдун орорду нйанэ овкил нэлбэкунэ, калтама.

Когда жарко бывает эвенки топят дымокур *самӈин* оленям, чтобы оводы их не кусали. Делают навесы, где прохладно бывает. Женщины лапы выделывают, шьют разные изделия. Иногда оленям большой навес делают – *калтама*. (см. с. 174–175)

Dulcha, Dullen Дулча, дуллэн	Springtime warmth from the sun in clear weather that causes the snow to melt (around + 10°C)
	Весной днем, когда сильно тепло становится из-за солнца – тепло, при котором снег тает

Dulcha occurs in spring: it designates the moment when the weather is getting much warmer after a long period of harsh cold. The Evenki hunt wild reindeer. (cf. Indigenous temperature measurements, seasonal chain, pp. 164–172)

Дулча таргачин нйэӈнэ бивки. Иӈин амардукин дулэвки. Нйамаллэкин бэйэл бэйувкил.

Дулча – такое весной бывает. После холода теплеет. Эвенки едут на охоту на диких оленей. (см. с. 164–172)

Kuraŋ, kuranen Кураӈ, куранэн	Heat wave in the summer (from +35°C) Пик жары летом (от +35°C)

Kuranen is the hottest period of the summer. Normally, it occurs over a considerable number of summer days. The smoke from forest fires causes the clouds to disappear. (cf. Evenki climatology: Indigenous temperature measurements, seasonnal chain, clouds typology)
In the studied regions, one of the most recent anomalies in the summer is the disappearance of a large number of hot days: they have been replaced by weeks of continuous rain (for instance in 2012, 2013, 2014, and 2016). Some of the effects of such changes on reindeer herding are presented in Lavrillier and Gabyshev 2017.

Кураӈнэкин сокунди окукучэвки. Кэтэдыван дйуҕанивэ окукучивки. Дуннэви дэгдэчэвки, нуныкун туксулвэ дйэпивки.

Куранэн – самый жаркие дни. Большая часть лета бывает жарким. Много дыма от пожаров, тучи исчезают. (см. с. 174–175, 164–172, 203–212)

Tamna Тамна	Fog, steam Туман, пар

Generic term for fog. Steam derives from the explosion of an icing blister *(bukte)* in the winter or the surface of a river in the summer. There is often fog (and thus poor visibility) after rain. (cf. also Clouds typology: tamnaksa)

Тамнакса гунвувувки уландйэракин тамнадйавки, иӈиниллэкин биралы тамнадйавки. Тыгда амардукин тамнавки. Экунда эвки ичэврэ.

Общее название тумана. Пар который выходит после взрыва наледного бугра пучения *(буктэ)*, или летом над рекой. Когда холодно в речках туманы бывают. После дождя туманит и видимость плохая. (см. с. 206–207)

Epkere Эпкэрэ	Air with smoke or steam that is difficult to breathe
	Воздух с дымом или с паром, которым трудно дышать

Epkere is air with smoke or steam which is hard to breathe. The smoke caused by forest fires and the steam caused by very harsh cold make it difficult to breathe.

Эпкэркун бирэкин нуныдук эридйами эпкэркун бивки эрукун эрида. Дуннэ дэгдэдйэрикин нуныдук нйанэ эпкэр бивки. Иӊиникундук нйанэ эпкэркун бивки.

Эпкэрэ – трудно дышать при дыме. При дыме от пожаров, или при сильном холоде трудно дышать.

Idia Идйа	Glacial or very cold air in the morning and evening (sunrise and sunset)
	Ледяной или холодный воздух утром и вечером (закат, рассвет)

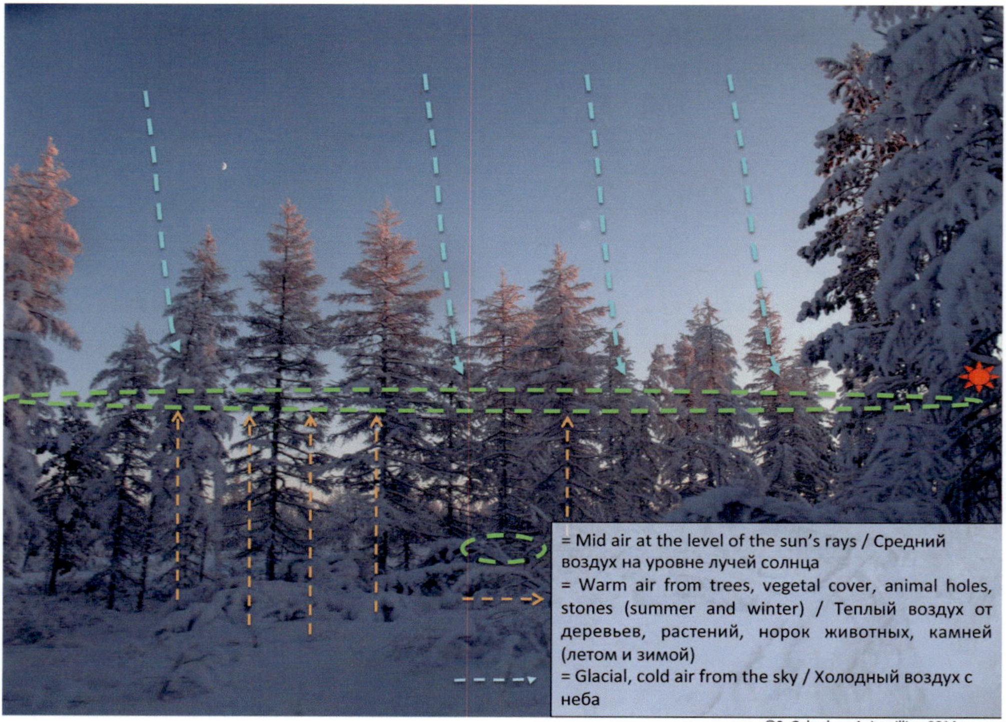

©S. Gabyshev, A. Lavrillier, 2014

Idia = Glacial or cold air in morning or evening. During SUNSET (winter) / _Идйа_ = холодный воздух утром и вечером.

Во время ЗАКАТА - Зимой

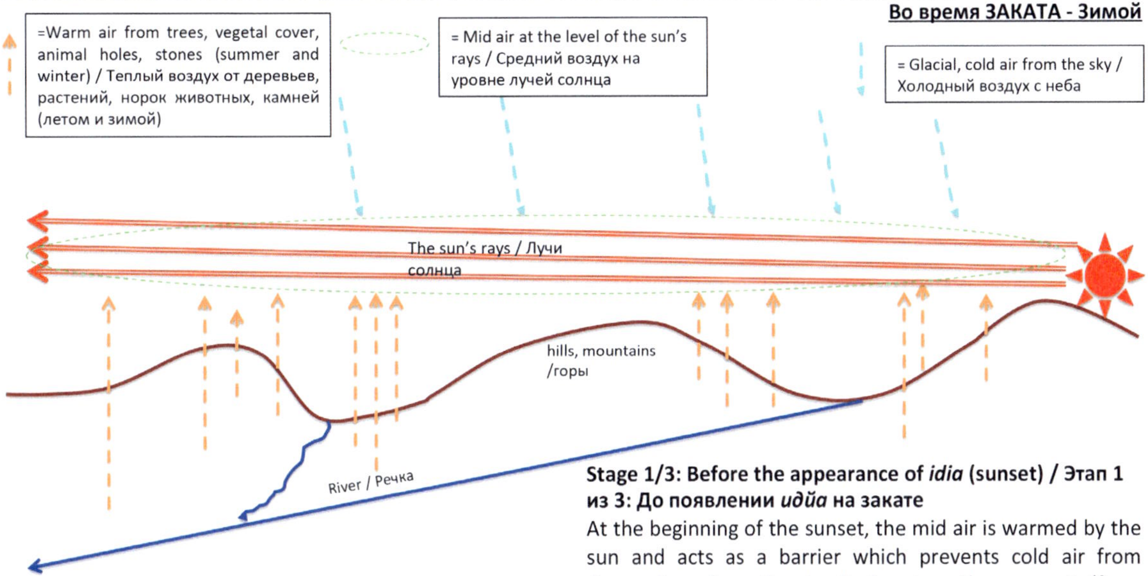

Stage 1/3: Before the appearance of *idia* (sunset) / Этап 1 из 3: До появлении *идйа* на закате

At the beginning of the sunset, the mid air is warmed by the sun and acts as a barrier which prevents cold air from descending from the sky. It also stops the warm air (from the ground) from rising into the sky. The presence of the mid air can be seen when rising fog hovers at this point in the air and rises no further.

Дылача тэҕэдйэрэкин, дулгу воздух дылачадук курэҕачин овки, тадук эвки опканэ эмур воздухту буҕадук эвупкандавки нйэскаки. Тыкадат дулгу воздух эвки бурэ нйама воздухту дуннэдук уҕирипканда буҕала. Тар дулгу воздух ичэвувки окин тамнакса уҕирипдйэрэкин - тар тамнакса дулинду намарадйэрэн.

На закате, средний воздух от лучей солнца служит как преграда, и не дает холодному воздуху с неба спуститься вниз. Также средний воздух не дает теплому воздуху от земли подняться в небо. Можно увидеть существование среднего воздуха, когда туман поднимается вверх и останавливается именно на среднем воздухе.

***Idia* = Glacial or cold air in morning or evening. During SUNSET (winter) / *Идйа* = холодный воздух утром и вечером.**
Во время ЗАКАТА - Зимой

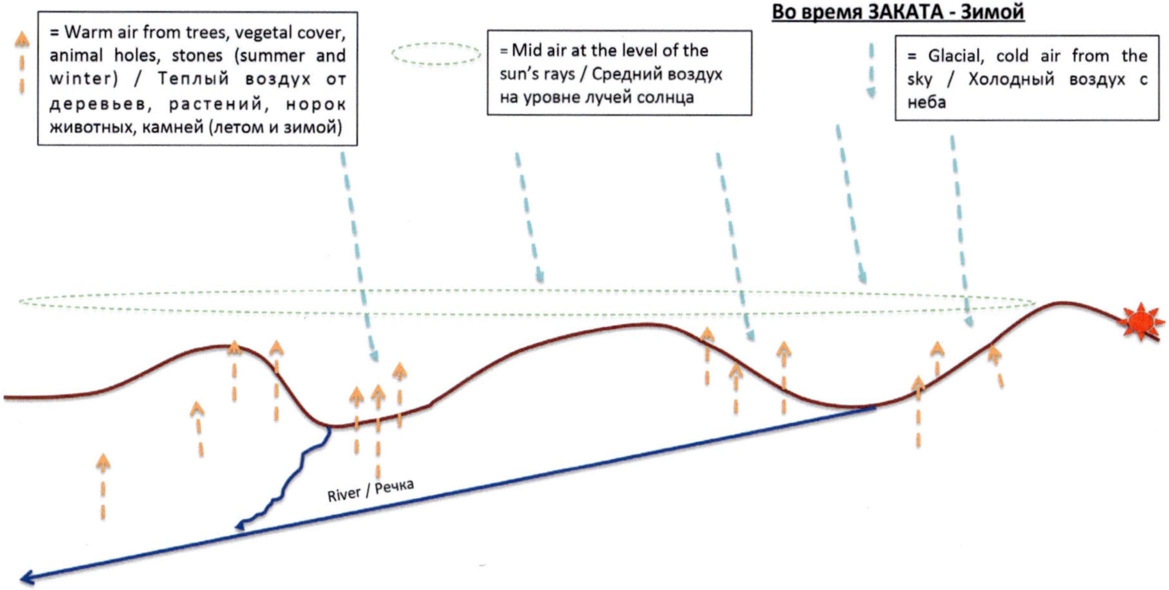

Stage 2/3: Appearance of *idia* during sunset / Этап 2 из 3: Появление *идйа* на закате
As the sun disappears below the horizon, the warm air disappears: this means the amount of glacial/cold air has increased. *Idia* descends according to the direction of the flow of the river.

Дылача тэђэчэрэкин, окин дылача тэђэвки, эмур воздух кађивки (нйама воздухпэ чумиканэвки). Кэтэтмэр эмур воздух овки. Идйа тыкивки дуннэла, тадук биралдули эйакуки урувки.

На закате, когда солнце уходит за горизонт, холодный воздух побеждает (теплый воздух исчезает), значит больше холодного воздуха станет, который, опускаясь на землю, спускается по реке.

Idia = Glacial or cold air in morning or evening. **BETWEEN SUNSET AND NIGHT (winter)**

Идйа = холодный воздух утром и вечером. **Во время СУМЕРЕК, ЗАКАТА – зимой**

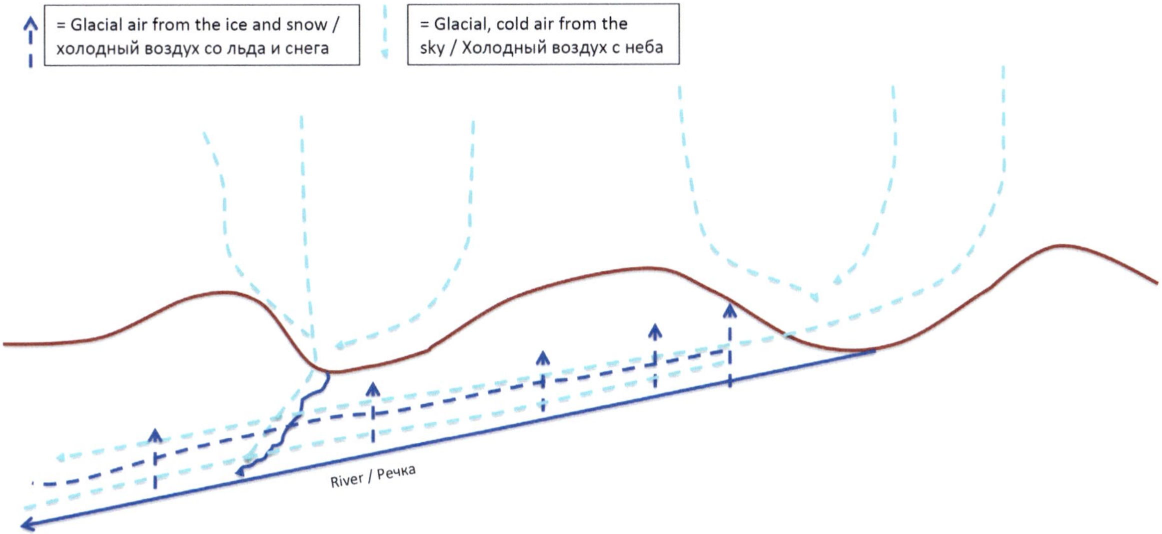

Stage 3/3: Development of *idia* after sunset / Этап 3 из 3: Развитие *идйа* после заката

Idia = When sun has only just disappeared, the glacial air descends, following the flow of the river; it thus attracts and carries away the glacial air from the ice and snow. Thus, two cold airstreams move along the river: one from the sky and the other from the ground (ice and snow).

Идйа = Окин дылача ачэн, эмур воздух буҕадук тыкивки дуннэла, тадук биралы урувки эйакуки, тандйэннэ эмурба дйукэлдук, иманналдук. Бирадулы эйаки дйур эмурба танэвки: 1. буҕадук, 2. дуннэдук (иманнал, дйукэ).

Идйа = на закате, когда солнца нет, холодный воздух с неба падает на землю и по речке уходит вниз, притягивая холод от льда или снега. По реке спускаются два типа холода - один с земли (лед и снег), другой с неба.

<u>*Idia* = Glacial or cold air in morning or evening.</u> **<u>During the SUNSET (winter). Topography</u>**

<u>*Идйа* = холодный воздух утром и вечером.</u> **<u>Во время ЗАКАТА – зимой. Топография</u>**

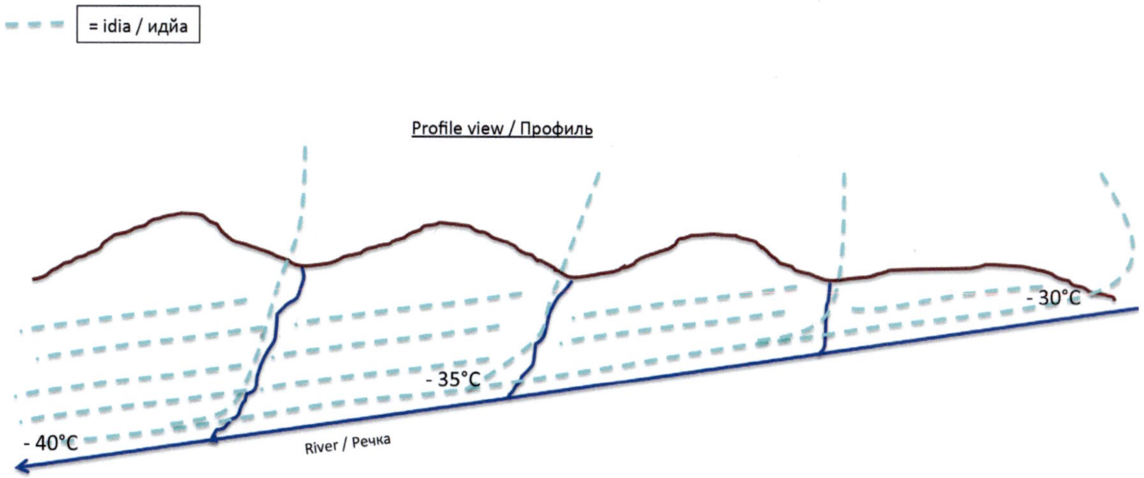

Idia – the cold air attracts the cold, especially from river ice. The more numerous the tributaries, the colder the valley will be. This is why it is much colder downstream than upstream: all the *idia* of the various tributaries merge and are further joined by the glacial cold from the river ice.

Идйа танэвки дйукэдук иҥинма бирадули. Кэтэтмэр биракар даптудйэрил бирала - иҥинитмэр. Олыһин эйаки иҥикун овки идйал бакалдырактын.

Идйа = холодный воздух, который притягивает холод к себе, особенно от льда рек. Чем больше притоков, тем холоднее в долине. Поэтому внизу по реке холоднее, чем вверху – ведь *идйа* соединяются с другими *идйа*-ми и плюс холод от льдов.

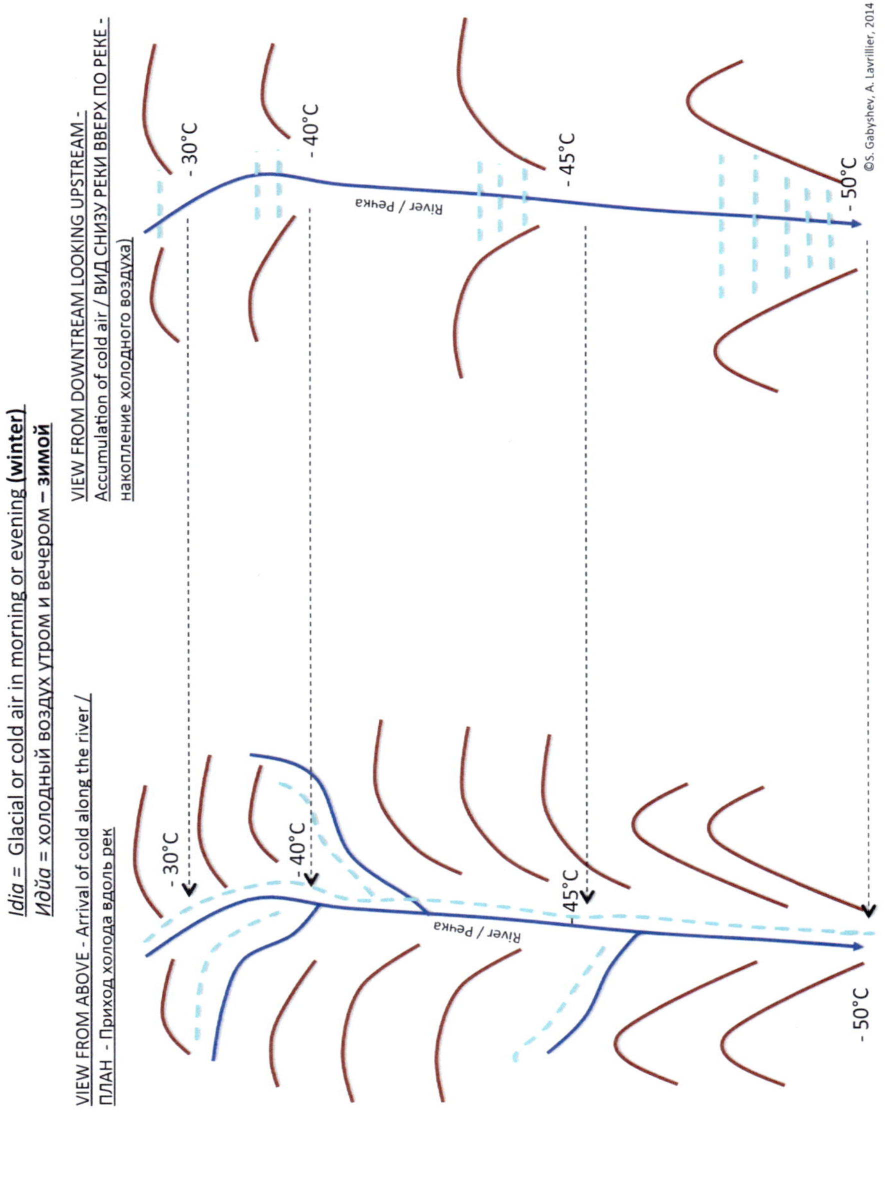
Idia = Glacial or cold air in morning or evening **(winter)**
Идйа = холодный воздух утром и вечером – **зимой**
VIEW FROM ABOVE - Arrival of cold along the river /
ПЛАН - Приход холода вдоль рек
- 30°C
- 40°C
-45°C
- 50°C
River / Речка
VIEW FROM DOWNTREAM LOOKING UPSTREAM -
Accumulation of cold air / ВИД СНИЗУ РЕКИ ВВЕРХ ПО РЕКЕ -
накопление холодного воздуха)
- 30°C
- 40°C
- 45°C
- 50°C
River / Речка
©S. Gabyshev, A. Lavrillier, 2014

<u>*Idia* = Glacial or cold air in morning or evening.</u> **<u>During SUNRISE (winter) /</u>**
<u>*Идйа* = холодный воздух утром и вечером.</u> **<u>Во время РАССВЕТА - зимой</u>**

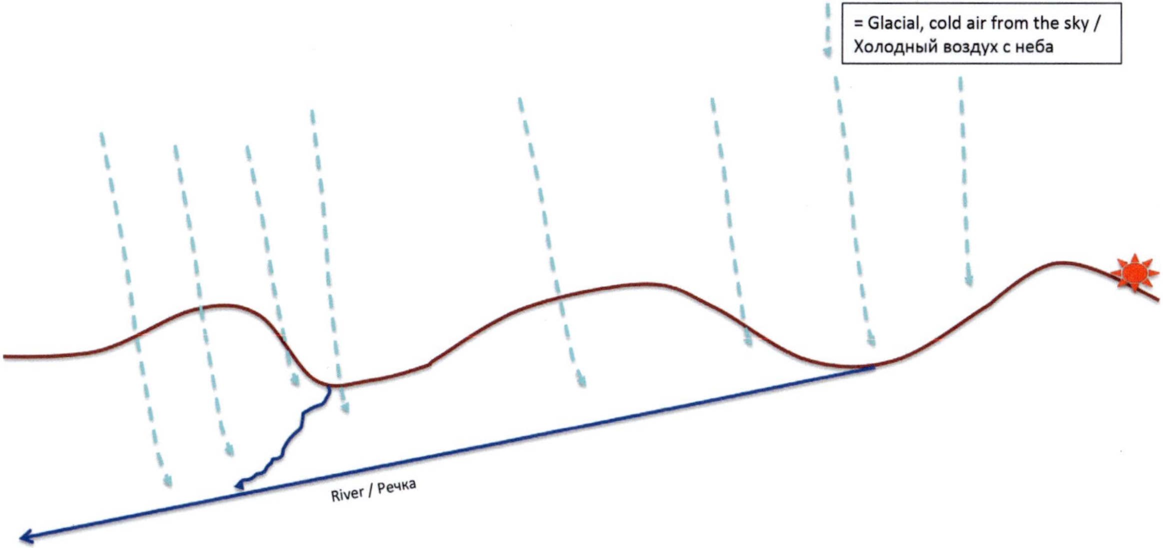

Stage 1/3: Very cold in the morning cf. *Idia* / Этап 1 из 3: Идйа утром очень холодный
At sunrise (when the sun has not yet risen), only cold/glacial air circulates, in particular along rivers and tributaries: this is because there is an accumulation of glacial air during the night (nothing warms it). *Idia* descends and goes downstream.
Ңарилдйэрэкин (дылача эдйэлин йурэ), эмурнун воздух ңэнэктэдйэвки, биралдули. Долбоныду эмур воздух кэтэлэвки (экунада эвки нйамалэвканэ воздухпэ). Идйа тыкивки дуннэла, тадук биралы урувки.
На рассвете, когда солнце еще не вышло из-за горизонта, только холодный воздух гуляет, особенно по рекам и притокам, потому что за ночь холодный воздух накапливается (ничто его не согревает). *Идйа* падает на землю и спускается по реке.

©S. Gabyshev, A. Lavrillier, 2014

Idia = Glacial or cold air in morning or evening. **During SUNRISE (winter)**
Идйа = холодный воздух утром и вечером. **Во время РАССВЕТА – зимой**

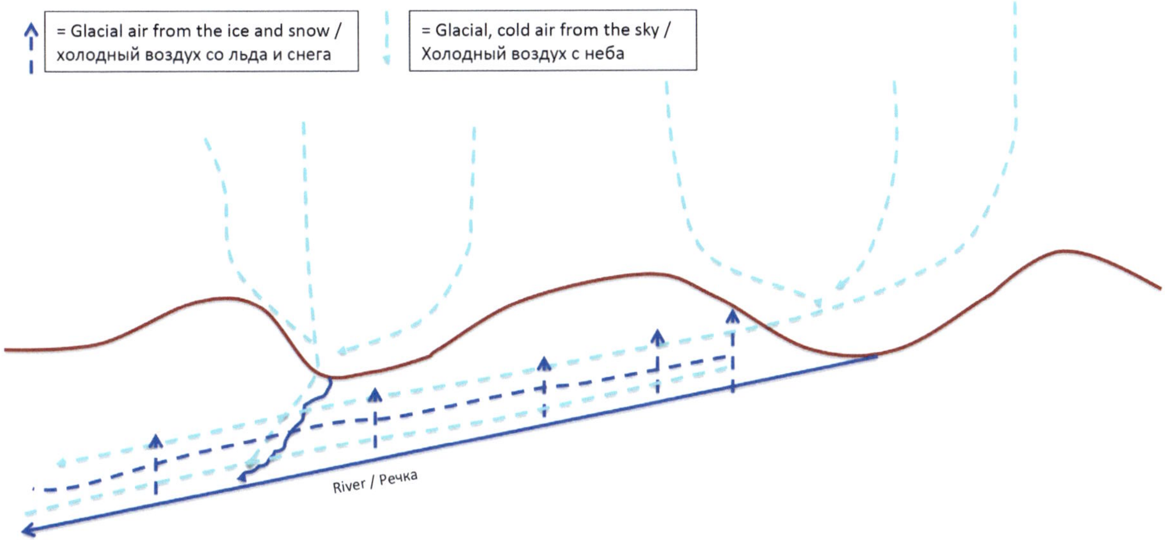

Stage 2/3: Development of *idia* after sunrise / Этап 2 из 3: Развитие *идйа* после расцвете
Just before sunrise, the glacial air descends and follows the course of the river, attracting the cold from the ice and snow as it does so. Thus, there are two cold airstreams moving down the river: one from the sky and the other from the ground (ice and snow).
Ңарилчалан (эдйэлин дылача йурэ) эмур воздух буђадук тыкивки дуннэла, тадук биралы урувки эйакуки, тандйэнэ эмурбэ, дйукалдук, иманнадук. Бирадулы эйаки дйур эмурба танэвки: 1. буђадук, 2. дуннэдук (иманнал, дйукэ).
Идйа = на рассвете, когда солнца нет, холодный воздух с неба падает на землю и по речке уходит вниз, притягивая холод от льда или снега. По реке спускаются два типа холода - один с земли (лед и снег), другой с неба.

©S. Gabyshev, A. Lavrillier, 2014

Idia = Glacial or cold air in morning or evening. **During SUNRISE (winter)**
Идйа = холодный воздух утром и вечером. **Во время РАССВЕТА – зимой**

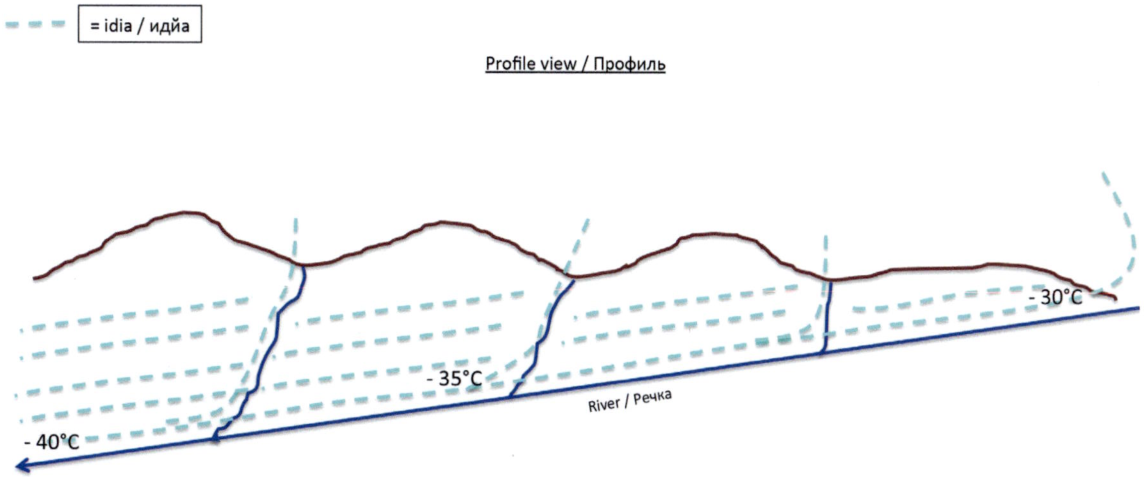

The cold air attracts the cold, especially from river ice. The more numerous the tributaries, the colder the valley will be. This is why it is much colder downstream than upstream: all the *idia* of the various tributaries merge and are further joined by the glacial cold from the river ice.

Идйа = эмур воздух танэдйэрэн эмурба мантыкиви, дйукэдук. Кэтэтмэр биракар даптудйэрил бирала – иӈитмар – идйал бакалдйэвкил уӈтул идйалнюн. Олыһин эйаки иӈикун овки идйал бакалдэрэктын.

Идйа = холодный воздух, который притягивает холод, особенно от льдов рек. Чем больше притоков, тем холоднее в долине. По этому внизу по реке, холоднее, чем вверху – ведь *идйа* соединяются с другими *идйа*-ми плюс холод от льдов.

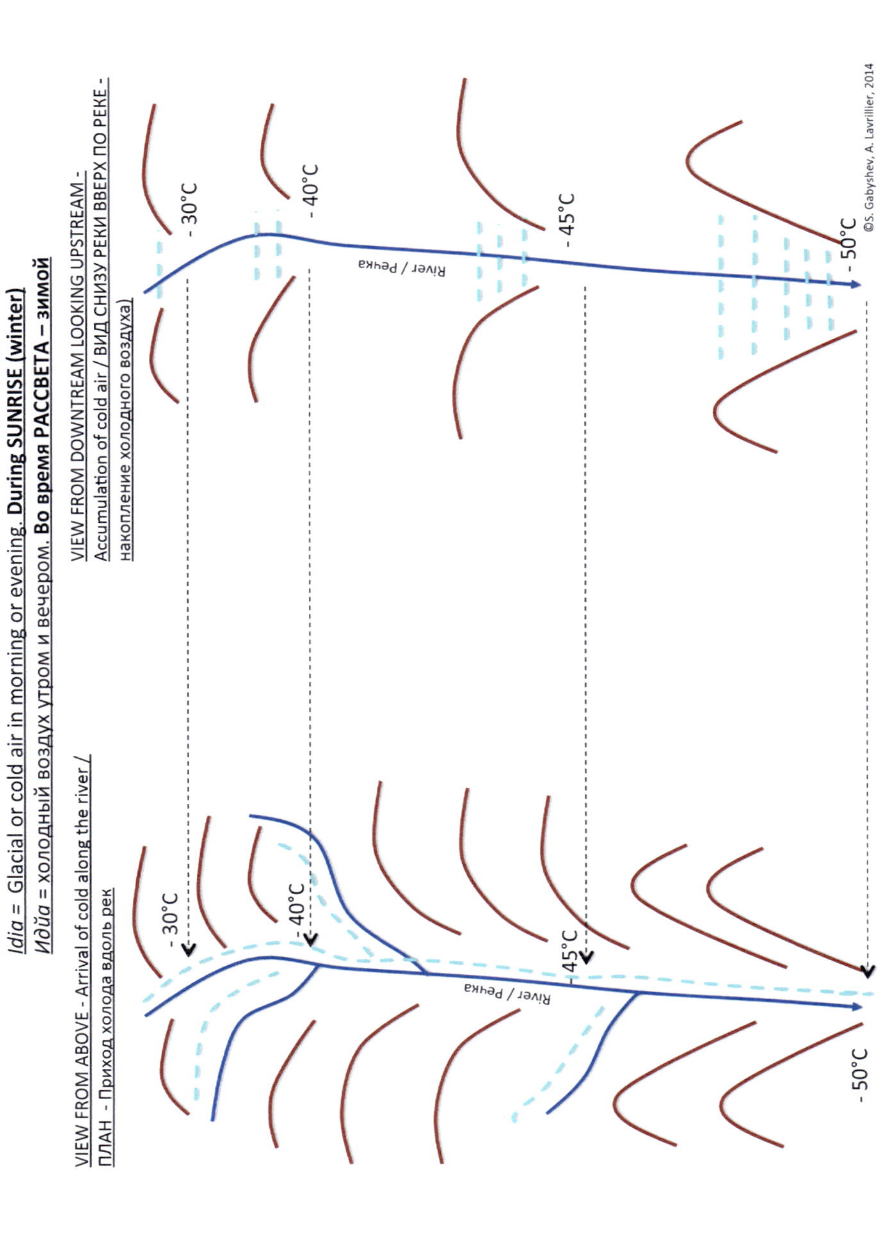
Idia = Glacial or cold air in morning or evening. During SUNRISE (winter)
Идйа = холодный воздух утром и вечером. Во время РАССВЕТА – зимой
VIEW FROM ABOVE - Arrival of cold along the river /
ПЛАН - Приход холода вдоль рек
VIEW FROM DOWNSTREAM LOOKING UPSTREAM -
Accumulation of cold air / ВИД СНИЗУ РЕКИ ВВЕРХ ПО РЕКЕ -
накопление холодного воздуха)
- 30°C
- 40°C
- 45°C
- 50°C
River / Речка
©S. Gabyshev, A. Lavrillier, 2014

Idia = Glacial or cold air in morning or evening. **During SUNRISE (winter) /** *Идйа* = холодный воздух утром и вечером.
Во время РАССВЕТА – зимой

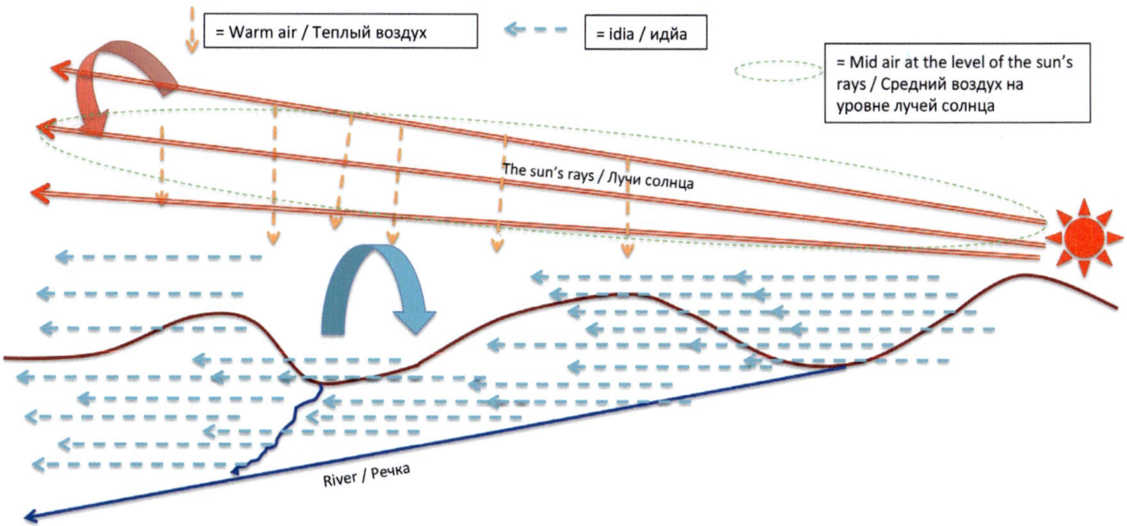

Stage 3/3 / Этап 3 из 3:
During sunrise, the mid air is warmed by the sun and acts as a barrier which prevents the glacial/cold air from rising. As a result, the glacial/cold air is pushed downwards: the temperature is lower than during the night, especially since this air joins the cold from the ground (rivers, snow). Thus, it is colder than during the night until the sun rises fully.
Дылача йудйэрэкин, буҕани эҥэһикун идйа овки нйэскаки урудйонэ, дуннэни идйа бакалдйэвкил. Дйур идйал – олыһин иҥинйэтмар овки.
На рассвете средний воздух от лучей солнца служит как преграда, которая придавливает и не дает нижнему *идйа* подняться к небу. В итоге верхний холод спускается вниз еще сильнее, чем ночью и соединяется с нижним холодным воздухом. Значит, еще холоднее становится, чем ночью, пока солнце не встало.

2.3.5 Snow and ice typology
by A. Lavrillier

From snow and ice

Continuing the chapter on Evenki climatology, we enter a very important section which categorises a set of precipitations (snow) and different types of ice according to the Evenki knowledge system. While both academic scientists and the Evenki treat snow and ice separately (the former is precipitation while the latter is a transformed state of ground water), we decided to analyse them together for several reasons. First, it became an obvious categorisation decision as we documented the snow types (see below). Second, the Evenki have complex knowledge about the physical transformation of the snow types into ice and the many interactions between snow falling from the air, snow lying on the ground, and ice. Since the Evenki seem to consider snow and ice as belonging to one and the same element (water), we decided not to separate these two chapters.

The importance of snows and ices

Imanna (иманна) is the generic term for snow in Evenki (*imannadiak, иманнадйак* is snowfall), but our years-long study found 27 occurrences for around 25 snow types in the Evenki snow typology, since some which correspond to particular terms or types appear twice a year (cf. Chronological table below). The snow is a crucial element in Evenki life and is omnipresent in their landscape. Southeastern Siberia is covered by snow for eight or nine months of the year (i.e. from 66 per cent to 75 cent of the year). From an indigenous point of view, snow is the Evenki's native element: it is their life, as expressed by S. N. Gabyshev:

> "Without the snow, we could not survive, because it would mean no hunting, no movement. The snow informs us about which human or animal groups or individuals have passed through our territory and when they did so.
> Without the snow, we would perish from the frost: the snow keeps us warm.
> Without the snow, the reindeer would not live, they would be sick. The snow keeps the master of the forest (bears) asleep, so it will not create problems; by this, it protects us and our reindeer during the longest part of the year.
> Thanks to the snow, we can find everything in our landscape – thanks to the tracks they leave, we can find reindeer, game, and people.
> The snow is very useful for the animals and the vegetal cover: it gives water to the animals and plants and is a long-lasting reserve of food for many animals. It irrigates the vegetal cover and protects the buds from the cold. It protects the forests and the vegetal cover against fires. It forces the birds to migrate and by

> this protects them from mass illnesses. The ice and snow together protect fish from humans, and animals from predators for several months.
> Snow for us is like a master-spirit because it gives us everything (food and water), it protects us and the natural environment. It treats nature, people, and animals by giving plants a strong medicinal effect.
> Without the snow, there would be a lot of stress, because without it hunting is very difficult, and we cannot survive without hunting.
> Without the snow, it would also be very difficult to predict the weather and animal movements. Snow is also crucial for maintaining a good balance between humans (especially non-natives) and animals. When the snow is deep, there are more animals because humans hunt with difficulty; in contrast, when the snow is shallower, humans live better because hunting is easier, but the animals become less numerous. So, the snow keeps a good balance between human and animals.
> The snow creates a good atmosphere, clean air, and a good mood among humans and animals. Snow makes humans, animals, and the vegetal cover stronger: people and animals become healthier, stronger, and hardier; it gives a strong medicinal effect to plants. The snow gives people and the environment resilience. In addition, snow and ice give us wisdom and knowledge. For instance, if you see a lot of small bird tracks in the snow, it means that the weather will be warm. The snow informs us about the present, past, and future states of everything: the vegetal cover, fauna, rivers, and weather.
> Of the ice, the Evenki say that it gives water, but it also represents an important part of their winter roads and gives them access to the opposite sides of rivers and fish."

In addition to such general understandings about the many functions of the snow and ice for the vegetal, animal, and human realms, the Evenki have a complex knowledge system that includes typologies. These contain not only knowledge ('know that', e.g. knowing something'), but also 'know-how' and an understanding of the 'physics of snow and ice'.

We will first present some elements of the Evenki physics of snow and ice with the system of snow depth measurements. Then a study of how the Evenki master the snow is provided. Afterwards, details regarding the specificity of extreme snow events and different categories of 'anomalies' is offered (cf. also Part III). With texts, pictures, and diagrams, the following section presents the knowledge system relating to the order of the appearance of snow types according to Evenki norms and their understanding of anomalies. It also considers related opportunities and threats. We then conduct a detailed analysis of Evenki snow physics (with a table showing the transformation of the physical states of the snow types) and native methods of analysing the snow cover. Detailed information is given about snow and ice types, including measurements of the size of flakes. Finally we present the typology of ice and its physics, opportunities, and threats for the community. In the final part of the book, the reader will find several case studies showing how these indigenous ice and snow typologies are used to study norms and anomalies and to analyse past and future important changes in the environment and climate.

An emic physics of snows and ices

The Evenki snow typology corresponds to a specific understanding of snow transformation, which can be conceived of as an emic 'physics of the snow and ice'. Our analyses of the snow typology demonstrate various kinds of snow are differentiated according to:

- the level of humidity,
- the size and form of the snowflake,
- the state of the flake (including icy flakes),
- the state of the snow cover itself (compact or friable; in a frozen or melted state, etc.),
- the colours (within the western 'white' category),
- its transformation by reindeer,
- the physical transformation it has to undergo,
- the depth,
- the noise it produces,
- its location (on the ground, in various layers within the snow cover, in the air, on the trees, and so on),
- the time it falls – each type of snow is expected during a specific period of the year with a specific order, which forms a norm. Of course, this norm admits a degree of variability; however, if the limits of this variability are crossed, the weather is consider to be disrupting this order and is defined as an anomaly according to Evenki ecological knowledge (Lavrillier et al. 2016).

Some specific snow types (like *sy, сы*, for instance) may be directly perceived as anomalies; after a certain limit, the presence of these types is considered an extreme event that will endanger reindeer herding (see below). For instance, such snow triggered the deaths of hundreds of reindeer in 2007 throughout Siberia.

In addition, herders and hunters are well aware that topography and the snow typology are interconnected. Herders use the various topographies of their nomadic lands to adapt to anomalies in the snow cover and ensure that they can find better pastures for the reindeer. The reader will find several examples in this book (cf. Topographic typology, Vegetal cover typology). The Evenki's ability to move and their knowledge about topographical peculiarities and the specific characteristics of the snow cover for each topographical type are essential tools for adapting to anomalies. During our research for the BRISK project, we documented and analysed several adaptive strategies that we will soon publish in other papers and books.

The snow typology can be considered a 'knowledge of specialists', somewhat like a scientific specialism within TEK. Even within the Evenki community, not all the Evenki know the complex terminology describing snow (this is true even for those who know the Evenki language well, like Evenki language specialists or specialists in oral literature, cf. Introduction). Indeed, even among the nomads, only middle-aged Evenki grasp the typology in its entirety.

The snow typology, like all the typologies in this book, is a link in a cognitive chain that allows the Evenki to use the natural environment and adapt to variations and anomalies.

The following table is the result of several years' work spent documenting the snow and ice typology. A nomad will never list all the typology at once or give a clear and concrete explanation of it. It is a type of knowledge that one must receive piece by piece. In addition, as far as Lavrillier noticed, such knowledge has a kind of 'temporality' (i.e. herders can remember the terms in use at a specific time while temporarily forgetting the terms for the following seasons). For instance, at the end of December 2013 during a working session in a tent, the Evenki present could not immediately recall the term for a specific snow that appears during the spring. They remembered it two days later.

It is also useless to work in Russian. For instance, it was initially impossible to gather names for the snow other than the generic *imanna, иманна,* when asking in Russian: 'What are the terms for snow?' The only way one can work is through the Evenki language. One must detect and track unknown terms in Evenki narratives about herding and hunting or follow herders in their everyday lives in the taiga and ask questions about the surroundings. One method was that Lavrillier walked alone in the forest and took a lot of pictures (topography, vegetal cover, snow cover layers, etc.), especially during the hunting period when the nomads are very busy. Then, in the evening, she asked the nomads for comments. However, some nomads, mostly S. Gabyshev and Vasilii Gabyshev, also took many pictures, either on their own initiative or when documenting indigenous knowledge (cf. Topographic typology: diagram Arbun solgu for instance) and environmental changes. For example, asking for such commentary about the snow layers allowed her to identify the terms for snow presented in this typology. In addition, the daily observations and documentation designed and completed by the herders (who also took a lot of pictures of the processes that occur during the spring and autumn) and their collective analysis helped a lot in terms of identification of these typologies.

We should note that these terms can change from one dialect to another. Several terms can exist for one item, even among groups that live very close to each other. When there are dialect differences, the name of the specific village is written in parentheses.

The types of snow are expected in specific periods, and the Evenki know that a certain variability from the norm is allowed, as we will see in the chapters of the book.

In one specific period, as is shown in the following diagrams, several types of snow coexist within the layers that make up the snow cover, rest on trees and bushes, or fall through the air.

According to our analysis, each type of snow is awaited because it has certain uses, be it for the human economy (and movement), the regeneration of the vegetal cover, or for the reindeer. As one can see in the following table, each type of snow enables (or threatens) various seasonal hunting and herding activities.

For example, snow types can produce more or less noise. This is important for hunting, since noisier snows will frighten away the game. Another example is the ability of a snow type to support a hunted sable, the dog hunting it, or the hunter (on sledge or on foot). Some types allow for snow roads to be established and maintained: without these, movement and survival in the taiga are simply impossible. A winter that is too warm will not offer the kind of cold that will freeze the surface of a snow road created the day before by herders. A frozen snow road is essential for supporting reindeer or snowmobiles at the mid-depth of the snow cover and avoiding an exhausting (or impossible) move. Anomalies with regards to road building appeared in the winter of 2014-15; many herders were unable to leave their camps or were stranded in the middle of the roads. (cf. Part III: diagram Snow roads)

An example occurred in 2014–15, one which created many problems with transportation. Be it by sledge, snowmobile, or ski, all transport became almost impossible and extremely dangerous, with an average temperature of -25/-50°C. Many herders/ hunters frequently got stuck in the middle of their trips: they had to spend hours digging reindeer sledges or snowmobiles out of the humid, porridge-like snow. A trip usually made in three hours was taking around 15-20 hours. This occurred because of the lack of strong cold (-45/- 55°C), which usually transforms the dry and soft snow *(duiukun, дуйукун)* into icy and seed-like flakes *(buldo, булдо)*. This snow type provides the necessary firmness for the establishment of snow roads (cf. diagram Buldo). In addition, an abnormally warm snowstorm deposited a deep layer of fluffy and slightly wet snow *(debdeme, дэбдэмэ* – a snow type usually found in the late spring or early autumn). When the harsh cold set in, it could not reach the inner layer of snow because of the *debdeme, дэбдэмэ* snow and thus it could not transform the inner layer into *buldo, булдо*: the deep snow cover created thermic isolation. (cf. Part III: diagram Anomalies in snow cover 2014–2015)

As in the winter of 2014–15, the lack of strong cold and the abnormal warming (even at night) meant the Evenki were blocked into their camps because they could not create snow roads: the strong cold at night normally freezes the snow on the surface of the roads, creating a hard layer of snow that allows the nomads to move for hunting and herding.

Snow depth: Measurements and uses

The Evenki have their own system for measuring snow depth: *elekin, элэкин* (up to the knee) is the ideal depth for herding and hunting activities; *suaban, суабан* (up to the ankle) is an anomaly in the winter; *arban, арбан* (up to the calf) and *sungta, суӈта* (up to the top of the leg) are considered manageable; and *suŋtakun, суӈтакун* (up to the torso) is considered too deep for hunting and herding activities.

The snow depth is a crucial factor for the access of the reindeer to grazing grounds. The Evenki lead their herds to places where there are bearable snow conditions. Here again being mobile is the most important tool for adapting. The Evenki use the different snow depths. Thin layers of snow *(suaban, суабан* and *arban, арбан, arbakun, арбакун)* are used in the autumn for sable hunting with dogs. In contrast, deep snow, *suŋta, суӈта,* is used to keep reindeer gathered in one area and ensure that predators are kept at a distance: wolves cannot cross areas where the snow is deep.

Anomalies in the snow depth can have harsh consequences. A snow cover which is too deep threatens sable hunting, transportation, and thus household purchasing power (Lavrillier et al. 2016).

The presence of a very thin layer of snow *(arban, арбан, arbakun, арбакун* or *suaban, суабан)* throughout the winter, like in 2011, leads to an extreme process in different domains. It was very cold and the reindeer moved very far in various directions. Thus, herders had problems with keeping the herd together and predators found it easy to attack it.

Sometimes, the Evenki see both the direct and indirect negative consequences of snow depth on the development of vegetal cover and the presence of fur and food game (for more details, see Lavrillier and Gabyshev 2017; Lavrillier et al. 2016: 118–119).

Arban, Arbakun imanna / Арбан, Арбакун иманна
Literally means shallow snow (up to the calf). The same system of measurement is used for rivers
Мелкий снег до колена. Эта система измерения употребляется также для измерения уровня реки

Elekin imanna (UN) – Enŋetli (Iengra) / Элэкин иманна (УН) Энӈэтли (по колено) (Иенгра)

Snow depth 'up to the knee' is ideal for herding and hunting

Уровень снега по колено – идеальный для охоты и оленеводства

Suŋta, Suŋtakun imanna / Суӈта, Суӈтакун иманна

The snow level (up to the torso) is considered too deep for hunting and herding

Глубокий снег до пояса

Mastering the snow

Snow road construction: In its natural state, snow in the forests is soft. A deep snow cover does not allow for easy movement; indeed, it makes moving around extremely difficult. For instance, a snow cover that is too deep or a snow type that is too soft causes many serious problems for transportation, among other things: it can even threaten the lives of the nomads. Being unable to move means death in the forest because it deprives nomads of the ability to gather reindeer, hunt, forage for food, travel to the village, and so on. The nomads themselves never explain it in such a way, but our study of nomadic practices and comparisons of the evolution of snow cover over several years have demonstrated that this is the case. This is why it is crucial for nomads to create a network of snow roads from the first snowfalls of late autumn and to maintain them throughout the snowy period (cf. Part III: diagram Snow roads: case study on anomaly with porridge-like snow). With such an aim, they will always travel on the same set of roads several times a month. This allows the road to harden, compacting the surface snow: after several trips, a kind of 'puff pastry' or laminated snow cover that will last until the very end of the snowy spring will be created. The other 'tool' the Evenki use during construction is the evening and morning frost *(eltan, элтан)*. This freezes and hardens the snow disturbed by the passage of sledges or snowmobiles. According to the Evenki, this morning and evening frost *(eltan, элтан)* plays a very important role in the physics of the snow, (cf. diagrams Processes of snow cover, Cheγa, Evenki snow physics, Clouds typology, Winds and Airs typology).

In addition, such roads are a crucial source of information, since each human group that goes along them leaves very useful information for others. Sledge or snowmobile tracks can be read by the Evenki, who deduce from them information about how many people went down the road, how heavy their baggage was, in which direction they went, and when during the day they travelled. By knowing where people have installed their camps, nomads can deduce, with a high level of accuracy, the identity of the person who went along the road. Gabyshev explains that 'snow gives us knowledge'. This is an additional argument for considering the snow to be part of the cultural landscape.

With climate change and warming, this 'technique' of mastering the snow with the frost has come under threat, as was the case in 2015.

The second kind of practice that proves that the Evenki 'master' the snow is the occasional use of a deep snow cover (in bad years or in a well-known landscape) as a herder. Once again, the nomads Lavrillier met do not explain it in such terms, but it is indeed a practice they deploy. Before going further into the explanation, let us note that there are two other elements of the natural environment 'used as a herder': some high mountains and horseflies. (cf. Topographic typology, Evenki calendar, Evenki climatology, clouds typology)

The principle at the basis of this practice seems to be that reindeer, just like humans, are lazy and thus prefer to walk in shallow and hard snow rather than in deep and soft snow. Sometimes, herders use deep and soft snow to herd their reindeer. Firstly, they might guide the reindeer to a landscape known for its deep snow (like mountains, for instance). This allows to keep them grazing in one delimited area: meanwhile, the Evenki can occupy themselves with hunting. In this case, it is crucial that the snow be very soft and the pastures very rich so the reindeer can graze fully (cf. Topographic typology; Vegetal cover typology). Secondly, the Evenki can sculpt circulation channels into the deep snow that they can open or close: they thus control the movements of the herd. Upon arriving in a landscape typically known for its deep snow cover (or during a year with deep snow conditions), herders sculpt the deep snow in two ways. Firstly, open channels are created by reindeer sledge or snowmobile: these guide the herd to the desired pastures. Secondly, they close off open channels from extant snow roads with a kind of deep snow door, thus preventing the reindeer from milling off (cf. diagram Snow as a herder).

In this case, they are not only using an element of the natural environment for human work (i.e. herding), but are also modifying it in order to help humans control the herd in their absence.

Regarding responses to extreme events, such practices **are more than adaptation** to a bad snow condition, since they **use features of the extreme event** in favour of humans. In our study of climate change and extreme events, we met several similar behaviours. For instance, during a terrible flood in a village, some nomads used the very high level of the river to travel by boat down to a mountain slope that was otherwise inaccessible in order to gather high-quality firewood.

The normal order of the appearance of snow types and the many functions of snow

Our analyses of the anthropological data from fieldwork and the daily observation tables produced by the nomads show some important elements in the functions of the different snow types:

- Snow types show tracks that give information about the main (regular) and strongest winds, temperatures, future weather patterns, and animal movements. Some herders await the installation of the snow cover with impatience because then, as they say, the 'taiga will become an open book'.
- As explained above and shown in the diagram 'the snow cover as a herder', a deep snow cover is often used to help the herder: it keeps the reindeer in one place, guides them in a specific direction, or protects them against predators (cf. diagram Snow as a herder).
- Some snow types are especially anticipated. Here are the examples of two snow types that both result from physical transformation when they are already on the ground (i.e. they are not snow types from the sky (cf. diagram Evenki physics of

the snow here below)). In November, the snow type *chuiur, чуйур* (hard surface snow) is really anticipated by the hunter: if it is made strong enough by the seasonal winds, it makes hunting easier, allowing the hunting dogs to run on the surface of the hard snow and catch more sable. It also makes transport easier if it is able to support reindeer sledges or snowmobiles on its surface, since they move much quicker on such snow than in soft and deep snow. Sometimes, the snow type *cheya, чэҕа* (scab-like snow created by successive warming and freezing) has the same functions. Another highly anticipated snow type in November and December is the *buldo, булдо* snow that appears thanks to the way in which harsh cold physically transforms various snow types. This snow type has several functions. It frees the vegetal cover from the ice or icy layer *(sy, сы)* that appears in the early autumn. In addition, by transforming the snowflakes (cf. diagram Buldo), the cold reduces the volume (and depth) of the snow cover. Finally, because of the large size and hard consistency of the flakes, the snow type can support people, reindeer, and snow mobiles above the ground (cf. diagram Buldo). This is essential for transport and preventing sinking when in a deep snow cover. (cf. Order of appearance of snow types)

- A harsh cold offers many more opportunities than warm weather. However, the climate changes that emerged in the 2000s have affected the normal increase and decrease of temperatures and has endangered many aspects of the environment. For instance, the evolution of the snow cover (see here below) and the vegetal cover have been affected, which has consequences for the animals which depend on this vegetal cover. By disturbing the normal evolution of the snow cover, climate changes are also endangering the mobility of nomads as well as their economic activities. As we can see in the table about the order of the appearance of snow types, the snows typical during warmer temperatures create many problems for transportation and food conservation. In addition, warming weakens humans confronting the cold. For instance, chilblains are more likely during periods of wet snow than during a harsh cold. In other words, the harsh cold prevents people from freezing. (cf. Order of appearance of snow types: from spring)

Other snow types are feared, like the appearance of *cheya, чэҕа* (a scab-like snow layer) in the autumn or the formation of *sy, сы,* an ice-snow layer embedding the vegetal cover. Sometimes, the consequences of the formation of one snow type are so numerous and interlinked that it has been crystallised by TEK into one concept. For instance, there is the concept of *sydiŋan, сыдиӈан.* The root of this term is *sy-, сы-,* the name of this snow type. A verbal ending has been added, so it can be literally translated as 'it will *sy*'. This causes a panic among nomads. The meaning of such a term-concept refers not only to the formation of this snow, but also to a set of interlinked consequences. The first time Lavrillier heard about this term-concept, it was translated by several nomads as: "*The domestic reindeer and game will move away from our usual areas of nomadisation in order to reach places where there is shallow and soft snow. We must move away from our usual areas to other places ASAP, otherwise it will be too late.*" This explanation and the name of the snow was only given when more information was requested: only after further explanation from even more informants did the entire process became clear to the anthropologist. So the logic chain of this term-concept is as follows. The snow type *sy, сы*, appears (because of several conditions shown in the diagram *sy*) and embeds the vegetal cover, including the pastures for wild and domestic reindeer. Upon grazing in such areas, wild and domestic reindeer lose weight very quickly: the weaker become sick and die. As a result of the *sy, сы* snow type, game and domestic reindeer will move great distances to get away from places touched by this phenomenon. The nomads would then be without their herd and game to hunt, which would be a catastrophe.

This concept is also interesting for understanding the Evenki conceptualisation of an 'extreme event'. Indeed, *sydiŋan, сыдиӈан* is a typical example of how nomads conceptualise extreme events, a chain of interactions involving the snow cover, vegetation, economics, and transportation. It is, in other words, a process that endangers the survival of the human group. (cf. Part III: Anomalies in the snow cover 2014–2015)

Snow and the notion of an extreme event

In the following table and the part about the physics of the snow, we have shown the order in which snow types normally appear and known anomalies according to Evenki ecological knowledge. We have observed or documented between 2012 and 2016 an increase of unprecedented anomalies. Some are indicated in this book and previously published papers (cf. Lavrillier et al. 2016; Lavrillier and Gabyshev 2017; diagrams Anomalies in snow cover 2014–2015 and 2013–2014, etc.), while others will be published elsewhere. Within the Evenki category of 'anomalies', we can distinguish the following: 1) A norm that takes in account a set of variations; 2) Known anomalies (already observed from year to year); and, 3) Unprecedented anomalies (never observed before) that cause panic. We have noted several of these elements during the BRISK project.

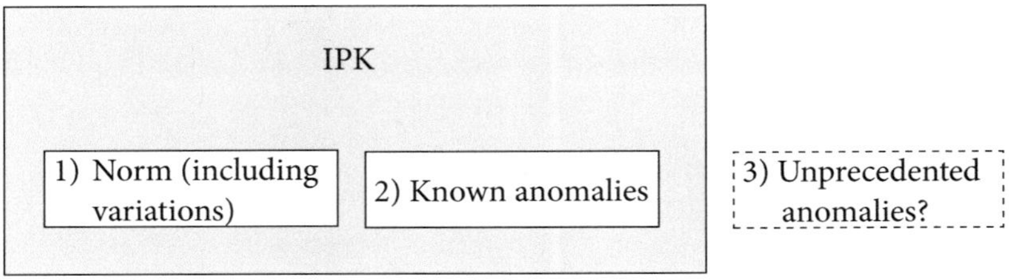

In our analysis, we noticed that 'anomalies' do not necessarily mean 'extreme event' or 'catastrophe' for the nomads if they do not endanger life or the economy. However, there is a difference between 'known anomalies' and 'unprecedented anomalies', since the latter create panic among people even if the event does not threaten them. The people are facing an unknown reaction of the natural environment that they cannot understand. This panic is expressed in exclamations in Evenki like 'nature is broken' *(Buya ukchapcha, Буҕа укчапча)*; 'I do not understand anything of the illogical function of this poor nature, *(Olok achem uidore on buyakun manak nekekutten, Олок ачэм уйдорэ он буҕакун манак нэкэкуттэн)*; 'our earth has been turned upside down' (*Dunnekun keltelgeren, Дуннэкун кэлтэлгэрэн*), and 'the natural environment is angry' *(Buya tukullen, Буҕа тукуллэн)*. The latter phrase suggests that the natural environment is angry at humans for 'bad treatment' (i.e. for all the extractive industries that 'wound' the earth and nature) (Lavrillier 2013, Rojo et al. 2016, Lavrillier and Gabyshev 2017). This type of discourse is omnipresent in the modern representation of pollution (PARCS project). The Evenki notions of climate and weather are detailed in the section on Evenki climatology.

Order of the appearance of snow types
by S. Gabyshev and A. Lavrillier

ORDER	EXPECTED PERIODS Периоды ожидания	NAMES Названия	DESCRIPTION ORIGIN / PLACE OF APPEARANCE / FORMATION описание и места появления	USES / PROBLEMS Функции / проблемы
1	September/ October Сентябрь/ Октябрь	Lebga, lyabga (UN) Лэбга, лйабга (УН)	First snow, which melts Первый снег, который тает. 'Sky' / 'небо'	Moving: Easy to walk Herding & hunting: Fresh tracks (for hunting and finding reindeer). Транспорт: легко ходить. Оленеводство и охота: свежие следы (оленей и дичь найти).
2	Beginning of October and in the spring	Imanna debdeme ulapkuhincha (UN) Иманна дэбдэмэ улапкуhинча (УН)	Fluffy snow, a bit wet and sticky. Пышный, маленько мокрый, липкий снег. 'Sky' / 'небо'	Moving: Hard to walk through, one cannot use a sledge yet. Herding: Reindeer can stray far in search of mushrooms (the smell of mushrooms circulates better because of the humidity). Hunting: Good for hunting because this snow makes no noise (wild reindeer and wood grouse). Транспорт: тяжело ходить по такому снегу, рано еще на нартах ездить. Оленеводство: олени могут далеко ходить за грибами (запах грибов хорошо чувствуется из-за влажности). Охота: Хорошо для охотника, потому что этот снег не шумный (дикий олень и глухарь).
2 bis	Beginning of October and in the spring Нач. Октября и весной	Bona (UN, Iengra) Бона (УН, Иенгра)	Hail Град 'Sky' / 'небо'	Cf. Evenki climatology, precipitations typology.

ORDER	EXPECTED PERIODS Периоды ожидания	NAMES Названия	DESCRIPTION ORIGIN / PLACE OF APPEARANCE / FORMATION описание и места появления	USES / PROBLEMS Функции / проблемы
3	During the first snow: in some places, it disappears (norm). If it does not disappear, it is an anomaly. Во время первых снегов: в некоторых местах исчезает (норма). Если *сы* не исчезает - это аномалия?	Sy (UN), Si (Iengra) Сы (УН) Си (Иенгра) Extreme event Экстремальное событие	An ice layer in which the vegetal cover is embedded. Appears from the first snow on the ground and then in some places disappears with the sun or because of the cold at the end of November (the norm). Appears in most places and does not disappear (anomaly). Норма: лед, охватывающий растительность местами (появляется от *лэбга* растаивающего), исчезает с солнцем или с холодом в конце ноября. Аномалия: везде и не исчезает за всю зиму. ‚Snow cover‘ / ‘снежный покров‘	<u>Moving</u>: No problem. <u>Herding</u>: The reindeer have no access to the vegetal cover (cf. diagram Sy). <u>Hunting</u>: No game in such places. <u>Vegetal cover</u>: Not accessible. <u>Транспорт</u>: Нет проблем. <u>Оленеводство</u>: олени не имеют доступ к растительному покрову (см. схемы, с. 300–304). <u>Охота</u>: нет дичи на таких местах. <u>Растительный покров</u>: не доступен.

ORDER	EXPECTED PERIODS Периоды ожидания	NAMES Названия	DESCRIPTION ORIGIN / PLACE OF APPEARANCE / FORMATION описание и места появления	USES / PROBLEMS Функции / проблемы
4	From the end of Oct – all winter С конца октября и всю зиму	Duiukun imanna (UN) Uŋtalama (Iengra) Дуйукун иманна (УН) Уӈталама (Иенгра)	Dry and soft snow Сухой мягкий снег 'Sky' / 'небо'	Moving: Very good, especially by sledge. Herding: Reindeer graze well; also used as a 'herder' by constructing a 'channel' for guiding reindeer where the herders want them to go. (cf. diagram Snow as a herder) Hunting: Dogs can catch sable easily (if not too deep). Vegetal cover: Good harvest (berries and pinus pumila cones). Транспорт: Очень хорошо, особенно на нартах. Оленеводство: олени хорошо кормуются; в глубоком снегу пастухи делают дорогу-канал, чтобы направить оленей туда, куда надо. (см. с. 346–347) Охота: если не слишком глубокий, собаки легко гоняют соболей. Растительный покров: хороший урожай (ягод и стланиковых шишек).

ORDER	EXPECTED PERIODS Периоды ожидания	NAMES Названия	DESCRIPTION ORIGIN / PLACE OF APPEARANCE / FORMATION описание и места появления	USES / PROBLEMS Функции / проблемы
5	Anomaly From Oct Аномалия с Октября	Cheγa (UN, Iengra) Чеђа (УН, Иенгра)	Scab-like snow layer that appears after melting and then the refrigeration of a pre-existing snow cover (a norm in springtime and an anomaly in autumn). Верхняя корка снега появляющаяся от таяния снега и потом замерзшая (весной это норма, а осенью это аномалия). 'Snow cover' / 'снежный покров'	<u>Moving</u>: Difficult to walk in (for humans and animals) if too deep. <u>Herding</u>: Reindeer have poor access to the lichen; (if in autumn) the reindeer go to the tops of mountains, where the snow is softer. <u>Hunting</u>: Dogs have problems catching sable, which run across the surface of the snow; the dogs wound their paws. <u>Vegetal cover</u>: Bad because the hard snow breaks the small branches (berries and pinus pumila). <u>Snow physics</u>: It does not allow for the normal physical transformation of the snow to occur. It does not allow the different layers of snow to be packed down as should occur through the thawing or the weight of the upper snow layer(s); so the depth of snow cover does not lessen as it should (cf. diagram Cheγa and additional information below).

<u>Транспорт</u>: Труднопроходимый (для людей и животных), если слишком глубокий.
<u>Оленеводство</u>: Оленям трудный доступ к корму; тогда олени идут на вершины гор, где снег мягче.
<u>Охота</u>: Собаки с трудом догоняют соболей, которые бегают по поверхности снега, собаки ранят себе лапы.
<u>Растительный покров</u>: Плохо, потому что снег ломает ветви и замораживает (ягоды, стланик).
<u>Физика снега</u>: Это не дает нормальной физической трансформации снега произойти как должно. Например, не позволяет разным слоям снега компрессироваться и оседать (благодаря таянию снега и весу верхнего слоя). Значит глубина снега не уменьшается как должно быть. (см. схемы, с. 319–323)

ORDER	EXPECTED PERIODS Периоды ожидания	NAMES Названия	DESCRIPTION ORIGIN / PLACE OF APPEARANCE / FORMATION описание и места появления	USES / PROBLEMS Функции / проблемы
6	Anomaly in the late autumn/ winter Аномалия в позднюю осень/зимой	Maŋa imanna Маӈа иманна	Hard snow – a layer of sticky winter snow which freezes and becomes extremely hard and compact (different from the other 'hard snows' *chuiur* or *cheya*). Твердый слой снега. Зимний липкий снег, который при заморозках становится плотным, как сыр (другое чем *чуйур* или *чеӡа*). 'Snow cover' / 'снежный покров'	Moving: Bad – the snow cover is too deep and hard to travel across. It does not allow the different layers of snow to be packed down as should occur through the thawing or the weight of the upper snow layer(s); so the depth of snow cover does not lessen as it should. Herding: Reindeer get tired very quickly when moving and have difficulties grazing through the snow. Hunting: Bad because it is hard to walk. Vegetal cover: Bad because it breaks small bushes of berries and pinus pumila cones. Транспорт: Плохо, потому что снежный покров слишком глубокий и трудно ехать по нему. Не позволяет разным слоям снега компрессироваться и садиться (благодаря таянию и весу верхнего слоя). Значит глубина снега не уменьшается как должно быть. Оленеводство: Олени устают быстро, когда ездят куда нибудь, и с трудом добывают корм под снегом. Охота: Трудно потому, что трудно ходить. Растительный покров: Плохо, потому что ломает маленькие кустарники ягод, также шишек.

ORDER	EXPECTED PERIODS Периоды ожидания	NAMES Названия	DESCRIPTION ORIGIN / PLACE OF APPEARANCE / FORMATION описание и места появления	USES / PROBLEMS Функции / проблемы
7	Throughout the snow period В течение всего снежного периода	Soŋor (UN) Keraha (Iengra) Соӈор (УН) Кэраhа (Иенгра)	A new, thin, and soft snow layer (during day or night) +/- 1.5-2 cm in depth Пороша 'Sky' / 'небо'	Moving: Good because it is slippery. Hunting & herding: very good, offers fresh tracks which are useful for finding reindeer and hunting. Транспорт: Хороший потому что скользкий. Оленеводство и охота: Очень хороший, потому что видно свежие следы, полезно для охоты и оленеводства.
7 bis	Throughout the snow period В течение всего снежного периода	Soŋordioron (UN) Соӈордйорон (УН)	Verb: smaller and invisible snow-flakes (like pollen), usually in the mor-ning and evening. All the sky is clear. (Smaller than *Kereyadieren*). It leaves a layer several millimetres in depth. Глагол: Очень мелкий снегопад (не видно что снег идет); обычно утром или вечером мелкой пыльцой (мельче, чем *кэраҕадйэрэн*). Слой снега составляет несколько мм. 'Sky' / 'небо'	Moving: The sledge slides easily, making the trip easy for the reindeer pulling the load. Hunting & herding: Very good, offers fresh tracks which are useful for finding reindeer and hunting. Транспорт: нарты хорошо скользят, оленям легче тащить груз. Оленеводство и охота: Очень хорошо, потому что видно свежие следы и легче найти домашних оленей, также как и диких оленей и дичь.

ORDER	EXPECTED PERIODS Периоды ожидания	NAMES Названия	DESCRIPTION ORIGIN / PLACE OF APPEARANCE / FORMATION описание и места появления	USES / PROBLEMS Функции / проблемы
7 ter	Throughout the snow period В течение всего снежного периода	Keraγadieren (UN) (cf. picture here below) Кэраҕадйэрэн (УН)	Verb: very light and fine snowflakes, which obscure visibility along the horizon (while the sky is clear). The moisture in the air is transformed into very fine snow-flakes. Очень мелкий снег падающий – мутная видимость от него идет по горизонту (а небо чистое). Влажность воздуха от мороза превращается в очень мелкие снежинки. 'Sky' / 'Небо'	Moving: It is a good snow type for moving around on because it is slippery. Hunting & herding: A very good weather type, since animals leave fresh tracks in it, making it easy to find reindeer and hunt. Транспорт: Хороший, потому что скользкий. Оленеводство и охота: Очень хорошо, потому что видно следы, позволяющие найти оленей и дичь.

Keraγadieren / Кэраҕадйэрэн © A. Lavrillier

ORDER	EXPECTED PERIODS Периоды ожидания	NAMES Названия	DESCRIPTION ORIGIN / PLACE OF APPEARANCE / FORMATION описание и места появления	USES / PROBLEMS Функции / проблемы
8	From November – all winter С ноября и всю зиму	Chuiur (UN) Koŋorok / koŋarak (Iengra) Чуйур (УН) Коӊорок/ коӊарак (Иенгра)	Hard surface layer of snow which forms after winds and blizzards (on mountain tops and in open spaces vulnerable to strong winds). Твердый наст после ветра и пурги (на возвышенности и в чистых и открытых местах, где гуляют сильные ветра). 'Snow cover' / 'снежный покров'	Moving: Good for humans (possible to walk and to ski across the surface of the hard snow). Herding: easy for reindeer to move; in some places, where there is a thin layer of snow, grazing is easier (wild and domestic reindeer). Hunting: The hunters await the formation of this snow (for hunting and transport purposes). Sometimes good for hunted animals when the hard layer is strong enough to support their flight from the hunter but not strong enough to support the hunting dog and the hunter. (cf. diagram Chuiur, Cheγa, Part III: Analysis of grazing pastures) Транспорт: Хорошо для людей (возможность ходить на лыжах по поверхности твердого снега). Оленеводство: Оленям легко ходить; в некоторых местах, где слой снега не толстый, легко диким и домашним оленям кормоваться. Охота: Охотники ждут появления этого снега (для охоты и транспорта). Иногда хорошо для дичи, которая может убежать от охотника по поверхности снега, тогда как слой снега недостаточно крепкий, чтобы выдержать вес охотника.

ORDER	EXPECTED PERIODS Периоды ожидания	NAMES Названия	DESCRIPTION ORIGIN / PLACE OF APPEARANCE / FORMATION описание и места появления	USES / PROBLEMS Функции / проблемы
9	All winter Всю зиму	Kumnuk Кумнук	Snow wads on the trees and bushes. Комки снега на деревьях, кустах. 'Snow cover (tree)' / 'снежный покров'	Hunting: Bad, no visibility. If these wads are no longer in the trees, it means that they have been blown off by the wind. If there is no wind, there is no *chuiur* snow: there is thus no reason to hunt in this small area. Indeed, the visibility is bad and the snow is too soft for skiing. Охота: Плохо, нет видимости. Если нет комков на деревьях, значит были ветра. Если ветра не было, значит твердый наст *чуйур* нету, поэтому не стоит в этой зоне охотиться, потому что видимость плохая и снег мягкий, лыжи проваливаются.

ORDER	EXPECTED PERIODS Периоды ожидания	NAMES Названия	DESCRIPTION ORIGIN / PLACE OF APPEARANCE / FORMATION описание и места появления	USES / PROBLEMS Функции / проблемы
10	End October to mid-Dec: throughout a cold winter Конец октября по декабрь: в течении холодной зимы	Buldo (UN) Siŋilgen, siγilgen (Iengra) Булдо (УН) Сиӊилгэн, сиҕилгэн (Иенгра) иманна дйукаҕачин доҕу	Appears during winter freezes: these transform the state of the snowflakes within the snow cover into icy and seed-like flakes. При сильных морозах снег превращается в снег-льдинки (зернообразные) внутри снежного покрова. ‚Snow cover' / 'Снежный покров'	Moving: Nomads await the formation of this snow. This allows the different layers of snow to be packed downwards thanks to the weight of the upper snow layer(s): this is as it should be. This reduces the depth of the snow cover, which facilitates the lives of the Evenki. Hunting: Good because it is easier to move in. Herding: Good for reindeer to graze: 1) because of the reduction of the snow depth 2) because the vegetal cover is freed from the snow and/or ice layer (cf. diagrams Buldo, Sy, or Cheγa and the entries under these names here). Транспорт: кочевники ждут появление такого снега. Он позволяет разным слоям снега компрессироваться и садиться (благодаря весу верхнего слоя). Значит, глубина снега уменьшается как должно быть. Это облегчает жизнь эвенков. Оленеводство и охота: Хорошо, олени хорошо кормуются: 1) потому что глубина снежного покрова уменьшается; 2) потому что растительный покров освобожден от снега и/или от слоя льда. (см. схемы, с. 315–318, 300–304, 385, 319–323, 287)

ORDER	EXPECTED PERIODS Периоды ожидания	NAMES Названия	DESCRIPTION ORIGIN / PLACE OF APPEARANCE / FORMATION описание и места появления	USES / PROBLEMS Функции / проблемы
11	End of February, March-April Конец февраля, март и апрель	Oigu imanna bagdama Ойгу иманна багдама	When it gets warmer, the snow on the surface of snow cover becomes 'whiter'. Когда теплеет, верхний слой снега очень белый и очень мягкий. 'Sky' / 'Небо'	Moving: Good for transport by sledge because of the slippery snow. In this period (springtime warming), the different layers of the snow are compressed by the thawing and the weight of the upper layers, thus reducing the depth of the snow cover. Easier to move through. Herding & hunting: Good because it transforms/ destroys abnormal snows *(cheya, magna imanna)*. Транспорт: Хорошо из-за скользкого снега. В этом периоде (весеннее потепление), разные слои снега компрессируются благодаря таянию некоторых слоев и весу верных слоев снега; это уменьшает толщину снежного покрова и позволяет легче передвигаться. Оленеводство и охота: Хорошо, потому что это трансформирует/ разрушает аномальные слои снега *(чеӡа, маңа иманна)*.

ORDER	EXPECTED PERIODS Периоды ожидания	NAMES Названия	DESCRIPTION ORIGIN / PLACE OF APPEARANCE / FORMATION описание и места появления	USES / PROBLEMS Функции / проблемы
12	Beg of March (around the 10th) Начало марта (прим. 10ого)	Tepama imanna Тэпамэ иманна	Damp, sticky snow that sticks to clothes. Влажный снег, который прилипает к одежде. 'Sky' / 'Небо'	Moving: This snow is a sign that springtime has arrived. Bad because clothes get wet and people freeze. Herding: Because of warmer temperatures, the reindeer are drowsy and run more slowly. Snow sticks to the sledge and it is difficult for reindeer to pull the sledge. Hunting: Good because this snow is not noisy. Транспорт: Этот снег является знаком, что весна наступила. Плохо потому что вещи намокают и люди мерзнут. Оленеводство: Из-за потепления, олени вялые и бегают медленнее. Снег липнет к нартам. Оленям тяжело тащить нарты. Охота: Хорошо, потому что этот снег не шумный.

ORDER	EXPECTED PERIODS Периоды ожидания	NAMES Названия	DESCRIPTION ORIGIN / PLACE OF APPEARANCE / FORMATION описание и места появления	USES / PROBLEMS Функции / проблемы
13	End March – April *Anomaly in autumn (cf. diagrams Anomalies in snow cover 2013–2014)* From May – within snow cover (due to snow physics, cf. *tygda imannanun*) Конец марта-апрель *Аномалия осенью (см. схемы, с. 400–415)* С мая, внутри снежного покрова (из-за физики снега, см. с. 284)	Teptaren imanna Тэптарэн иманна	White, sticky, and damp snow – snowflakes larger than in winter (different from uniaksa). Белый липкий, влажный снег – снежинки покрупнее, чем зимой (другой чем унйакса). 'Sky' / 'Небо'	<u>Moving</u>: Bad because sledges are wet and snow sticks to them; the reindeers are lazy. Unpleasant dampness. <u>Herding</u>: Reindeer graze well (shallower snow cover) <u>Hunting</u>: Wild reindeer become lazy (and run less) (cf. Part III: Analysis of grazing pastures, November). <u>Транспорт</u>: Плохо потому что нарты мокрые и снег липнет к ним; олени вялые. Неприятная влажность. <u>Оленеводство</u>: Олени хорошо кормуются (неглубокий снежный покров). <u>Охота</u>: Дикие олени становятся вялыми (меньше бегают). (см. схемы, с. 378–399)

ORDER	EXPECTED PERIODS Периоды ожидания	NAMES Названия	DESCRIPTION ORIGIN / PLACE OF APPEARANCE / FORMATION описание и места появления	USES / PROBLEMS Функции / проблемы
14	March Март	Uŋniariktadieren Uŋnia Уӈнйарик-тадйэрэн Уӈнйа	Short snowfalls of fluffy snow (or rain showers or light hail). + cf. Evenki climatology – variable strong winds lead to this type of snow (rain or light hail). + muddy weather from spring snow or rain.[1] Кратковременные снегопады (пышный густой снег или дожди, ливень или мелкий град). + см. Evenki climatology – часто меняющиеся сильные ветра, ведущие к кратковременным осадкам (снегу, дождю, граду). + мутная погода во время весенних снегопадов. 'Sky' / 'Небо'	Moving: Bad because the visibility is limited; snow sticks to clothes. Herding: Good for blood circulation; the reindeer predict this weather by 'playing'; they graze easily because the snow is softer (from the warm humidity). Hunting: If a lot of snow falls, it is bad for hunting. Impossible to hunt, the tracks are unreadable. Транспорт: Плохо потому что видимость плохая; снег липнет к одежде. Оленеводство: Хорошо для кровообращения; олени предсказывают погоду «играя»; они пасутся хорошо, потому что снег помягче (из-за тепла и высокой влажности). Охота: Если много снега падает – это плохо для охоты: невозможно охотиться, ведь следы невозможно интерпретировать.

1 Let us note that this term is mentioned together with the entry about *bona* hail (cf. precipitation).

ORDER	EXPECTED PERIODS Периоды ожидания	NAMES Названия	DESCRIPTION ORIGIN / PLACE OF APPEARANCE / FORMATION описание и места появления	USES / PROBLEMS Функции / проблемы
15	20 March 20 ого марта	Uniaksa Uniadiaran унйакса, унйадйаран	Wet snow which is in the process of melting. It looks like snow but is wet. Влажный снег в начале процесса таяния. 'Sky' / 'Небо'	Moving: No problem for moving. Makes clothes and items slightly wet. People can transport food and good reserves to storage houses. Herding: Reindeer graze well (shallower snow cover). Hunting: No hunting except for self-use. Транспорт: Нет проблем для транспорта. Этот снег слегка намачивает вещи и одежду. Люди могут транспортировать продукты и товары в амбары. Оленеводство: Олени хорошо кормуются (снежный покров не глубокий). Охота: не охотятся или только для собственного употребления.

ORDER	EXPECTED PERIODS Периоды ожидания	NAMES Названия	DESCRIPTION ORIGIN / PLACE OF APPEARANCE / FORMATION описание и места появления	USES / PROBLEMS Функции / проблемы
16	From April to the melting away of all snow С апреля до полного таяния снега *Anomaly* From Oct Аномалия с Октября	Cheγa (UN, Iengra) Чеҕа (УН, Иенгра)	Scab-like snow layer that appears after melting and the refrigeration of a pre-existing snow cover (a norm in springtime and an anomaly in autumn). Верхняя корка снега, появляющаяся от таяния снега и потом замораживания (весной это норма, а осенью это аномалия). 'Snow cover' / 'Снежный покров'	*Very different from the abnormal autumn* cheγa, *because much more strong and resistant.* <u>Moving</u>: Good, on the surface of the snow. <u>Herding</u>: Good for grazing when there is a thin snow cover. <u>Hunting</u>: End of hunting because of the noise of the snow (cf. diagram Cheγa). <u>Vegetal cover</u>: this prevents berries and horsetail from freezing. *Очень разные специфики чем осенний* чэҕа, *намного крепче и плотнее.* <u>Транспорт</u>: Хорошо на поверхности снежной корки. <u>Оленеводство</u>: Хорошо для пастбища, если тонкий слой снега. <u>Охота</u>: Конец охоты из-за очень шумного снега Растительный покров: Защищает пушицу, ягоды от мороза. (см. схемы, с. 319–323)

ORDER	EXPECTED PERIODS Периоды ожидания	NAMES Названия	DESCRIPTION ORIGIN / PLACE OF APPEARANCE / FORMATION описание и места появления	USES / PROBLEMS Функции / проблемы
17	April-May (also Oct) Апрель-май (тоже октябрь)	Imanna debdeme ulapkuhincha Иманна дэбдэмэ улапкуhинча	Fresh and fluffy snow, a bit wet and sticky. Свежий, пышный, маленько мокрый, липкий снег. 'Sky' / 'Небо'	<u>Moving</u>: Hard to walk through, but good for sledges because it is slippery if it is not too deep. Bad for snowmobiles because it sticks to the tracks. <u>Herding</u>: Reindeer stray far in search of mushrooms (the smell of mushroom circulates better because of the humidity). <u>Hunting</u>: Good for hunting because this snow produces no noise (wild reindeer and wood grouse). <u>Транспорт</u>: Труднопроходимый, но удобный для нарт, потому что скользкий (только если не слишком глубокий); плохо для снегоходов, потому что прилипает к тракам. <u>Оленеводство</u>: Олени далеко уходят в поисках грибов (запах хорошо ощущается из-за влажности). <u>Охота</u>: Хорошо для охоты потому что снег не шумный (дикие олени и глухари).

ORDER	EXPECTED PERIODS Периоды ожидания	NAMES Названия	DESCRIPTION ORIGIN / PLACE OF APPEARANCE / FORMATION описание и места появления	USES / PROBLEMS Функции / проблемы
18	From April to the entire snow melting С апреля до полного таяния снега	Ulapkun imanna Улапкун иманна	Wet snow Мокрый снег 'Sky' and 'snow cover' / 'Небо и снежный покров'	Moving: Bad (sledges are wet and heavy). Snow is melting. Hunting: No hunting during this period except in times of scarcity (and then only males because of the reproductive period). (cf. Evenki calendar) Herding: Good because the reindeer search for mushrooms (in the snow cover from the previous autumn) and walk easily. Транспорт: Плохо: нарты становятся мокрыми и тяжелыми; снег тает. Охота: Нет охоты во время этого периода, только по необходимости (например, из-за недостатка продуктов) и тогда только самцов убивают, из-за отела. (см. с. 43–59) Оленеводство: Хорошо, потому что олени ищут грибы, которые остались в снегу с прошлой осени и передвигаются легко.

ORDER	EXPECTED PERIODS Периоды ожидания	NAMES Названия	DESCRIPTION ORIGIN / PLACE OF APPEARANCE / FORMATION описание и места появления	USES / PROBLEMS Функции / проблемы
19	End of April, May Rarely, in the autumn Конец апреля, мая Редко осенью	Tygda /dygda imannanum Тыгда / Дыгда иманнанун	Rain with snow Дождь со снегом 'Sky' / 'Небо'	Moving: The worst snow for transport and people: everything gets wet and freezes in the evening. Snow physics: this snow causes the snow already on the ground to turn into *teparkun imanna* (wet snow with water). Herding: Important to survey newly born calves that can freeze (get sick or die) in such snow. (cf. additional information below) Транспорт: Это худший снег для транспорта и людей: все намокает днем и замерзает к вечеру. Физика снега: Из-за этого снега снежный покров, который уже лежит на земле, превращается в *тэпаркун иманна* (мокрый снег с водой). Оленеводство и охота: важно следить за только что родившимися телятами, которые могут замерзнуть (заболеть и умереть) в таком снегу. (см. дополнительная информация здесь ниже)

ORDER	EXPECTED PERIODS Периоды ожидания	NAMES Названия	DESCRIPTION ORIGIN / PLACE OF APPEARANCE / FORMATION описание и места появления	USES / PROBLEMS Функции / проблемы
20	Beg of May – June Начало мая-июнь	Teparkun imanna Тэпаркун иманна	Wet snow with water and a porridge-like texture (along the road and across the ground and ice). Мокрый снег с водой, как каша (по дороге, на земле и на льду). 'Snow cover' / 'Снежный покров'	Moving: Bad because water makes all items and clothes wet – no nomadisation is possible at this time. People look for dry places to live. Herding: The reindeer are calving and require attention. Reindeer graze well (places without snow – *iliakak*). Hunting: No hunting. Транспорт: Плохо, потому что вода мочит вещи – кочевки стали невозможными. Люди ищут сухие места, чтобы установить стоянку. Оленеводство: Олени телятся и нуждаются во внимании. Олени хорошо пасутся (из-за мест без снега – *илйакак*). Охота: нет охоты.

O R D E R	EXPECTED PERIODS Периоды ожидания	NAMES Названия	DESCRIPTION ORIGIN / PLACE OF APPEARANCE / FORMATION описание и места появления	USES / PROBLEMS Функции / проблемы
21	May (around 10 th) Май (прим. 10 ого)	Murbucha imanna Мурбуча иманна	When water flows under the snow. Когда под снегом вода. 'Snow cover' / 'Снежный покров'	<u>Moving</u>: Bad because the sledge cannot glide over the ground, mostly covered by water. Even small steams become entire rivers: it becomes impossible to cross to the opposite bank. The water level raises considerably. <u>Herding</u>: Bad but the reindeer go to the driest places. <u>Hunting</u>: Bad because of noisy movements in the water and snow. Difficult to move through (regardless of whether deep or not). In the streams and lakes, water appears from melted snow. This means that very soon the small rivers will melt and begin flowing – the melting process will be accelerated. (cf. diagram Cheγa) <u>Транспорт</u>: Плохо, потому что нарты не могут катиться по земле, большинство покрытой водой. Даже маленькие ручейки становятся большими реками – невозможно перейти на ту сторону, вода сильно поднимается. <u>Оленеводство</u>: Плохо, но олени идут в более сухие места. <u>Охота</u>: Плохо из-за шумных движений по снегу и воды. Трудно проходить или проехать насквозь (будь это мелкие или глубокие). В озерах и ручейках появляется вода от снега. Это значит, что очень скоро мелкие реки начнут таять и будут течь – процесс таяния будет быстрее происходить. (см. схемы, с. 319–323)

ORDER	EXPECTED PERIODS Периоды ожидания	NAMES Названия	DESCRIPTION ORIGIN / PLACE OF APPEARANCE / FORMATION описание и места появления	USES / PROBLEMS Функции / проблемы
22	May (around 12-15 th) Май (12-15 прим)	Iliakak Илйакак	Round places where there is no snow Места, где снега нету (растаял) 'Snow cover' / 'Снежный покров'	Moving: Bad because nomads cannot move, although they can walk easily because there is almost no snow left. Herding: Good because reindeer graze well and the herders know where the reindeer are (in places without snow). Hunting: Good because of silent movement; rare hunting males wild reindeer and wood grouse. Транспорт: Плохо, потому что кочевники не могут кочевать, зато им легче ходить – почти снега нету. Оленеводство: Хорошо, потому что олени хорошо кормуются и оленеводы знают где их найти (в местах, где нет снега) Охота: Хорошо, потому что движения не слышны; редкая охота на самцов диких оленей и глухари кормуются.

ORDER	EXPECTED PERIODS Периоды ожидания	NAMES Названия	DESCRIPTION ORIGIN / PLACE OF APPEARANCE / FORMATION описание и места появления	USES / PROBLEMS Функции / проблемы
23	May (around 15 th) Май (прим. 15 ого)	Imanna chum-dieren Иманна чумдьерэн	Snow dissolves and slowly vanishes снег разлагается, исчезает, растворяется 'Snow cover' / 'Снежный покров'	Moving: Bad. CHANGE OF TRANSPORT FROM SLEDGE TO PACKS (backs of reindeer). Large rivers become too wide to cross. Good for walking (because the land is frozen). Herding: Good because fresh *nirgakta* grass (cf. Vegetal cover typology) grows (they go to *kever*; cf. topographic typology). Calves are more independent and less fragile. Hunting: Good because the flowing rivers make noise (covering the sound of the hunter). Транспорт: Плохо. ПЕРЕХОД ОТ НАРТОВОГО ТРАНСПОРТА НА ВЬЮЧНЫЙ (на спине оленей). Широкие реки стали слишком большие, чтобы их переходили. Хорошо для ходьбы (потому что земля еще мерзлая). Оленеводство: Хорошо из-за свежей травы пушица *(ниргактэ)* (см. с. 138), которая растет на кочкарниках *(кэвэр)* (см. с. 60, 111). Телята более самостоятельные и здоровье крепче. Охота: Хорошо, потому что реки сильно шумят и это заглушает шум от охотника.

O R D E R	EXPECTED PERIODS Периоды ожидания	NAMES Названия	DESCRIPTION ORIGIN / PLACE OF APPEARANCE / FORMATION описание и места появления	USES / PROBLEMS Функции / проблемы
24	End of May Конец мая	Imanna eiaicha Иманна эйаича	The melted snow flows away Снег (растаявший) уплывший 'Snow cover' / 'Снежный покров'	NO SNOW LEFT Moving: Good for moving on foot. Herding: Bad because predators are everywhere, but reindeer graze well. Herders use all their time watching the herd. Hunting: Difficult to find game (because there are no tracks; there is no snow for them to leave tracks in). Транспорт: Хорошо для движения пешком. Оленеводство: Плохо, потому что хищники везде, но олени хорошо кормуются. Оленеводы проводят все время, осматривая стадо. Охота: Трудно найти диких оленей (ведь следов почти нет без снега, и не могут быть на земле, потому что мерзлая).

The physics of the snow

by A. Lavrillier

Evenki conceptualise the order of the appearance of different snow types, as shown in the previous chronological table. Even if they are not separated expressly into two different categories by the nomads, within this list of snow types we can distinguish between snow types that fall from the sky (designated here as *'sky snow type'* and indicated by 'sky' in the table) and the other types that appear from the moment the snow types start to constitute the snow cover on the ground or on the trees (as '*snow cover snow type*', 'snow cover' in the table).

If we take distance ourselves somewhat from the text and pay attention to the methodology of analysis used in this book, we can see that the anthropologist has added a classification ('sky snow types' and 'snow cover snow types') which does not explicitly exist among the Evenki. This interlocking or intermingling of classifications from indigenous and anthropological knowledge and analysis is a sign of 'co-production'.

I mean by this that certain snow types fall from the sky, while others appear only due to physical transformations within the snow cover. I expressed this knowledge (gathered from Evenki knowledge about the physical transformation of the snow) in the following table. We can see that the possibilities for the transformation of snow types are limited and that the Evenki conceive of these transformation as being dependent on air temperature (but also on ground temperature, as we can see in the many proposed diagrams about snow and ice), on the succession of warm and cold airs, on winds, and on different types of cold (episodic evening or morning cold; the winter permanent cold; a cold produced by a mixture of rain and snow) (see below for more details).

We note also that some types from the sky are transformed into types of snow cover, but also that some 'snow cover types' can be transformed into another 'snow cover type'. For instance, *sy, сы*, (a typical snow cover type) can be transformed into *buldo, булдо* (another snow cover type). (cf. diagram, Evenki physics of the snow)

In the various diagrams proposed in this book, we can see that some snow types can coexist at the same time.

Indigenous methods of analysis of the snow cover

In the different diagrams, we see that the Evenki always analyse the composition of the snow cover and its different layers, identifying the dates of snowfall they correspond to (cf. various diagrams related to grazing). Analysing a snow cover composition, they distinguish the lower layers *(ergu imanna, эргу иманна)*, the middle layers *(dulgu imanna, дулгу иманна)*, and the upper layers *(oigu imanna, ойгу иманна)*, a 'layer that touches the air', and *uyiskaki imanna, уӈискаки иманна,* a 'layer on the upper middle layer').

They not only identify the transformation of one snow type to another, but also the equivalence of snowflakes. They know how one snowflake from one snow type compares to snowflake of another snow type (or, in other words, the equivalence between snowflakes of different snow types) and the variety of flakes within one snow type. (cf. diagram Buldo)

When analysing the physics of the transformation of one snow type into another, they also **take into account the effects** of ground temperature, air temperatures, the time scale of the freezing process (regular or abrupt) on the ground and in the snow layers, the pressure from the upper layer of snow, and the role of thermic isolation played by a thick layer of dry and soft snow *(duiukun, дуйукун)*. (cf. diagrams Buldo, Sy, Part III: Anomalies snow 2014–2015)

During this analysis, they also take into account the effects of additional precipitations (different snow types, rain, and hail) on the established snow cover, the results of the combination of precipitation and air temperature, and the effects of the combined interaction between vegetal cover, topography, and the wind (cf. diagram chuiur). In addition, nomads conceptualise the specific effects of the angles at which the rays of the sun hit various local topographic types and its consequences for the physics of the snow and the presence or absence of snow types. This knowledge is detailed enough to include nuances regarding the time of the year and the seasonal axis of the sun in relation to the Earth (translated in the nomadic expression as 'the sun's position in the sky'). (cf. diagram, Cheγa)

During our study from 2013 to 2016, we noticed that the Evenki constantly analyse the snow cover and its evolution – be it normal and abnormal, be there an extreme event or not. As our fieldwork shows, nomads are in a permanent state of observing and analysing the snow cover (among other thing). This detailed and continual analysis allows nomads to identify many variations each year, seeing large and small anomalies together with normal features. Thus, it seems impossible to determine whether one year is entirely good or entirely bad. So, research about 'bad' and 'good years' for reindeer herding is probably not detailed enough: it would be more accurate to talk about various bad aspects for each year (cf. part III: diagram Inter-annual comparisons).

In addition, we have noted that the nomads both analyse existing facts and make forecasts about the possible evolution of the snow cover over the coming weeks and months: indeed, they even make predictions for the coming years. According to these forecasts or predictions, they estimate the consequences for hunting, gathering, herding, the evolution of the vegetal cover, and sometimes even landscape transformation. (cf. part III: diagrams, Anomalies in snow cover 2013–2014; Landscape transformation)

In their daily analysis of the snow cover, the nomads distinguish the physical state of the snow as it falls from the sky and its physical transformation(s) from the moment it reaches the ground or joins the extant snow cover. Upon seeing a specific snow type in the snow cover, nomads can identify when it fell or/and from which snow type it was formed. (cf. Part III: diagrams Analysis of grazing pastures)

In their analysis of normal and abnormal processes in the snow cover and ice, the Evenki also take into account the circulation of the air both in the air itself (cf. part III: diagrams Anomalies in the snow cover 2014–2015, Anomalies in the snow cover 2013–2014), from the ground (cf. part III: diagram Anomalies in snow cover 2013–2014), and underground circulation through plants and their roots. (cf. Part III: Ice anomalies)

Regarding the vocabulary used to talk in Evenki about the transformation of the snow, they mostly use two generic verbs and a set of verbs attached to each snow type. The most used is 'it turns into' *(ovki, овки, odieren, одйэрэн)*, but they also say 'snow type X eats snow type Y'. For instance, this is said about the transformation of soft and dry snow *(duiukun, дуйукун)* into icy flakes *(buldo, булдо)* ('the *buldo* snow type is eating the *duiukun* snow type', in Evenki – *Buldo diepdievki duiukun imannave, Булдо дйэпдйэвки дуйукун иманнавэ).* They also use this phrase to describe the interactions between other elements and the snow types. For instance, the Evenki say that *salgyn, салгын* air eats the snow (i.e., when this air appears, the snow cover is reduced without a thawing process). This occurred in 2016 when a lot of snow fell all at once in the autumn: with abnormal warming, *salgyn* air appeared and the deep snow cover almost disappeared without any obvious melting taking place. This expression is also deployed to discuss the interaction between water and snow: the Evenki say that water 'eats' snow (for instance, along some river banks). The other verbs about snow type formation are based on the names of snow types. For instance, *duiukun imanna* ***chuiurcha**, дуйукун иманна чуйурча* (i.e. 'the *duiukun* snow *chuiur-ed*') means that *duiukun, дуйукун*, snow turned into *chuiur, чуйур* snow. (cf. Winds and airs typology, Snow and ice typology; diagram Ulan-Bukte; Part III: Ice anomalies)

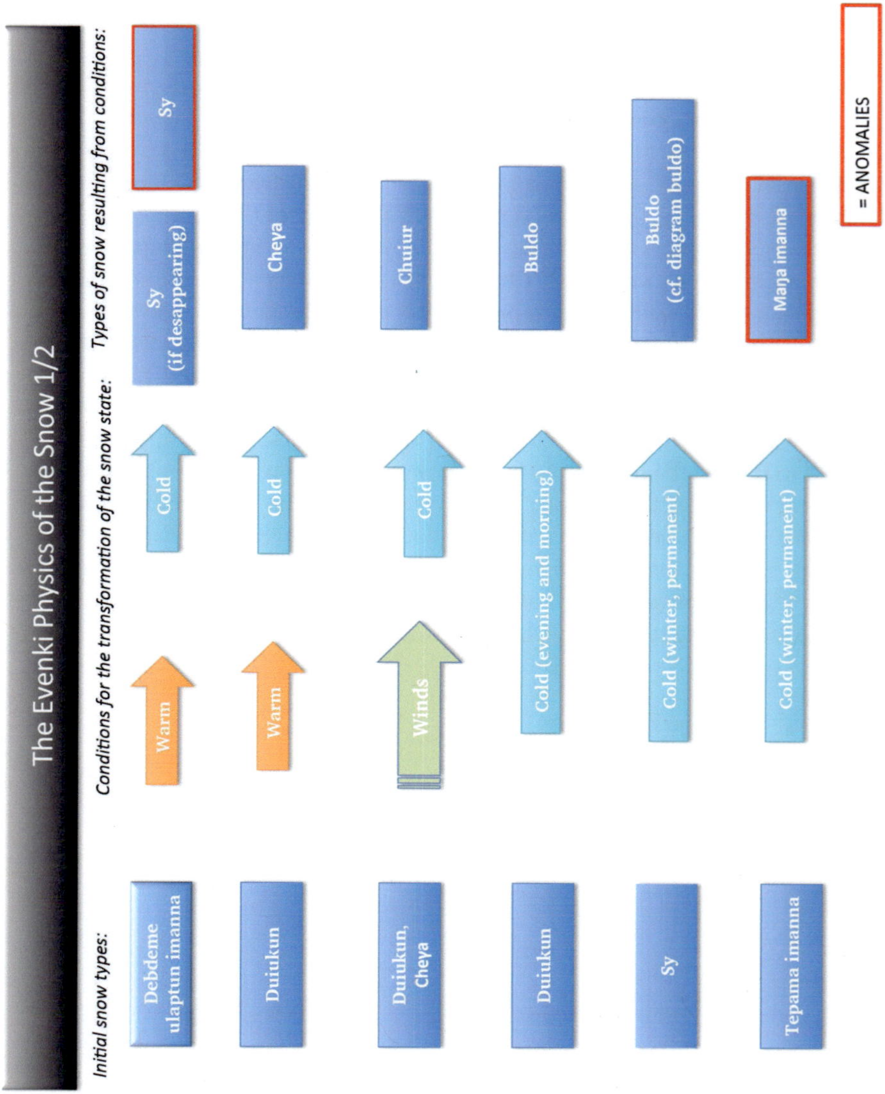
The Evenki Physics of the Snow 1/2
Initial snow types:
Conditions for the transformation of the snow state:
Types of snow resulting from conditions:
Debdeme ulaptun imanna
Warm
Cold
Sy (if desappearing)
Sy
Duiukun
Warm
Cold
Cheya
Duiukun, Cheya
Winds
Cold
Chuiur
Duiukun
Cold (evening and morning)
Buldo
Sy
Cold (winter, permanent)
Buldo (cf. diagram buldo)
Tepama imanna
Cold (winter, permanent)
Maŋa imanna
= ANOMALIES

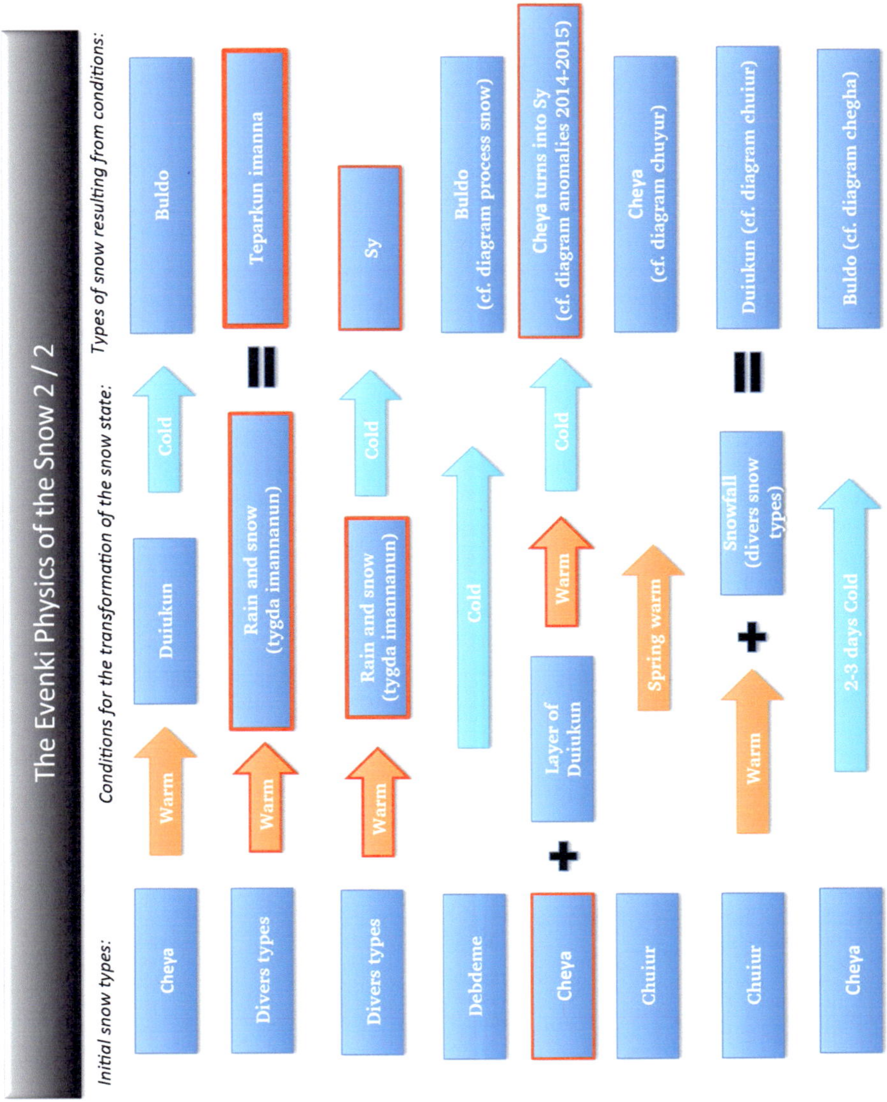
The Evenki Physics of the Snow 2 / 2
Initial snow types:
Conditions for the transformation of the snow state:
Types of snow resulting from conditions:
Cheya
Warm
Duiukun
Cold
Buldo
Divers types
Warm
Rain and snow (tygda imannanun)
=
Teparkun imanna
Divers types
Warm
Rain and snow (tygda imannanun)
Cold
Sy
Debdeme
Cold
Buldo (cf. diagram process snow)
Cheya
+
Layer of Duiukun
Warm
Cold
Cheya turns into Sy (cf. diagram anomalies 2014-2015)
Chuiur
Spring warm
Cheya (cf. diagram chuyur)
Chuiur
Warm
+
Snowfall (divers snow types)
=
Duiukun (cf. diagram chuiur)
Cheya
2-3 days Cold
Buldo (cf. diagram chegha)

Additional information about snow types
by S. Gabyshev and A. Lavrillier

Tygda imannanun (dygda) / Тыгда иманнанун (Дыгда)	Rain with snow at the end of April, May or, more rarely, in the autumn
	Дождь со снегом, ожидается в конце апреля, мая и осенью, но редко

Moving: This is the worst kind of snow for transportation: everything, including people, gets wet and then freezes in the evening. The snow sticks to the sledge, which then becomes heavy and difficult to drive.
Snow physics: This snow makes the snow already on the ground wet.
Herding: It is important to keep an eye on newly born calves, since such snow can make them freeze, become ill, or die.
Hunting: This is very bad snow for hunting wild reindeer because it sticks to the hunter's clothes: these clothes then freeze.
We note here that the warmer weather which creates such snow conditions is certainly considered a threat by the Evenki. So, very cold conditions offer more opportunities for protecting humans against freezing than warmer ones. Such rain or snow conditions are becoming more and more frequent in western Siberia and northern Russia (cf. Bartsch et al. 2010).

Таргачинди иманнами тэпарми бивки, иманна овки. Таргачинду бэйуктэдйэми эруми, тэткэ доӊкотодйовки. Таргачирду эӊнэкан балдывуллэкин ичэччэвкил эдан эӊнэкан доӊкоторо. Таргачимиду эруми нулгидйэда, сыргала намарадйамивкил иманнами, бэйэл и тэткэл улапчавкил, кэђа доӊкотовкил.

Такой снег плохой, потому что влажным снегом становится. По такому снегу очень неудобно охотиться на диких оленей, ведь одежда мокрой становится, потом замерзает на человеке. При таком снеге во время отела надо за оленятами хорошо присматривать, чтобы они не замерзали. При таком снеге очень тяжело кочевать: снег липнет к нартам, которые становится тяжелыми, вещи и одежда мокнут, а к вечеру все замерзает.

Sy (UN), Si (Iengra) / Сы (УН) Си (Иенгра)	Hard snow with ice that appears on the ground from the autumn onwards Твердый снег со льдом, который на земле лежит с осени

Hard snow with ice that appears on the ground from the autumn onwards. It appears after the first snow (which melts under the sun): it freezes into ice during the evening frost and then creates a crust within which the vegetation is embedded. This is really bad for the reindeer, since they cannot reach the pastures. *Sy* is often found on the slopes of mountains and hills that receive little sunlight: it is almost completely absent on slopes where the sun shines a lot (this melts *sy* snow).
However, the *sy* snow type is much more complex in reality since there are three categories: there is *sy* from the sun (see above), *sy* from rain on snow, and *sy* from the transformation of the *cheya* snow type. The latter two are highly anomalous. (cf. diagram Sy)

Твердый снег со льдом, который на земле лежит с осени. Образуется осенью, когда первый снег падает, потом тает на солнце и от мороза замерзает и корка образуется, охватывая растительность. Такое плохо бывает для оленей – они не могут добыть корм. Такое постоянно образуется на склонах гор, где солнца мало. Где солнце постоянно/долго бывает, такая корка не образуется надольго, поскольку она тает на солнце.

Но, *сы* сложный тип снега в практике – есть три категории: этот *сы* от солнца (см. высшее), и среди сильных аномалий, *сы* – от дождей после снега и *сы*- от трансформации типа снега *чэҕа.* (см. схемы, с. 300–304)

Buldo (UN), Siŋilgen, siɣilgen (Iengra) Булдо (УН) / Сиҥилгэн, сиҕилгэн (Иенгра) иманна дйукагһачин догһу	Icy and seed-like flakes (for details, see chronological table above) снег-льдинки

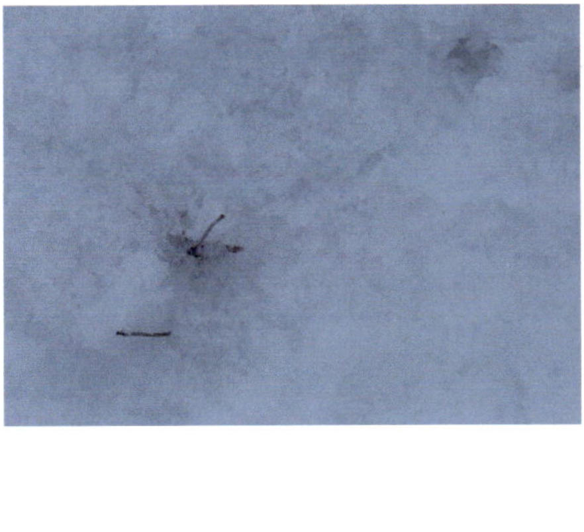

Kumnuk / Кумнук

This snow type indicates that the wind has not blown for a long time

Говорит о том, что долго не было ветра

Kumnuk tykcha / Кумнук тыкча

The wads have fallen

комки снега упали

This indicates that there has been wind; it is now possible to hunt because of high levels of visibility

Значит, ветер был и можно охотиться – видимость хорошая стала

Songor (UN)
Keraha (Iengra) /
Сонгор (УН)
Кэраhа (Иенгра)

A new, thin, and soft snow layer (during day or night)

Пороша

Oigu imanna bagdama / Ойгу иманна багдама	Upper layer of the snow in March: extremely soft (like flour), very thin snowflakes Верхний слой снега в март, слишком мягкий, как мука, мелкие снежинки.

The upper layer of the snow in March: extremely soft (like flour), very thin snowflakes. In February, when it is not too cold, this snow is visible on top of the deep snow: it is very white and soft. The appearance of this white snow means that it is definitely warmer and spring is approaching.

Верхний слой снега в марте, слишком мягкий, как мука, мелкие снежинки. В феврале, когда холода нет, *ойгу иманна* (верхний слой снега) становится заметней: более белым и мягким. Значит уже значительно теплеет – первая весна на носу.

Cheɣa (UN, Iengra) / Чеӡа (УН, Иенгра)	A hard scab-like layer of snow (wet snow that has frozen) Твердый наст (мокрый, а потом замерзший снег)

A hard scab-like layer of snow (wet snow that has frozen). Walking on such snow creates a lot of noise, making it difficult to hunt. In addition, such snow supports the weight of sable but not of hunters or dogs: thus, they cannot catch the sable. No hunting is possible in such conditions. Normally, this type of snow appears only in spring, but it has now been appearing in autumn for several years. It has started to appear because of the considerable temperature jumps, which are a recent phenomena. This snow triggers a lot of problems, like in 2015–2016 when it spoiled hunting and thus the income for the entire year.

Твердый наст (мокрый, а потом замерзший снег). Шумный, когда по нему ходят, для охоты плохо. Еще такой снег соболя выдерживает, а охотника и собак нет, и оба проваливаются. Любая охота невозможна при таком снеге. В принципе этот снег появляется только весной, но теперь это наблюдается уже несколько лет осенью. Такая аномалия происходит от больших перепадов температуры, которого раньше не было. Это вызывает разные проблемы, как в 2015–2016 – наша осенняя охота была вся испорчена.

Nuances of additional details

The size of the flakes

In addition, the specific characteristics of snowflakes can be expressed by attaching suffixes to normal Evenki words.

Verbs	English translation	Russian translation
Imanna+ktadieren Иманна+ктадйэрэн	Light snow is falling	Мелкий снег падает
Imanna+dieren Иманна+дйэрэн	Moderate snow is falling	Средний снег падает
Imanna+kutten Иманна+куттэн	Heavy snow is falling	Сильный снег падает

The intensity of the snowfall

Imannadiiatyn – Иманнадийатын	It is snowing very heavily Очень сильно снег идет

When the snowfall is very dense, people stay at home. It is impossible to hunt wild reindeer and visibility is practically zero.

Иманнадийатэкин бэйэл дйуду тэђэчэвкил, эңнэнны бэйурэ, экунада этанны ичэрэ.

Когда очень сильный снег идет, люди дома сидят, невозможно охотиться на дикого оленя и все равно ничего не видно.

The temperatures during snowfall

Imannadieren niamadiene/ Иманнадйэрэн нйамадйэнэ	When it snows in the winter at relatively warm temperatures Это когда зимой снег падает в теплую погоду

The snow sticks to the trees, forming clumps of snow.
Иманна намарадйэрэн ирйактэлтыки – тадук кумнук овки.
Снег липнет к деревьям и становится комками снега *кумнук*.

Other items

Tain, tair / Таин, таир	Puddles from melting snow Лужа от снега

When snow melts, a lot of puddles appear. This water is clean, so one can take drinking water from them. The puddles flow towards rivers.

Иманна унэкин таир овкил. Му айакикин – мулинңэрум. Таир эйандйэрэ биралтыки.

Когда снег тает, много луж становится. Вода чистая и оттуда воду берем для питья. Лужи текут в реку.

Snow transformed by reindeer

Ir / Ир

Places where wild or domestic reindeer have dug at the snow in order to graze

Места, где олени (дикие и домашние) кормуются

With their hooves, the reindeer turn the snow upside down to dig holes. During hunting, Evenki study the state of the snow in this *ir* and deduce how many reindeer there were and when they left.
This kind of logic chain is crucially important for hunting and reindeer herding, since it helps the nomads to identify the possible position(s) of domestic or wild reindeer (Lavrillier 2005–2006).

Орор̌-да, бэйур-да кукчанди иманнавэ оңковкил ир овки. Бэйуктэдйэнэ саңнынны экуды иманна ады сэӡин бича, окир бинэвэтын.

Домашние и дикие олени копытом снег переворачивали и ямы делали. На охоте эвенки исследуют состояние снега в *up* и делают выводы о количестве оленей и времени их пребывания.

Ekir / экир	Snow trampled by reindeer Снег, истоптанный оленями

Ekir are found in camp areas: one can easily walk on them because the snow has been trampled hard and flat by reindeer. When one arrives in a camp, the Evenki look at the *ekir* and know when the reindeer came to lick the salt. If the camp is empty, the nomads can know when the people left by looking at the state of the snow on *ekir*.

Урикичирду экир бивкил, экирду айакикин ңэнэктэдйэми иманна экипча бивки. Урикиттыки эмэми, саңнынны окир орор удйумучал, тадук нйан саңнынны окин бэйэл нулгичал.

Экир бывают на территории стоянок, по ним хорошо ходить, потому что снег истоптан оленями. Когда на стоянку приезжают, эвенки, глядя на *экир*, знают, когда олени приходили лизать соль, или если стоянка пустая, будут знать, когда люди укочевали.

Elun / Элун	Snow along the road, usually deep but untouched by reindeer Снег обычно глубокий, не тронутый оленями, вдоль дороги или в других местах

Along nomad roads covered by snow or away from the usual nomadic roads, one can find untouched snow through which the reindeer must blaze a trail. Only some reindeer can go through *elun*: they must be strong, possess great endurance, and be well trained. It is especially difficult for the first reindeer in the caravan. This is why it is important for the Evenki to 'create' and maintain snow roads at the beginning of the snow season and to train the guiding reindeer in the sledge harness properly.

һокто ачэндули сунҭакун иманна бирэкин элундули, бэрэчит ңэнэвки. Эңэһикун, айат татча бэрэчит ңэнэпдйэвки элгэңалбэ. Бэрэтчитту ңэнэдйэми элундули дэрувсэкун. Олыһин эвэнкил татыђадйэлвкил бэрэчикилбэ.

Элун – Вдоль дороги по глубокому снегу или по новой дороге, нетронутый снег по которому олени должны пробить дорогу. Только сильные и выносливые и хорошо обученные упряжные олени могут проехать по *элун*. Особенно тяжело первому оленю в караване. Поэтому для эвенков важно создать снежные дороги с начала сезона и хорошо выбирать и обучать упряжных оленей.

Snow conclusion
by A. Lavrillier

As a conclusion regarding snows, we can say from an anthropological point of view that snow plays several roles for the Evenki. It is a surface for transportation that most of the time facilitates movement. It is also an insulating material for tents, but also from the frozen ground, and between the air and the ground (cf. diagrams of snow types and anomalies). It also allows nomads to cover up the carcasses of hunted game until the next day, when the hunter will take it home: this protects the meat from predators and *salgyn, салгын* air. The snow also offers protection against the predators. Some snow types are also very good 'allies' for hunting (for instance, *chuiur, чуйур*). It also conditions the development of the vegetal realm, which, in turn, determines the quality of the pastures for domestic reindeer and the presence (or absence) of hunted species in specific areas. So, in two different ways the snow can help or hinder the nomadic economy. It also acts as a surface for reading information about the movements of game and domestic reindeer. It allows and reinforces social relationships by informing the nomads about the presence and movements of other humans.

The typology of ices

by A. Lavrillier and S. Gabyshev

As we have seen in the different diagrams, the Evenki conceive of snow and ice as two states of one element (water). In addition, they can identify many of the interactions between snow and ice. In particular, they know the effect of a deep snow cover on the lower snow cover, as we have seen (cf. part III: diagrams Anomalies in snow covers 2014–2015, Anomalies in snow covers 2013–2014), and on river ice (cf. diagrams Ulan bukte, Part III: Ice Anomalies). Many parts of this knowledge are almost impossible to express clearly with a simple text, so we will use a lot of diagrams.

The physics of the ice

The Evenki perceive the elements of the ice as integrated into one system of physical transformation. For instance, they do not conceptualise *ulan, улан* and *bukte, буктэ* as separate but as being interlinked throughout the process of water transformation in its entirety. The elements of the ice are seen as a part of a physical process where various elements interact.

Indeed, the physics of the ice, as per Evenki TEK, is well demonstrated in the diagrams related to *ulan, улан* and *bukte, буктэ*. As is the case for the snow cover, the Evenki also analyse the states of the ice. In their analyses of ice physics, they take into account the interplay between the ice and topographic features (for example, *emker, эмкэр* and *amnunna, амнунна*) and the interactions between snow and ice (for instance, an overly deep snow cover does not allow the ice to get thicker). They also understand the interactions between water, the outside air temperature, snow, and ice. Finally, they conceptualise the creation and circulation of 'air' contained in the water during the freezing process. (cf. diagram Ulan bukte, Part III: Ice anomalies)

Sirikte / Сириктэ
Crack in the ice
Трещина на льду

Along the cracks, it is easy to split the ice and bring ice blocks home to provide water. Within the water, air accumulates and pushes up against the river's ice layer, creating a crack. Some of these cracks are very small, while others are very large (cf. picture). Under the crack, there is a quiet stretch of river with fish (a small hole in the river basin). When the *ulan* flow, the nomads prefer to travel along the ice cracks because there is no water, only dry snow. (cf. diagram Ulan, bukte)

Сириктэду дйукэниду дйукэ айат йолдорговки, дйукэлидйэнэ муйэ оңнэнны. Таргачирду уригдар бивкил, олло бивки. Таргачирдулы уландйэрэкин сириктэ даҕалын эвки уланэ, нулгидйэми айа иманначи бивки.

По трещине лед хорошо раскалывается. Эвенки берут лед для питья. Внутри воды воздух накапливается, надувает и раскалывает лед и образуется трещина. Бывают очень длинные или короткие. Под ним бывает *(уригдан)* с рыбой, т.е. яма в бассейне реки. Когда идут наледи, то кочевники предпочитают кочевать вдоль трещины, потому что возле трещины не бывает воды, а сухой снег. (см. схемы, с. 348–367)

Ulan / Улан

A natural hole in the river ice through which water flows from under the ice to the surface. After flowing, it freezes. Ulan designates only the melted parts of these holes; naled.

Талая наледь – выходящая вода из-подо льда и выступающая на поверхность льда.

Ulan are very important for transportation by reindeer and (especially) by snowmobile. An *ulan* can be a significant obstacle: it can block people on the middle of the road, as happened during our expedition in 2014, when an *ulan* weakened the ice even in the harsh cold. It is impossible to know if the water is too deep and therefore it is impossible to estimate whether the sledge/snowmobile will move across it or whether it will sink into the water. It is not always possible to pull the transport out of glacial water. Sometimes, one must improvise a camp for the night in the middle of the frozen forest. In other cases, one must walk soaked through for kilometres until one reaches an occupied camp with warmth and dry clothes available. Anyway one must come back afterwards to break the ice and pull out the transport. *Usually, Evenki find a way out of such situations thanks to their inventiveness and resilience, although it can take hours of tremendous effort. This is why, when travelling by snowmobile, the Evenki always travel with a partner, a tent, and a wood burning stove.* Ulan *can abruptly appear and disappear in the course of two or three days, so the Evenki always ask about the* ulan *situation when nomads travel through their camps. It may surprise Western minds that* ulan *can deliver huge amounts of water above the ice, especially during harsher colds. The Evenki know the places where* ulan *can potentially appear. They can also benefit from* ulan, *since they can get drinking water for nearby camps.* (cf. diagrams Ulan, bukte, Part III: Ideal camp)

Улакурдулы эвэнкил нулгидйэми сыргалба чэпивкил улапивканэвкил эрукунди ӈэнэвувки. Таргачикурбэ эвенкил айилдйэвкил. Ӈэнэдйэми мамарилбэ анӈумачэӈнэрум саммэн иду улан биһин – нулгимӈат-у, эӈат-у? Буранди ӈэнэдйэми нйаны анӈумачиӈнэрум, буран эмиски чэпэвки улакурду. Палаткачи онда ӈэнэктэда нада.

Улан имеют важнейшее значение для кочевников, особенно для транспорта, на оленях и особенно на снегоходах. *Улан* могут представлять большие препятствия в дороге, вплоть до того, что блокируют людей в середине дороги, как это произошло с нашей экспедицией в 2014 г, ослабляя лед рек даже при лютой зиме. Тогда, и заранее нельзя знать, либо не очень глубокая вода и можно проехать, либо хуже – можно утопить нарту или снегоход. Не всегда можно в тот же день вытащить снегоход, иногда приходится устраивать ночлег в середине замороженной тайги, или идти мокрыми километры до стоянки, где есть люди, чтобы греться и сушиться, а потом вернуться, выдолбить транспорт из льда. *Но обычно после нескольких часов мощных усилий, вымокшие эвенки всегда выходят из положения, благодаря их удивительной выносливости и изобретательности. И поэтому эвенки предпочитают при езде на буране возить с собой палатку и печку.*

Уланы *могут внезапно появляться и исчезать за несколько дней, так что эвенки постоянно пытаются узнать об их состоянии от проезжающих людей. Как бы удивительно не было для европейского ума,* улан *выливая огромные массы вод по льду рек при самых лютых морозах. Места потенциального появления* улан *эвенки хорошо знают. Пользу тоже берут от* улан *– стоянки возле них иногда устраивают, чтобы из них воду брать (а не лед таять).*

Связанный с ним феномен буктэ *(наледный бугор пучения) имеет отношение с циклом жизни рыб.* (см. схемы, с. 348–367, 370–371)

Bukte / Буктэ

Icing blister

Наледный бугор пучения

When *bukte* do not appear, *ulan* will also not appear in large river basins *(amnunna)*. When Evenki nomadise, they try to ensure that the ice does not crack in order to prevent the sledge/snowmobile from falling into the glacial water. During the very harsh cold, *bukte* explode and throw huge blocks of ice into the air. Under *bukte*, there are fish. When *bukte* appear, the ice on the rivers is thick. From 2014 to 2016, *bukte* were not formed at all: this worried the nomads, since without *butke* it is hard to find fish. In the same years, the nomads were worried about the abnormal thinness of the ice: many were sinking through it. (cf. diagram Ulan, bukte; Part III: Ice anomalies)

After documenting Evenki indigenous knowledge about bukte, ulan, *and* sirikte, *we searched for a Russian translation and discovered that this phenomenon was studied in the Amur region and Yakutia by V. G. Petrov in 1928 (Alekseev 2001). It is interesting that this 'discovery' was made by a Russian geographer, who then decided to attach his name to it: however, the Evenki guides he worked with had known about this phenomenon for very long time. This is a very good illustration of the question about the belonging and authorship of knowledge relating to a natural phenomenon: does it belongs to the indigenous people that has collectively known about the phenomenon for centuries or to the scientific researcher who 'discovered it' and wrote about it?*

Буктэ эрэкин бирэ, амуннал эрэкин уланэ, нулгидйэми дйукэ эмтэргэвки. Чэпиӊат бэйэ сырагалби оротой. Буктэ эрэдун олло окинда бывки. Буктэдйэрэкин дйукэ дэрам бивки.

Когда не формируются буктэ, наледи *улан* не бывают на *амунна* (широкий бассейн реки). Когда эвенки кочуют, опасаются, чтобы лед не треснул, чтобы не утопить нарты и оленей. При сильном холоде *буктэ* взрываются, бросая огромные глыбы льда. Под *буктэ* всегда находится рыба. Когда *буктэ* формируются, то лед всегда толстый бывает вообще по всей реке. С 2014 г. по 2016 *буктэ* не формировались, что вызвало волнение кочевников из-за того, что если *буктэ* нету, то рыбу трудно найти. Одновременно волновались, что лед рек слишком тонкий и многие проваливались. (см. схемы, с. 348–367, 428–437)

Univun / Унивун

Naturally unfrozen part of the river, in some parts of some rivers

Натуральная талая часть льда в определенных местах некоторых рек

Water and the current are visible in some places along rivers with *univun*. On the edges of such *univun*, the ice is very thin. Because of the peculiar characteristics of this water, it does not freeze. Sometimes algae are visible. The Evenki prefer to avoid these places since they do not want to drown. However, such places are useful for gathering water for drinking, cooking, and washing: one does not have to bother melting snow. Sometimes, fish spend the winter in such rivers. (cf. univun, diagram Ulan, bukte)

Унивур бивкил адылду бира дулиндун эйандйэвки дйуҕаниҕэчин. Дйапкалын дйукэ нэмкукан бивки. Нулгидйэми аҕилэнанны эммэн чэпэрэ сыргалви. Адылдун олло туҕэвки. Мулэдйэми айа бивки.

Вода и течение видны местами по таким рекам. Лед по краям – очень тонкий слой над водой. Это из-за свойства этой воды, она не замерзает. Иногда местами водоросли виднеются. Эвенки предпочитают избегать такие места, чтобы не провалиться. Такие места удобные, чтобы воду набрать для питья, варки и мытья, чтобы лед не таять. Иногда рыба зимует там. (см. унивун, схемы, с. 348–367)

Kileŋ / Килэӊ
Extremely slippery part of the river ice without snow, like a mirror
Скользкая часть льда на реке – совсем без снега, как зеркало.

Kileŋ appear from frozen *ulan*. Where ulan appear regularly, there are usually *kileŋ*. This is hell for transportation because the reindeer slip and fall over. In snowmobiles, it is impossible to go fast because the sledge sways from side to side. Sometimes, the sledge breaks and both reindeer and humans can be seriously wounded. The Evenki avoid going through such places. (cf. diagram Ulan, bukte)

Улан доӊкотовки балдакикун овки – килэӊ. Улаӊаттун улан бивки килэӊ оммэн. Килэӊдули эруми нулгидйэда орор балдакачэвкил энэл сыргавэ танэ нулгидйэми аҕиливкил.

Килэӊ появляются от замерзших улан. Где постоянно улан бывают, обычно килэӊ выходят. Для транспорта кошмар, олени скользят так сильно, что не могут удержаться на ногах. На снегоходе невозможно разогнаться, потому что нарты сильно катятся по бокам, иногда ломаясь. Бывало, что олени и люди калечились. Эвенки опасаются через такие места ехать. (см. схемы, с. 348–367)

Koŋkurok / Коӈкурок

Empty space between ice and water or between ice and the ground in several layers

Пустолед – пустота между водой и льдом, или между льдом и землей, несколькими слоями.

When a river freezes, the surface layer freezes into a thin layer. When the water level under the ice lowers, the surface layer freezes, thereby creating a second sheet of ice. Between these layers of ice, there are empty spaces. This endangers nomadisation, since reindeer can sink. Sometimes, the Evenki avoid such places. (cf. diagram Ulan, bukte)

Бира доӈкотодйэрэкин дйукэ овки нэмкукан, тадук бира арбарэкин нйан доӈкотодйэрэкин. Тар дйукэл сйэгдилйадутын коӈкурок овки. Нулгидйэми эруми, орон эмтэргэвувки. Адылдун аӈилинэнны.

В процессе замерзания реки, сначала формируется тонкий слой льда на поверхности реки. Когда падает уровень воды, опять замерзает новый слой, но уже ниже. Между этими слоями льдов образуется пустолед. При кочевке опасно, ведь олени проваливаются и могут сильно пораниться. Иногда кочевники обходят такие места. (см. схемы, с. 348–367)

Iriaktal saŋipchal (UN) Uŋkakta, Uŋka (Iengra)
Ирйактал саӈипчал (УН) / Уӈкакта, Уӈка (Иенгра)
The trees are covered by hoar frost
Деревья инеем покрылись

If one sees trees covered by hoar frost when travelling, it means that there is water: there will be a very visible *ulan*. Having noticed such sign, the nomads prefer to set up camp there.

Ӈэнэдйэми ирйактэл саӈипчал бирэкин саӈныны таду му биһин. Ирйактэ саӈивувки, улан йудэрэкин тар со сама бивки. Нулгидйэми ичэми таргачинду уриӈнэнны даҕадун.

Если, проезжая, увидишь как деревья инеем покрылись, то значит там вода есть, наледь там выходит, это очень заметно. Кочевники увидев такой знак предпочитают установить табор рядом.

Oniovun, onior / Онйовун, Ониор
Patterns made by the hoar frost on the ground, ice, or windows in the morning
Рисунки, созданные инеем на земле при холоде, на льду или на окне

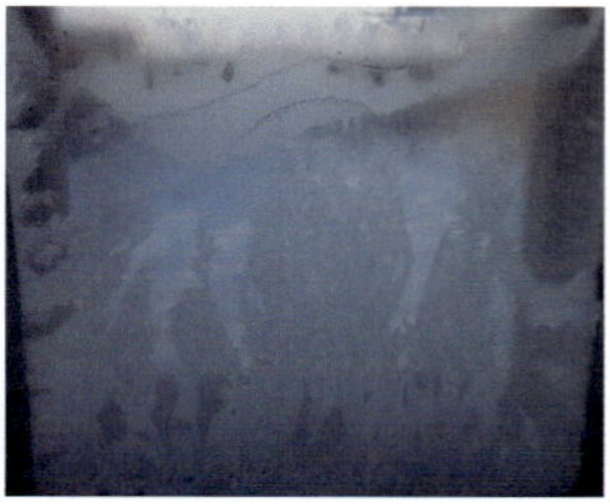

If one sees hoar frost patterns on the ground, one knows that the frost will remain for three to five days. This means it is better to stay at home.

Бэйэ ичэччэми онйовунмэ саӈныны горокунэ иӈинидиҕан диҕиллэйэ тунӈаллэйэ иӈинидиҕан. Дйудуви авалда нада.

Увидев рисунки инея на земле, знаем, что морозы будут стоять три-пять дней. Значит лучше работать дома.

Eiim / Эйим	Places, where the fish spend the winter under ice Места, где рыба зимует подо льдом

We will not describe this because we do not want other peoples to exhaust our fish reserves.
In accordance with desires of the Evenki nomads working with us, there is some information we will not provide. The Evenki were especially worried about giving non-natives too many keys for reading their environment and finding game, fish, and berries.

Тара этарэп тэдэврэ, адатын уңту бэйэл манадиңарэ.
Это не будем описывать, чтобы чужие люди, браконьеры не кончали рыбу.

The snow type SY - three categories: from rain on snow, from the transformation of the *cheya* snow type, and from successive phases of melting and freezing.

Тип снега СЫ - 3 категории: от дождя на снегу, от трансформации снега *чеҕа*, от солнца и холода

SY (embedding icy layer) from rain on snow / СЫ (ледяной слой) – от дождя на снегу

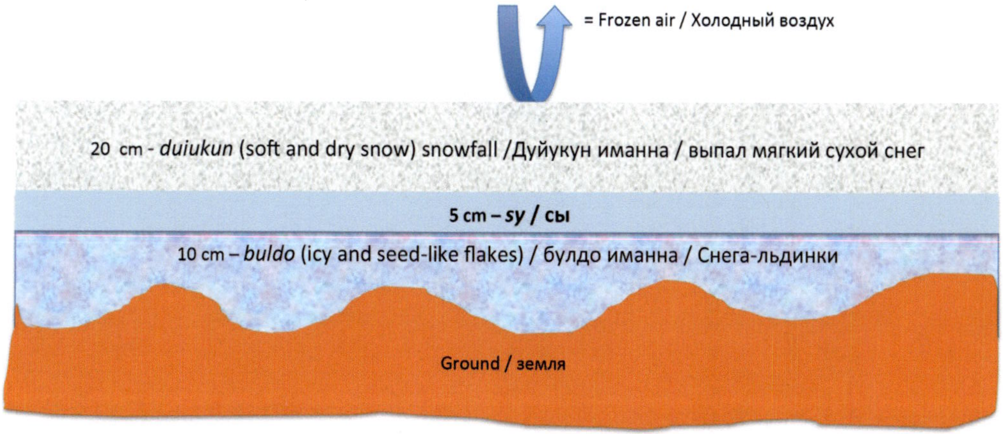

1 - If snowfall is followed by rainfall, a layer of *sy* appears upon the first frost. With frost under the *sy*, the snow turns into *buldo* snow (icy and seed-like flakes).

Иманнарэкин, тадук тыгдалдинан, элтанэкин сы овки. Элтанэкин умнат сы эрэдун булдо иманна овки.

Если снег падает, а потом дождь идет, то после этого при морозе становится *сы*. А под *сы* сразу снег превращается *булдо*.

2 – After the formation of *sy*, if a thick layer (cm 20) of snow falls and is followed by an abrupt frost, the *sy* layer will not thaw. Indeed, the thick upper snow layer will not allow the frost in the air to crumble the *sy*.

Сы амардукин кэтэкунмэ иманнарэкин, тадук иҥинйэктэллэкин сыкун тыкадат эманмувки, эвки чумнэ.

Если, после формирования *сы*, выпадет снег сразу см 20 и потом резко похолодает, то большой слой *сы* так и останется – толстый слой снега создает изоляцию и не дает *сы* крошиться.

The snow type SY - three categories: from rain on snow, from the transformation of the *cheγa* snow type, and from successive phases of melting and freezing.

Тип снега СЫ - 3 категории: от дождя на снегу, от трансформации снега *чеҕа*, от солнца и холода

SY from the physical transformation of the *cheγa* snow type /
СЫ от физической трансформации снега *чэҕа* (cf. Anomaly in snow cover 2014-2015)

= Lichen / лйавуктэ / ягель

If the snow melts, *cheγa* snow appears (a scab-like layer). If a thick layer of fresh snow falls onto it over several days, this *cheγa* snow will be transformed into *sy* during the first frost. The layer consists of hard snow and around 5-7 cm of *sy*. During the winter 2014-15, there was a surprisingly thick layer of *sy:* it could support a person. The reindeer grazing in it were completely soaked and froze. Thus the reindeer could not graze – this was an extreme weather event (cf. diagrams Anomalies in snow cover 2014-2015).

Иманна ундйэрэкин, чэҕа овки, адыллалывэл иманна тыкивки, тадук иӈинйэктэллэкин сыкун овки, иманна чэҕа - сыкун овки. Маӈа иманна сынйун – 5-7 см. Туҕэниду 2014-2015, оӈко улапиҕа, му орэн, тадук доӈкотовки.

Если снег начинает таять, появляется тип снега *чэҕа* (корка от таяния и замерзания) и если сверху несколько дней подряд снег идет, как только морозы наступят, этот снег *чэҕа* трансформируется в снег *сы*. Это твердый снег с *сы* – 5-7 см. В 2014-2015гг зимой был удивительно толстый слой *сы,* что люди ходили по поверхности, не проваливаясь. Олений корм намок, а потом весь замерз вместе с *сы*. Олени не могли кормоваться – и это стало таежной экстремальной ситуацией (cf. diagrams Anomalies in snow cover 2014-2015).

The snow type SY - three categories: from rain on snow, from the transformation of the *cheɣa* snow type, and from successive phases of melting and freezing.

Тип снега СЫ - 3 категории: от дождя на снегу, от трансформации снега *чеҕа*, от солнца и холода

SY from sun 1/3 / СЫ от солнца 1/3

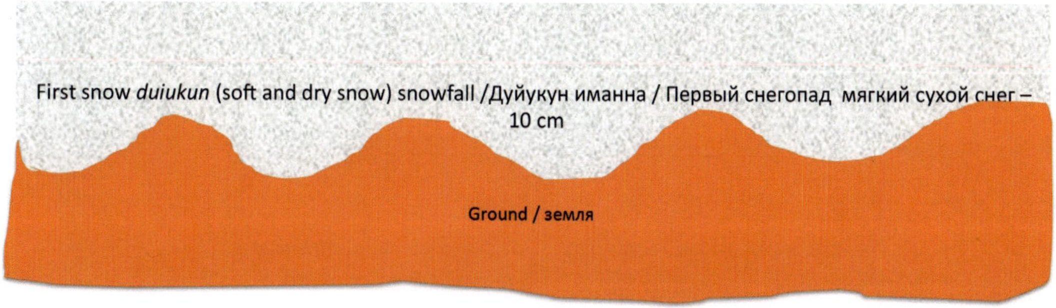

The snow type SY - three categories: from rain on snow, from the transformation of the *cheγa* snow type, and from successive phases of melting and freezing.

Тип снега СЫ - 3 категории: от дождя на снегу, от трансформации снега *чеҕа*, от солнца и холода

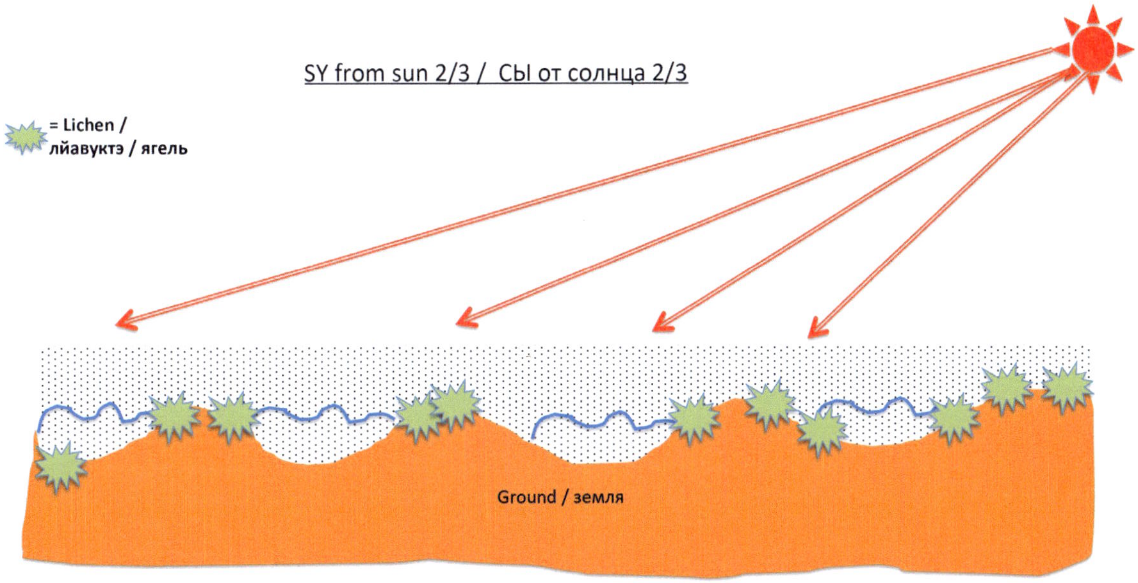

Due to the warmth from the sun, the *duiukun* snow turns into wet *ulapkun imanna* snow: the water flows downwards.
Когда днем солнце греет, то верхний слой снега превращается в мокрый снег *улапкун иманна*, и вода стекает вниз.

The snow type SY - three categories: from rain on snow, from the transformation of the *cheγa* snow type, and from successive phases of melting and freezing.

Тип снега СЫ - 3 категории: от дождя на снегу, от трансформации снега *чеҕа*, от солнца и холода

SY from sun 3/3 / СЫ от солнца 3/3

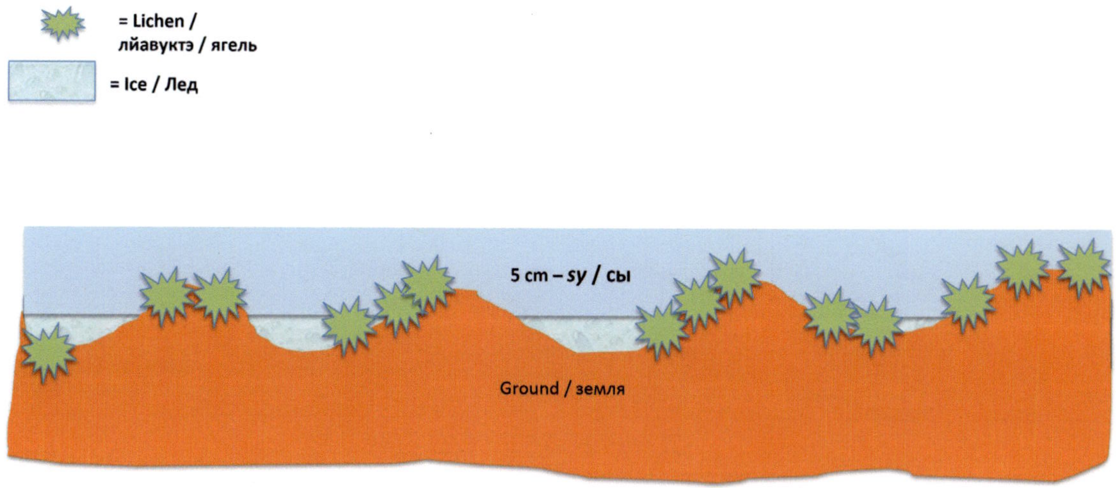

. ....in the evening, the frost turns the water into ice, i.e. *Sy*.

....то вечером при морозе вода в лед превращается – т.е. в *сы*.

CHUIUR snow type (hard surface layer on top of the snow) – 1st step (October, November) – before the appearance of *chuiur*

Тип снега ЧУЙУР (твердый наст на поверхности снега) - 1 этап (октябрь, ноябрь) – до создании чуйура

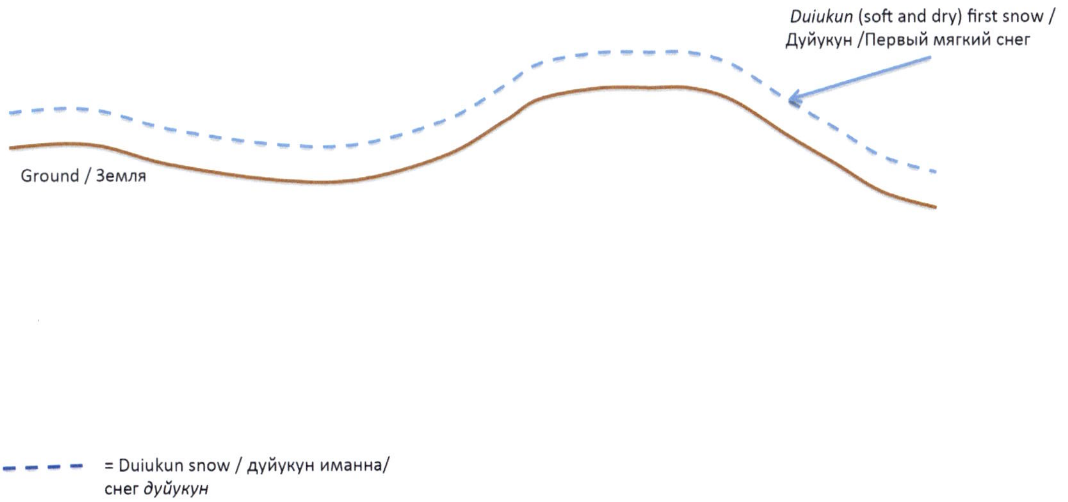

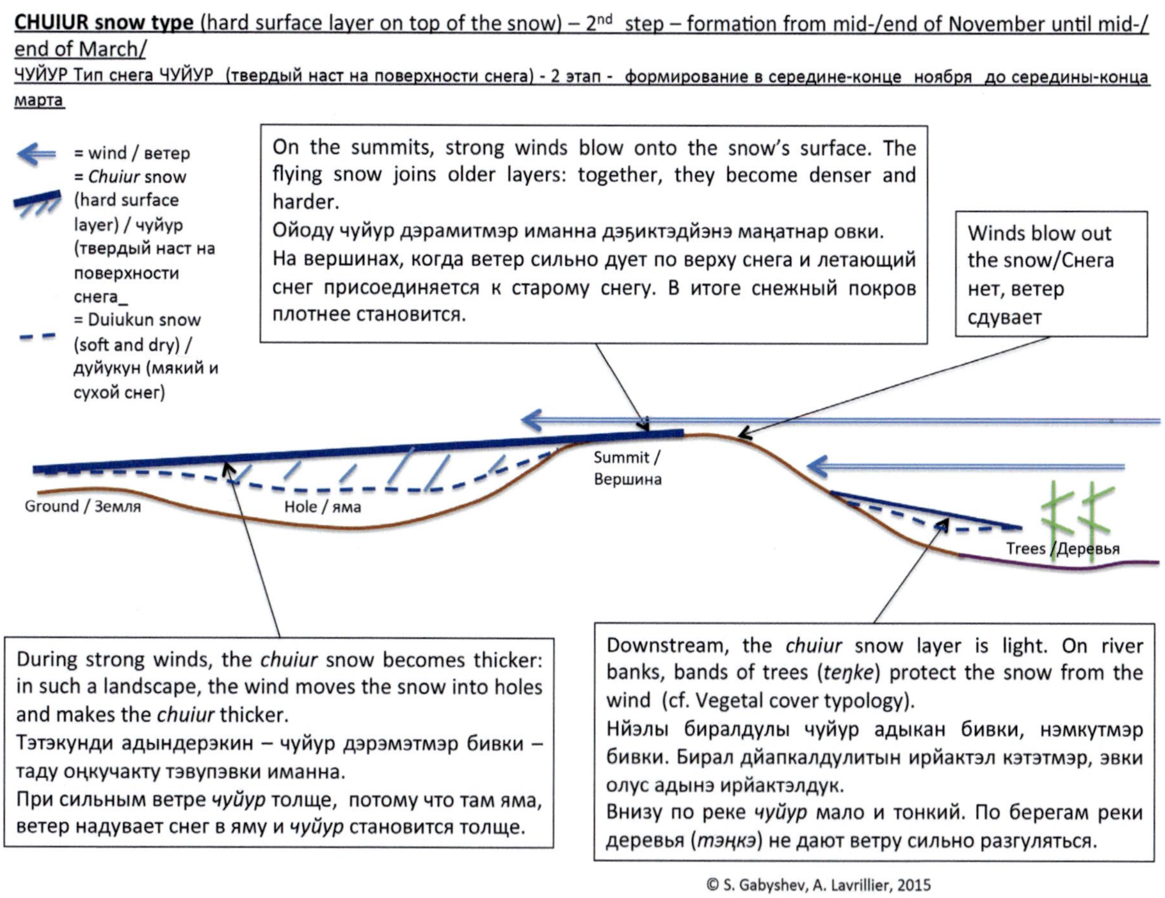
CHUIUR snow type (hard surface layer on top of the snow) – 2nd step – formation from mid-/end of November until mid-/end of March/
ЧУЙУР Тип снега ЧУЙУР (твердый наст на поверхности снега) - 2 этап - формирование в середине-конце ноября до середины-конца марта
= wind / ветер
= Chuiur snow (hard surface layer) / чуйур (твердый наст на поверхности снега_
= Duiukun snow (soft and dry) / дуйукун (мякий и сухой снег)
On the summits, strong winds blow onto the snow's surface. The flying snow joins older layers: together, they become denser and harder.
Ойоду чуйур дэрамитмэр иманна дэђиктэдйэнэ маңатнар овки.
На вершинах, когда ветер сильно дует по верху снега и летающий снег присоединяется к старому снегу. В итоге снежный покров плотнее становится.
Winds blow out the snow/Снега нет, ветер сдувает
Summit / Вершина
Ground / Земля
Hole / яма
Trees /Деревья
During strong winds, the chuiur snow becomes thicker: in such a landscape, the wind moves the snow into holes and makes the chuiur thicker.
Тэтэкунди адындерэкин – чуйур дэрэмэтмэр бивки – таду оңкучакту тэвупэвки иманна.
При сильным ветре чуйур толще, потому что там яма, ветер надувает снег в яму и чуйур становится толще.
Downstream, the chuiur snow layer is light. On river banks, bands of trees (teŋke) protect the snow from the wind (cf. Vegetal cover typology).
Нйэлы биралдулы чуйур адыкан бивки, нэмкутмэр бивки. Бирал дйапкалдулитын ирйактэл кэтэтмэр, эвки олус адынэ ирйактэлдук.
Внизу по реке чуйур мало и тонкий. По берегам реки деревья (тэңкэ) не дают ветру сильно разгуляться.
© S. Gabyshev, A. Lavrillier, 2015

CHUIUR snow type (hard surface layer on the top of snow) – 2st step – formation from mid-/end of November until mid-/end of March/
ЧУЙУР Тип снега ЧУЙУР (твердый наст на поверхности снега) - 2 этап - формирование в середине-конце ноября до середины-конца марта

During a snowfall in warm conditions, the *chuiur snow* is transformed into fresh *duiukun snow*. If the wind blows again, this *duiukun* snow will turn into *chuiur* snow.
From mid-/end of March, depending on the place, *chuiur* does not form anymore and transforms into *cheya* snow (a scab-like snow layer).

Нйамаду иманнадйэрэкин, чуйур чумивки тар иманнадук, дуйукун иманна овки. Нйан адынэкин – нйан тыкадат чуйур одиӊан.

Чуйур трансформируется при снегопаде в теплую погоду и становится *дуйукун*. Если ветер задует, то дуйукун снова станет чуйур.
С середины-конца марта в зависимости от местонахождения, *чуйур* уже не формируется и превращается в снег *чеҕа*.

Typical *chuiur* – a hard surface layer of snow created by wind. It can support a lot of weight and is very resistant.

Типичный *чуйур* – твердый наст снега, созданный ветром. Выдерживает тяжести: оленей, людей и тяжелые нарты.

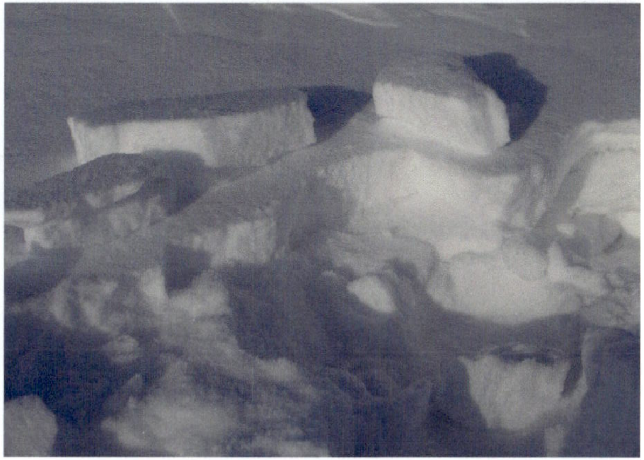

© S. Gabyshev, A. Lavrillier, 2015

On summits: a very thick and strong *chuiur* snow layer that can support a lot of weight
На вершинах: очень толстый и крепкий слой *чуйур* – выдерживает очень тяжелые нагрузки

Profile view, cf. Next picture /
Профиль, см. следующую фото

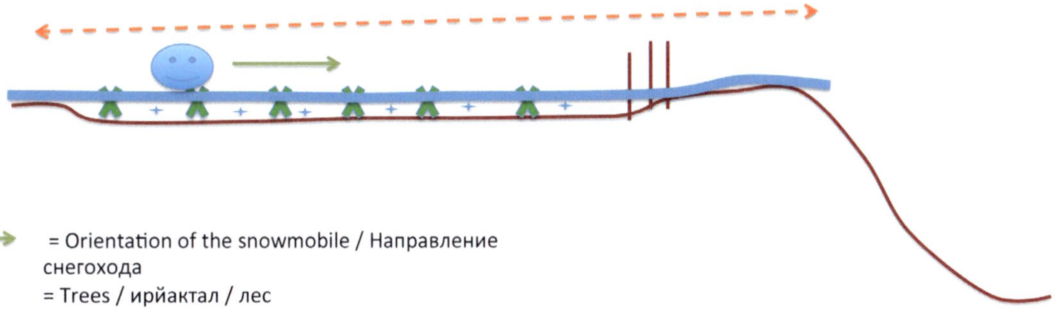

= Orientation of the snowmobile / Направление снегохода

= Trees / ирйактал / лес

= *Pinus pumila* / стланик / болгиктэ

= People on snowmobiles with food and goods / Люди на снегоходе с грузом +/- 600 кг

= Hard surface layer of *chuiur* snow / Чуйур

= Icy seed-like snow *buldo* / Снег-льдинки *булдо*

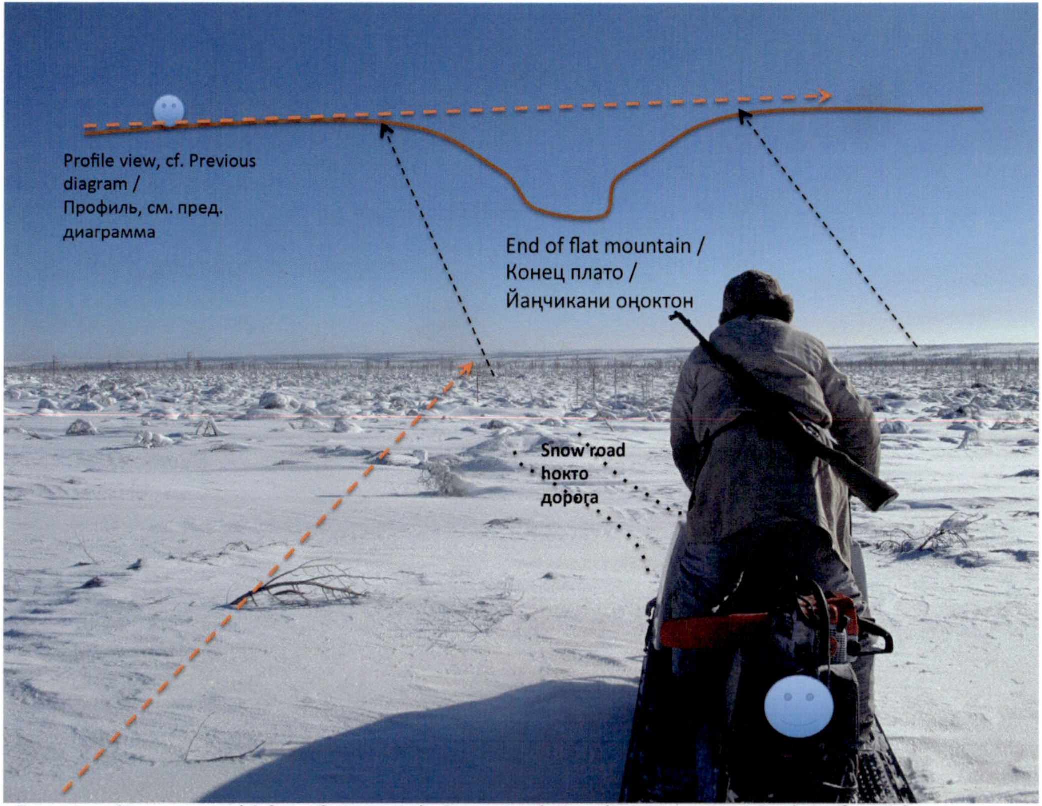

On summits: a very thick and strong *chuiur* snow layer that can support a lot of weight
На вершинах: очень толстый и крепкий слой *чуйур* – выдерживает очень тяжелые нагрузки

A thick *chuiur* layer created by the wind.

Чуйуркун оча адындук.

Ветра создали толстый слой снега *чуйур.*

On summits: a very thick and strong *chuiur* snow layer embeds the vegetal cover (*pinus pumila*)
На вершинах: очень толстый и крепкий слой *чуйур* – охватывает и заваливает стланики

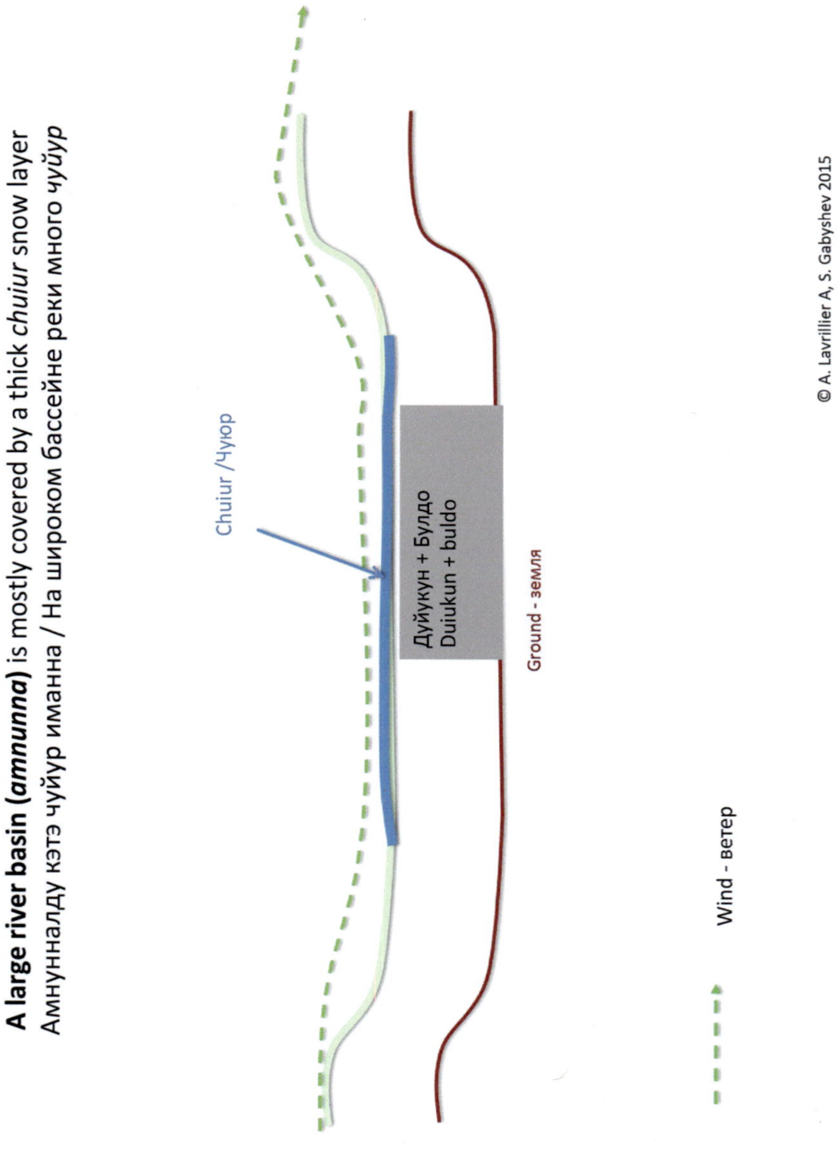
A large river basin (amnunna) is mostly covered by a thick chuiur snow layer
Амнуналду кэтэ чуйур иманна / На широком бассейне реки много чуйур
Chuiur /Чуюр
Дуйукун + Булдо
Duiukun + buldo
Ground - земля
Wind - ветер
© A. Lavrillier A, S. Gabyshev 2015

Chuiur in a large river basin.
Чуйур на широком бассейне реки.

Cf. Next diagram

Roads and a large river basin: *Chuiur* after a storm /
Дороги и широкие бассейны реки: *чуйур* после пурги

The snow road was erased / һокто унӈэча / Дорогу замело.

The large river basin (*amnunna*), is covered by a thick *chuiur* snow layer / Амнуннаду чуйуркуча/ Широкий бассейн реки после пурги покрылся снегом чуйур.

After a storm, one must 're-build' snow roads. *Chuiur snow* covers the surface of the roads and erases them.

Адын амардукин һоктово нйан һоктовувки. Иманна ойодун чуйур оча.

После пурги дорогу приходится делать по новому. Поверхность снега *чуйур* стало.

BULDO snow type, 1st step (October) in the normal installation of the snow cover (-5°-20°C)
Тип снега БУЛДО. 1 этап (Октябрь) при нормальном установлении снега (-5°-20°C)

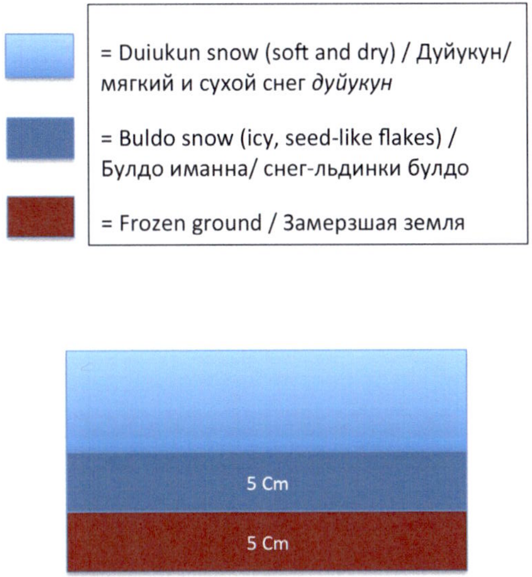

Upon the arrival of the first soft and dry snow (*duiukun*), the ground freezes up to 5 cm in depth during the morning and evening frost (*eltan*). The snow then turns into *buldo* snow because of the cold emitted by the ground (the icy flakes are around 2 mlm in diameter). During the day, when it becomes slightly warmer, the snow is dry and soft (*duiukun*): in the evening, the frost changes it into *buldo* snow.

Элэкэс иманнарэкин, дуйукун иманна иманнарэкин, элтанакин дуннэ доҥотовки, 5 см доҥоторон, булдо иманна овки дуннэдук доҥотоловки (иманна дйукукарэчир олэвкил 2 милиметрал овки). Инэниду нйамадйэрэкин иманна дуйукун бивки. Кэҕа элтанэкин долбонывэ иҥинирэкин – булдо иманна олэвки.

При первом мягком снеге с утренним и вечерним морозами земля замерзает на 5 см глубины, и снег (*дуйукун*) превращается в снег *булдо,* замерзая от земли (снежинки как маленькие льдинки - 2 мл диаметра). Днем, когда теплеет, снег сухой и мягкий (*дуйукун*), но вечером с морозом превращается в снег *булдо*.

(Cf. also diagram cheya snow type)

BULDO snow type, 2nd step (end October-mid November) in the normal installation of the snow cover

Тип снега БУЛДО, 2 ЭТАП (конец октября-середина ноября) при нормальном установлении снега

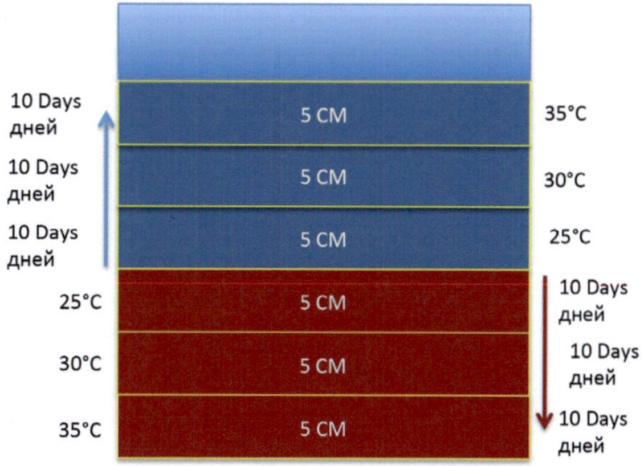

= Direction of freezing

= *Duiukun* snow (soft and dry) (layer and flakes) / Дуйукун/ мягкий и сухой снег *дуйукун* (слой и снежинка)

= Buldo snow (icy, seed-like flakes) (layer and flakes) / Булдо иманна/ снег-льдинки булдо (слой и снежинка)

= Frozen ground / Замерзшая земля

As snow falls and installs the snow cover, the ground freezes: the amount of *buldo* snow increases with the frost. The lower layer is composed of big *buldo* flakes (5mm), the middle layer of medium-sized flakes (3mm), and the upper layer of flakes measuring 2mm. On top of the snow cover, the transformation of the *duiukun* into *buldo* continues ('*buldo* eats *duiukun*').

Айат аракукан иманнадйэрэкин, дуннэ айат доӈкотодйэвки, булдо иманна иӈиӈду кэтэтмэр бивки. Эргу булдо – булдокун овки (5 млм), дулгу булдо – 3 мм, ойгу булдо – 2 мм овки. Уҕискаки булдо опканэвки иманнавэ.

По мере того, как снег постепенно падает, устанавливается, земля постепенно замерзает, снег типа *булдо* становится все больше с морозом. Нижний слой снега становится *булдо* с большими снежинками (5 мм); а в середине снежинки *булдо* размером 3 мм, а верхний *булдо* стал 2 мм. На верху снежного покрова *булдо* в процессе превращения *дуйукун* в *булдо* (*булдо* «сьедает снег *дуйукун*»).

Snow flakes equations / эквиваленты в снежинках:

flakes (*duiukun*) = 1 flake *buldo*

= 3 mm

= 5 mm flake. With the pressure of the snow layer, the *buldo* flakes merge together / Иманна тэҕэдйэрэкин, иманна булдо булдонун намаралдэдйавкил/ Давя, снежинки *булдо* соединяются между собой.

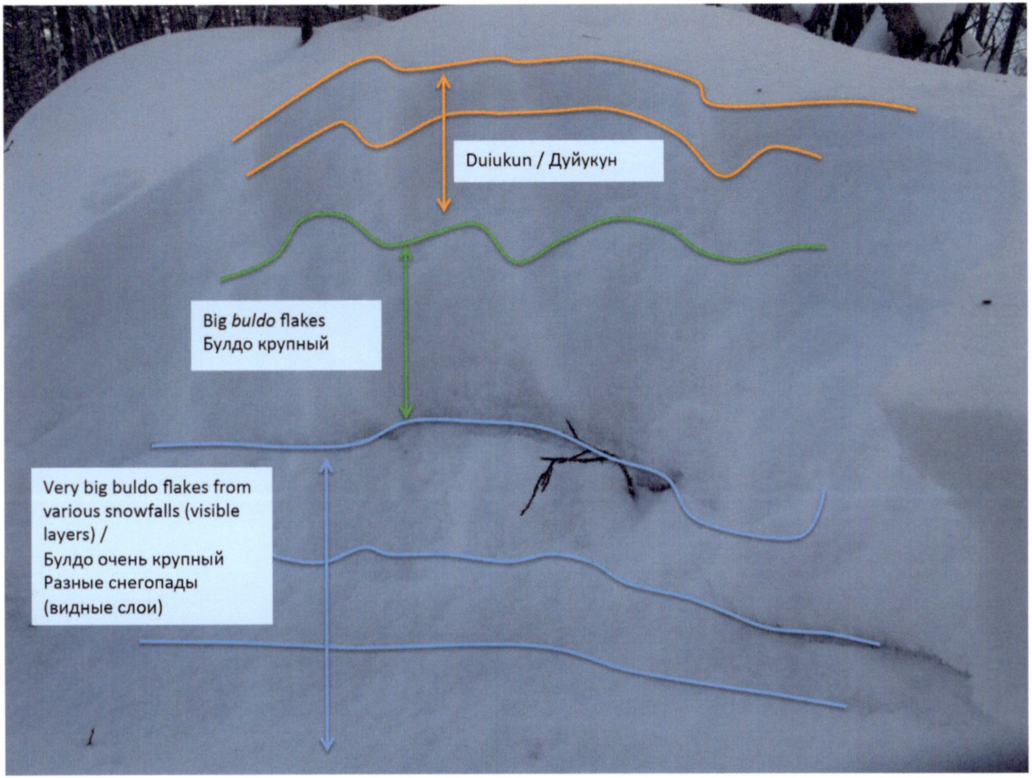

Along rivers, the amount of *buldo* is much greater because of the cold air *idia* (cf. Evenki climatology, diagram Idia).
Вдоль рек *булдо* больше бывает – это от холода, (cf. Evenki climatology, diagram Idia).
Бирал дйапкалын иӊинитмар бивки идйадук – булдо кэтэтмэр иманнаду. (31 Januray 2015)

BULDO snow type: supports people, reindeer, and snowmobiles above the ground's surface
Тип снега БУЛДО: Поддерживает людей, оленей и снегоход над уровнем земли

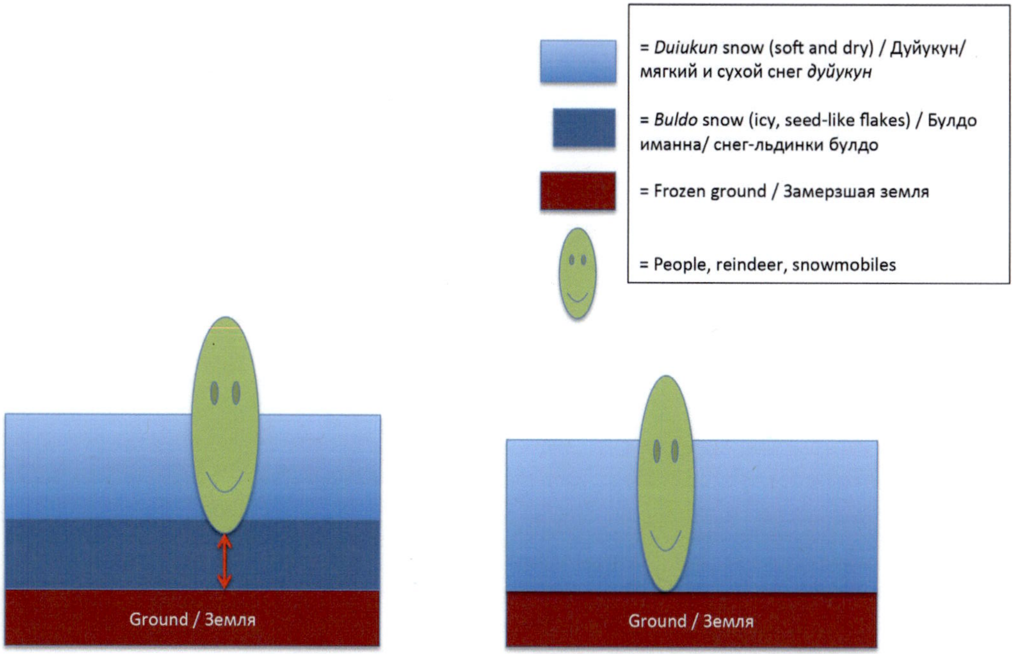

© S. Gabyshev, A. Lavrillier, 2015

CHEγA snow type (Scab-like snow layer from successive warming and freezing) – 1 st step (**March**)

Типа снега ЧЭҔА (корка снега от потепления и замерзания) - 1ый этап (март)

1 cm *cheya* – As soon as the snow starts thawing, the evening cold (*eltan*) freezes it again, making the *cheya* thin / Элэкэс унйалдйаран, кэҕа элтанэкин чэҕан нэмкутмэр / Как только снег начинает таять, вечером замерзает и *чэҕа* очень тонкий.

5 cm *cheya* – On mountain slopes (*elleŋe*), the sun has a strong warming effect. A considerable amount of snow thaws: during the evening and morning frosts (*eltan*), it freezes and creates a thick layer of *cheya* / Эллэнңэду дылача сот дылачавки. Иманна сокундит унйавки, тадук элтанэкин доңкотовки чэҕан дэрэмэтмэр бивки/ На склоне с боку горы (*эллэңэ*) солнце сильно греет, снег сильно тает, а при утреннем и вечернем морозах (*элтан*) замерзает и создает толстый слой снега *чэҕа*.

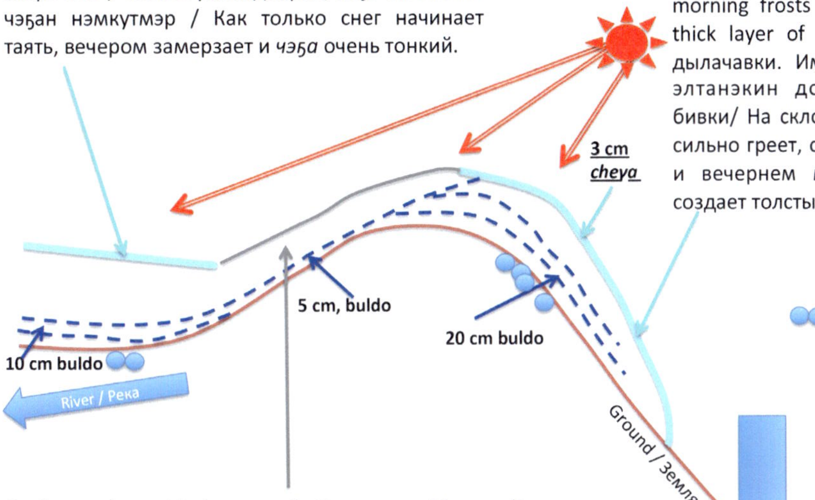

= Dampness from the thawing snow goes downwards during the morning and evening frosts: the snow turns into *buldo*, since this snow type attracts moisture / Влажность от таяния снега идет вниз, а при вечернем и утреннем замерзании становится снег *булдо*.

As the sun is unable to warm it, the snow neither melts nor refreezes; thus, there is no *cheya* snow, just dry and soft *duiukun* snow / Талы дылача эвки дылачарэ – иманна эвки унйарэ, эвки доңкоторо – таду чэҕа эвки бирэ. Дуйукун иманна / Туда солнце не светит и снег не тает и не замерзает, значит *чэҕа* нет, там только снег сухой и мягкий *дуйукун*.

CHEYA snow type (Scab-like snow layer from successive warming and freezing) – 1st step (**March**)
Типа снега ЧЭҔА (корка снега от потепления и замерзания) - 1ый этап (март)

Diagram keys'

= sun rays / лучи солнца

= *Buldo* snow (icy, seed-like flakes) / Булдо иманна/ снег-льдинки булдо

= *Duiukun* snow (soft and dry) / Дуйукун/ мягкий и сухой снег *дуйукун*

= *Cheya* snow (icy and seed-like flakes) /Чэҕа/ корка снега от потепления и замерзания

CURRENT ANOMALIES RELATED TO THIS SNOW TYPE:

Cheya is a snow type typical for the spring, but it can appear in autumn during anomalies. According to nomadic observations provided within the framework of the BRISK project, the *cheya* snow type appeared systematically in the autumns of 2013, 2014, and 2015 because of abrupt and unexpected warming in the middle of the autumn.

Чэҕа является типом снега типичного для весеннего процесса. Если *чэҕа* бывает осенью в конце сентября по середину октября, то это считается аномалией. По таежным наблюдениям по проекту БРИСКа, в 2013 2014 и 2015 систематически *чэҕа* формируется осенью из-за резких потеплений в середине осени.

CHEγA snow type (Scab-like snow layer from successive warming and freezing) – 2 nd step (**April**)

Типа снега ЧЭЂА (корка снега от потепления и замерзания) – 2ой этап (апрель)

5 cm *cheya* - The sun shines strongly, increasing the temperature: a substantial amount of snow thaws (and freezes with *eltan* cold), making a thick layer of *cheya* / Иманна унйадйакуттэн, дылача дылачадйэрэн, нйаматмар оча - таду чэђа дэрэмэтмэр (тольше)/ Солнце сильно светит, очень тепло стало, снег сильно тает и замерзает с *элтан* – там толстый слой *чэђа*.

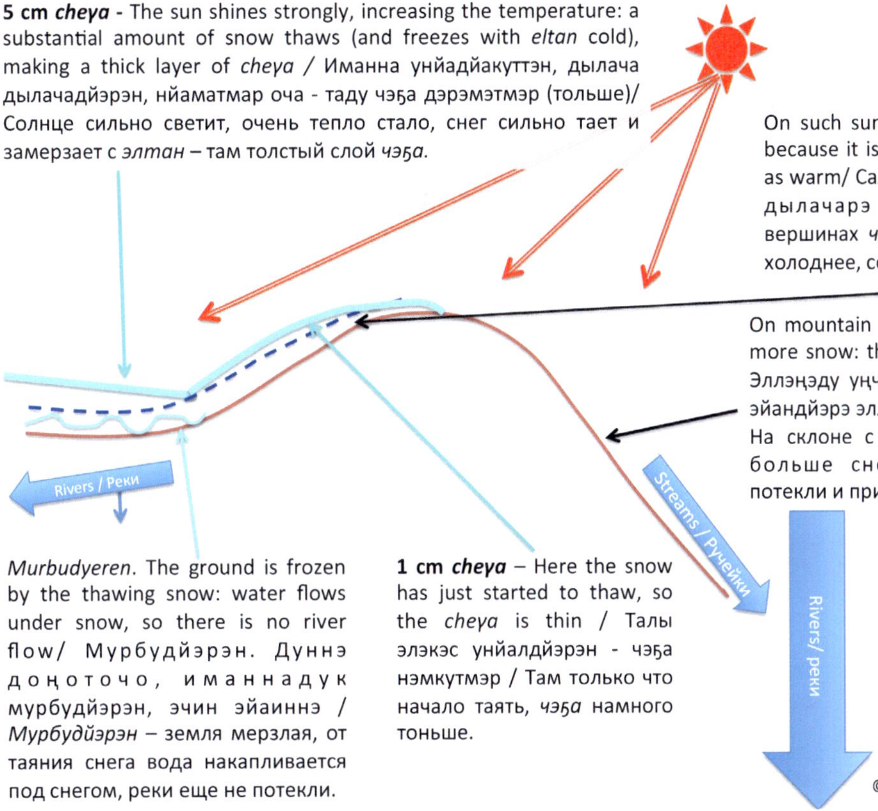

On such summits, there is less *cheya* because it is colder and the sun is not as warm/ Салгындйэрэн, дылача эвки дылачарэ / На не солнечных вершинах *чэђа* меньше, потому что холоднее, солнце меньше греет.

On mountain slopes (*elleŋe*), there is no more snow: the streams flow and flood/ Эллэӈэду уӈча, иманна ачэн. Биракар эйандйэрэ эллэӈэду, биракун мудача / На склоне с боку горы (*эллэӈэ*), нет больше снега. Мелкие ручейки потекли и прибывают.

Murbudyeren. The ground is frozen by the thawing snow: water flows under snow, so there is no river flow/ Мурбудйэрэн. Дуннэ доӈоточо, иманнадук мурбудйэрэн, эчин эйаиннэ / *Мурбудйэрэн* – земля мерзлая, от таяния снега вода накапливается под снегом, реки еще не потекли.

1 cm *cheya* – Here the snow has just started to thaw, so the *cheya* is thin / Талы элэкэс унйалдйэрэн - чэђа нэмкутмэр / Там только что начало таять, *чэђа* намного тоньше.

CHEγA snow type (Scab-like snow layer from successive warming and freezing) – 3 rd step (**May**)

Типа снега ЧЭҔА (корка снега от потепления и замерзания) – 3ий этап (май)

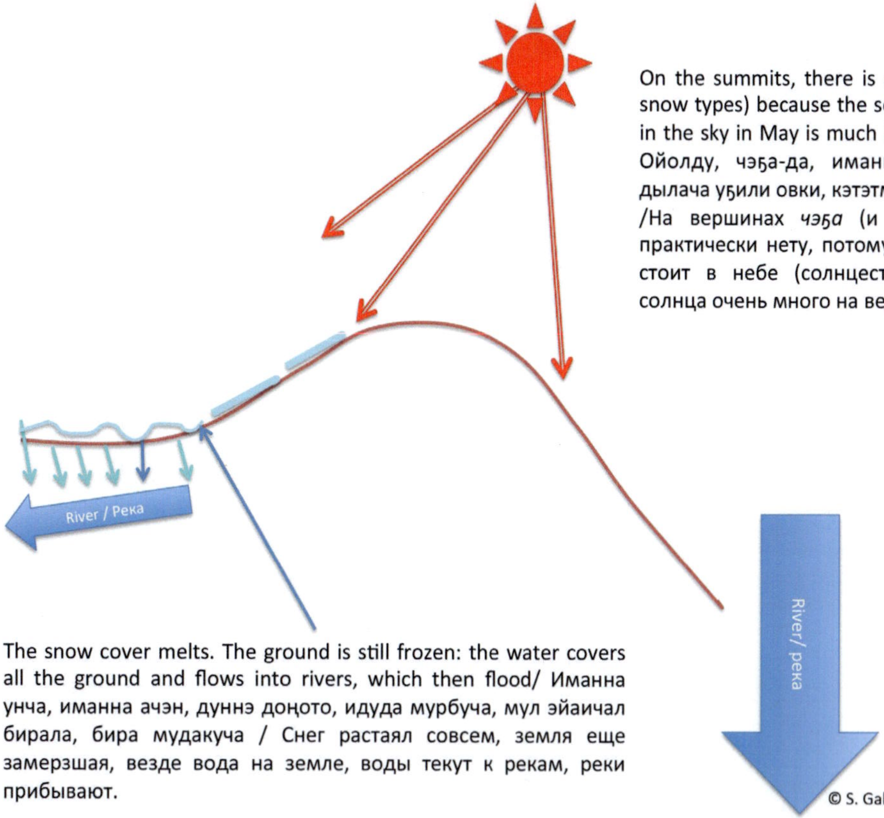

On the summits, there is almost no *cheγa* (or other snow types) because the seasonal position of the sun in the sky in May is much higher: it gets very warm / Ойолду, чэҕа-да, иманна-да адыкукан: майду дылача уҕили овки, кэтэтмэрбэ дылачавки.
/На вершинах *чэҕа* (и вообще любого снега) практически нету, потому что в мае солнце выше стоит в небе (солнцестояние выше весной) и солнца очень много на вершинах.

The snow cover melts. The ground is still frozen: the water covers all the ground and flows into rivers, which then flood/ Иманна унча, иманна ачэн, дуннэ доӈото, идуда мурбуча, мул эйаичал бирала, бира мудакуча / Снег растаял совсем, земля еще замерзшая, везде вода на земле, воды текут к рекам, реки прибывают.

CHEγA snow type (Scab-like snow layer from successive warming and freezing) – **Transformation into *buldo* snow type** (icy and seed-like flakes)

Типа снега ЧЭӠА (корка снега от потепления и замерзания) – Трансформация в тип снега *булдо* (снег-льдинки)

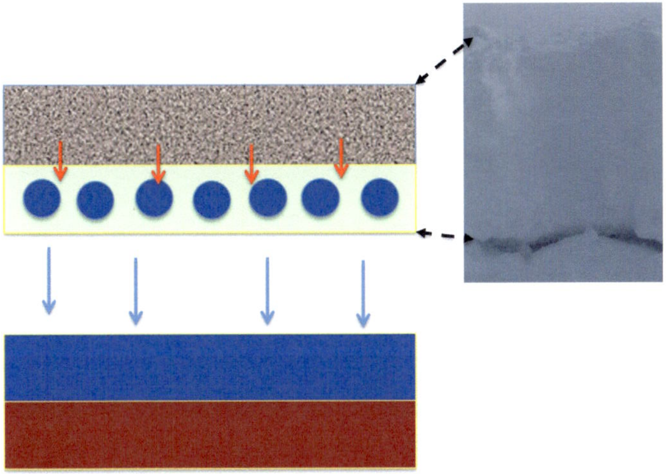

= The *cheγa* snow moves towards the ground if there is cold weather for three or four days consecutively. As soon as the *cheγa* freezes due to cold weather, it crumbles away while also turning into *buldo* and moving downwards to the ground. This can happen at any moment of the year and in all landscape types.

Чэӡа иманна тыкивки эргиски иӊинйэдйэрэкин илалайэ-у, диӡилэйэ-у.Чэӡадук булдо иманна овки, тадук тыкивки эргискаки, чэӡа доӊкоторокин иӊиӊду булдо ултамилэвки эргискакси тыкилэвки, окинда, идуда.

Снег *чэӡа* падает вниз, если холодно будет 3 или 4 дня подряд. Как только *чэӡа* замерзает от мороза, он рассыпается в *булдо* и падает вниз. Это в любой момент и при любом типе ландшафта.

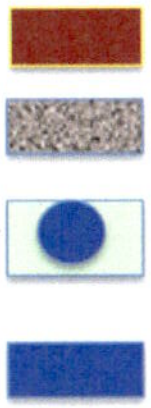

= Ground / земля
= *Cheγa* snow (icy and seed-like flakes) /Чэӡа/ корка снега от потепления и замерзания
= Empty space and old *buldo* snow falling and crumbling away / Булдо иманна тыкивки ултамдйэвки / Пустота и старое *булдо*, который сыпется
= *Buldo* snow (icy, seed-like flakes) falling down / Булдо иманна тыкивки эргиски/ снег-льдинки булдо падающие вниз

The installation of the snow cover – Evenki knowledge of norms // <u>September – before installation</u>
Формирование снежного покрова – Норма согласно эвенкийскому знанию // <u>Сентябрь до снежного покрова</u>

Before the ground freezes, the snow type *lebga* falls. Under the warmth of the sun, this snow melts. Before the *lebga* snow melts, the Evenki hunt wild reindeer and elk.

Эдйэлин дуннэ доңоторо – лэбгавки. Дылачарэкин, тар лэбга унэвки. Таргачинду лэбгаду эдйэлин унэ, бэйунэ, коңнокуйэ арычиңнанны / Пока земля не замерзла падает снег типа *лэбга*, который потом тает от тепла солнца. Пока этот снег не растаял, эвенки охотятся на диких оленей и лосей.

Onior – the morning and evening frosts cover the surface of the ground with hoar frost / *Онйор*, элтанча - дуннэ ойолын онйор одйаран / Утренний и вечерний мороз рисует иней на земле

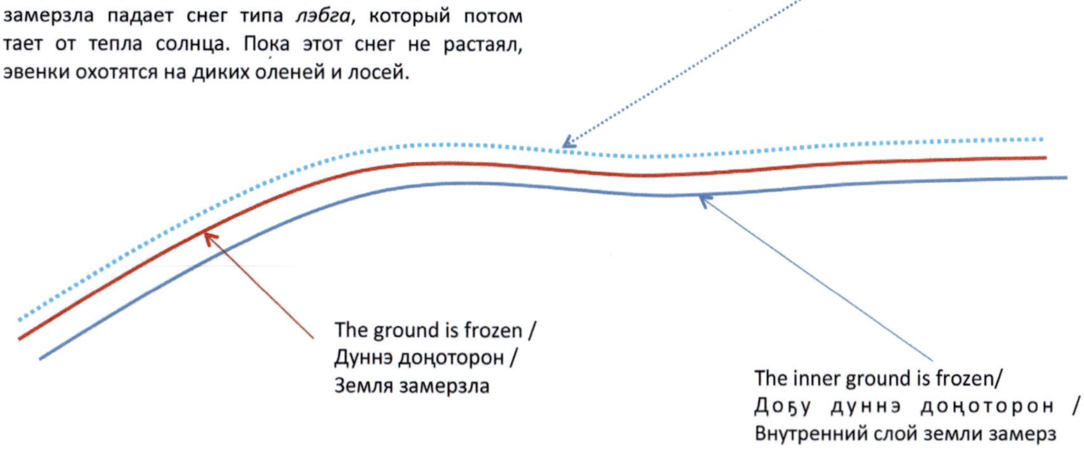

1st step: from (+/-) – 15°C / 1 этап: от (+/-) -15 C

The installation of the snow cover – Evenki knowledge of norms // <u>October A – installation</u>

Формирование снежного покрова – Норма согласно эвенкийскому знанию // Октябрь А инсталляция снежного покрова

1 – Snowfall: Fluffy, a bit wet, and sticky (*debdeme ulapkuhincha*): sable hunting with dogs starts / Иманнача дэбдэмэ улапкуһича – андаһидйавкил нинакиндит. Адылдун бонаһинэвки / Снег падает, пышный мокроватый, большими снежинками, иногда мелкий град падает – Сезон охоты соболя начинается собаками / 15 cm

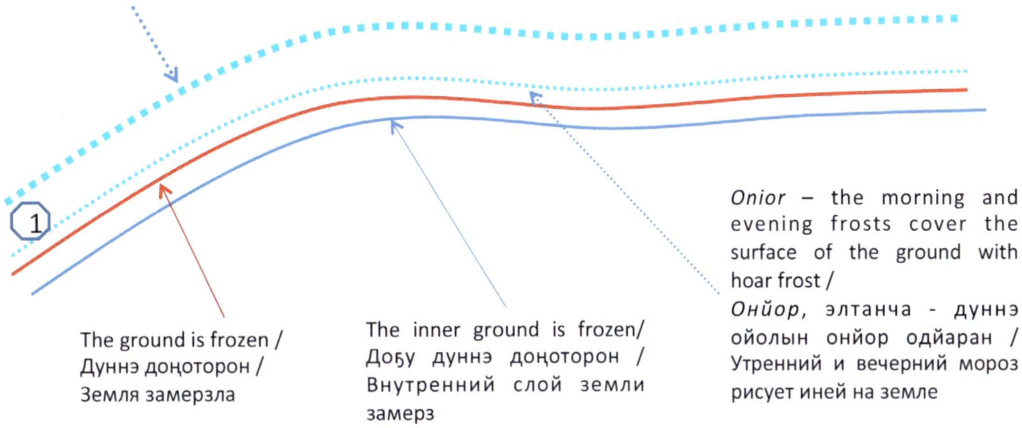

2nd step: from (+/-) – 25°C / 2 этап: от (+/-) -25 C

Snow cover installation – Evenki knowledge of norms // <u>October B – installation</u>

Формирование снежного покрова – Норма согласно эвенкийскому знанию // Октябрь Б инсталляция снежного покрова

1 - Because of the frost, the first snow fall turns into *buldo* (small icy and seed-like flakes) / Первай дэбдэмэ иманна – иӊиндук булдо овки / Первый слой снега от мороза превращается в мелкие крупинки *булдо*

2 – Snowfall: soft snow (*duiukun*) falls – the sable sink into the snow and their legs cannot reach the ground. Dogs thus catch them easily / Иманна дуйукукан, андаhил айат чэпадйэрэ, алганын дуннэла эчэл исчарэ, нинакин имакунди боконовки / Падает мягкий сухой снег *дуйукун*. Соболь хорошо проваливается, ноги не достают до земли, собаки его быстро догоняют / 20 cm

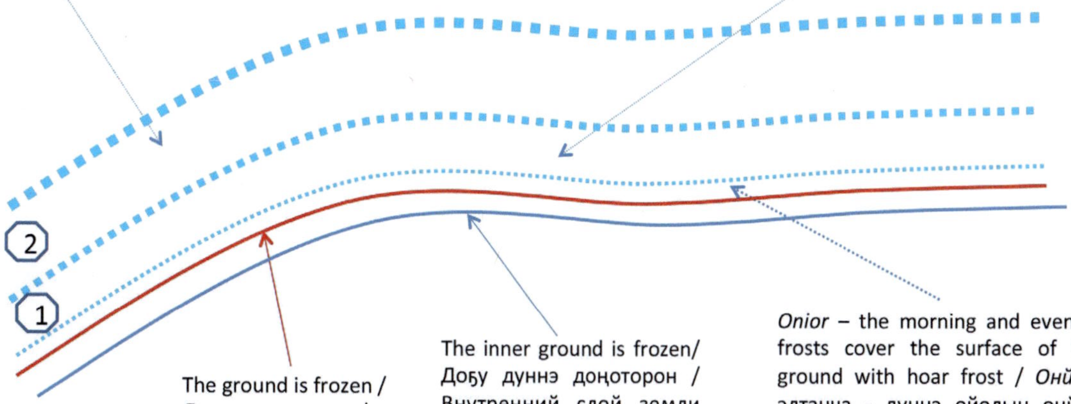

The ground is frozen / Дуннэ доӊоторон / Земля замерзла

The inner ground is frozen/ Доӻу дуннэ доӊоторон / Внутренний слой земли замерз

Onior – the morning and evening frosts cover the surface of the ground with hoar frost / *Онйор*, элтанча - дуннэ ойолын онйор одйаран / Утренний и вечерний мороз рисует иней на земле

3rd step: from (+/-) – 25, -30°C / 3 этап: от (+/-) -25, -30 C

The reindeer search for mushrooms / Олени ищут грибы и уходят далеко

Snow cover installation – Evenki knowledge of norms // <u>Before 15 November</u>

Формирование снежного покрова – Норма согласно эвенкийскому знанию // До Ноября 15ого

3 – Snowfall: soft snow (*duiukun*) falls – the sable sink into the snow and their legs cannot reach the ground. Dogs thus catch them easily / Иманна дуйукукан, андаһил айат чэпэдйэрэ, алганын дуннэла эчэл исчарэ, нинакин имакунди боконовки / Падает мягкий сухой снег *дуйукун*. Соболь хорошо проваливается, ноги не достают до земли, собаки его быстро догоняют / 20 cm

1 – The first snow that turned into *buldo* further transforms into big icy flakes, which are pressed down by the weight of the upper layers of snow / Первый слой снега, который превратился в мелкие крупинки *булдо,* переходит в стадию большие крупинки. Слой стаптывается от тяжести cm 10 (cf. diagram Buldo)

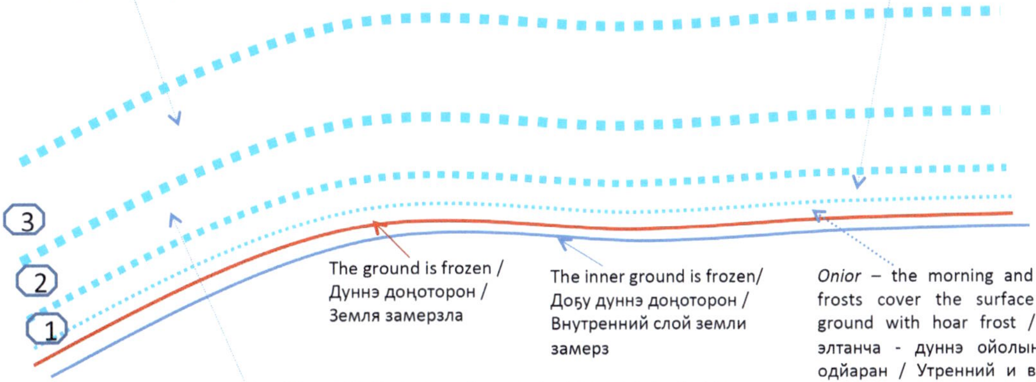

Onior – the morning and evening frosts cover the surface of the ground with hoar frost / Онйор, элтанча - дуннэ ойолын онйор одйаран / Утренний и вечерний мороз рисует иней на земле

2 – The middle layer of snow is turned into *buldo* (icy and seed-like flakes) by the frost / Средний слой снега от мороза только превращается в крупинки *булдо* / cm 10

4th step: from (+/-) – 30°C / 4 этап: от (+/-) -30 C: The reindeer stay in one place: they cannot smell mushrooms under the snow / Олени не ищут грибы, не чуют их под снегом

Snow cover installation process – Evenki knowledge Norm // 15-30 November
Формирование снежного покрова – Норма согласно эвенкийскому знанию // 15-30 ого ноября

4 – Snowfall: soft snow (*duiukun*) fall – sable sink in snow, legs do not reach the ground, dog catch sable easily / Иманна дуйукукан, андаһил айат чэпадйэрэ, алганын дуннэла эчэл исчарэ, нинакин имакунди боконовки / Падает мягкий сухой снег *дуйукун*. Соболь хорошо проваливается, ноги не достают до земли, собаки его быстро догоняют / 20 cm

1 - 2 – The first snow layers that turned into *buldo* further transforms into big icy flakes, which are pressed down by the weight of the upper layers of snow / Нижние слои снега *(эргу иманна)* сливаются в одно, превращаются в крупные льдинки (*булдо*) и стаптываются от тяжести / cm 20

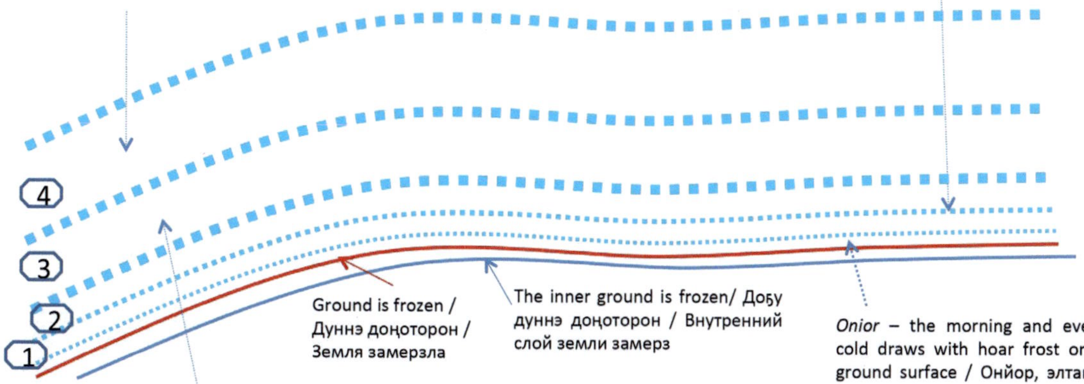

3 – Medium layer of snow is turning into icy and seed-like flakes *buldo* because of frost / Средний слой снега от мороза только превращается в крупинки *булдо* / cm 15

Onior – the morning and evening cold draws with hoar frost on the ground surface / Онйор, элтанча - дуннэ ойолын онйор одйаран / Утренний и вечерний мороз рисует иней на земле

5th step: from (+/-) – 30°C / 5 этап: от (+/-) -30 C: repeated process of *buldo* creation and increase; reindeer stay in one place / Процесс потом повторяется, *булдо иманна* все больше и больше; Олени далеко не уходят

The disappearance of the snow cover – Evenki knowledge of norms // <u>From March</u> <u>(changes from ground and air)</u>
Таяние снежного покрова – Норма согласно эвенкийскому знанию // <u>С марта (процесс изменение снизу и сверху)</u>

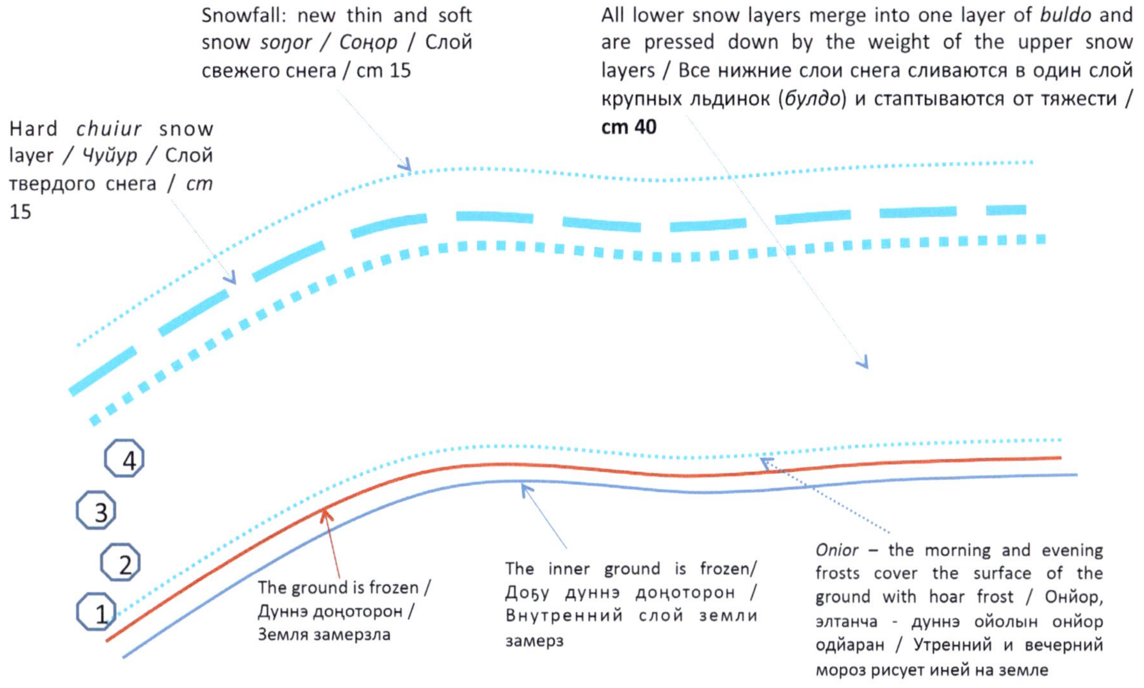

6th step: from (+/-) – 18, -20°C / 6 этап: от (+/-) – 18, -20°C

The disappearance of the snow cover– Evenki knowledge of norms // <u>April</u> <u>(changes from ground and air)</u>
Таяние снежного покрова – Норма согласно эвенкийскому знанию // <u>Апрель (процесс изменения снизу и сверху)</u>

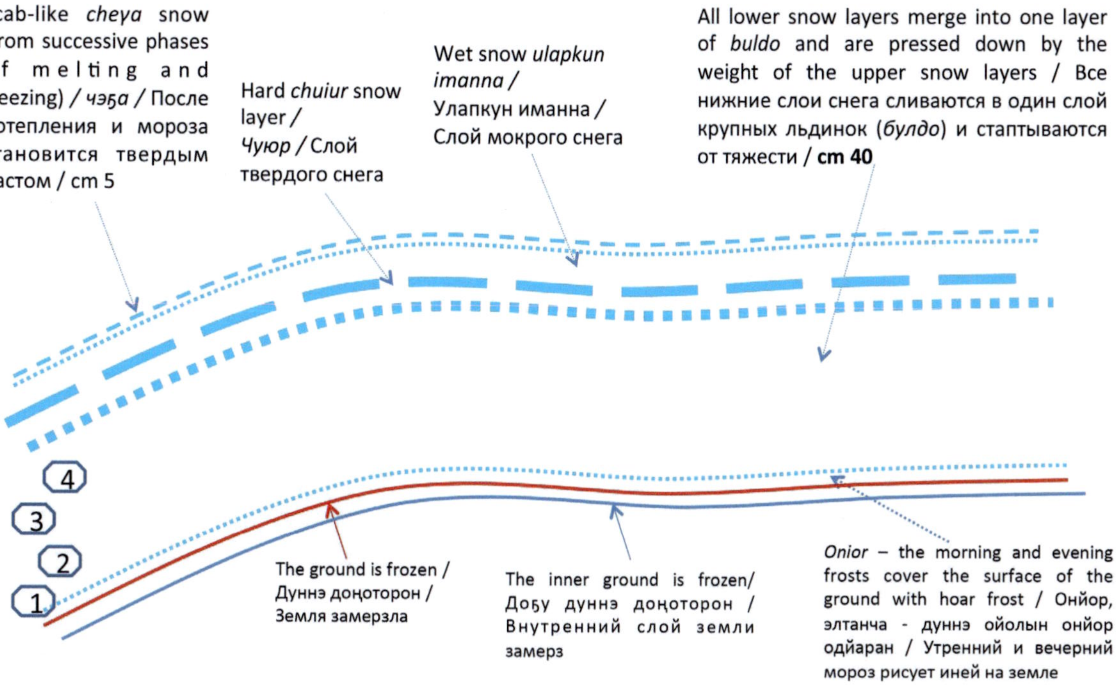

7th step: from (+/-) – 5, -9°C / 7 этап: от (+/-) – 5, -9°C: reindeer in one place

The disappearance of the snow cover – Evenki knowledge of norms // <u>from 5 May</u> <u>(changes from ground and air)</u>
Таяние снежного покрова – Норма согласно эвенкийскому знанию // <u>с мая 5 ого (процесс изменения снизу и сверху)</u>

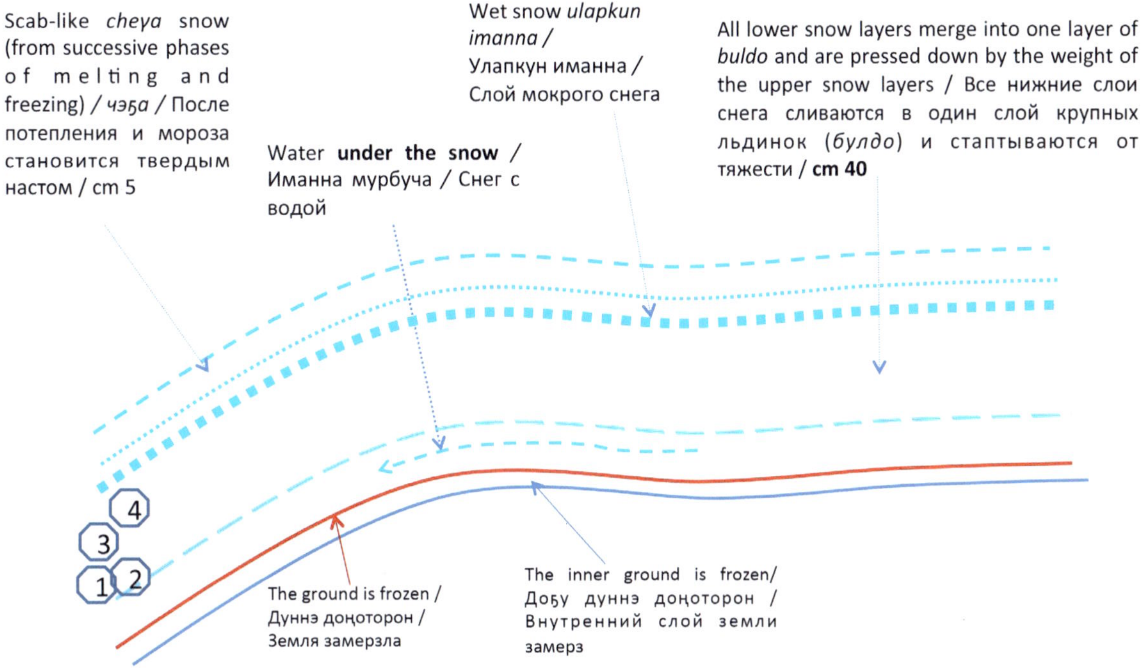

8th step: from (+/-) + 4°C / 8 этап: от (+/-) + 4°C: reindeer go away to places without snow (*iliakak*) / Илйакаҕилйа уручол орор / Олени далеко уходят туда, где снега нет

The disappearance of the snow cover – Evenki knowledge of norms // <u>from 10 May</u>
Таяние снежного покрова – Норма согласно эвенкийскому знанию // <u>с мая 10 ого</u>

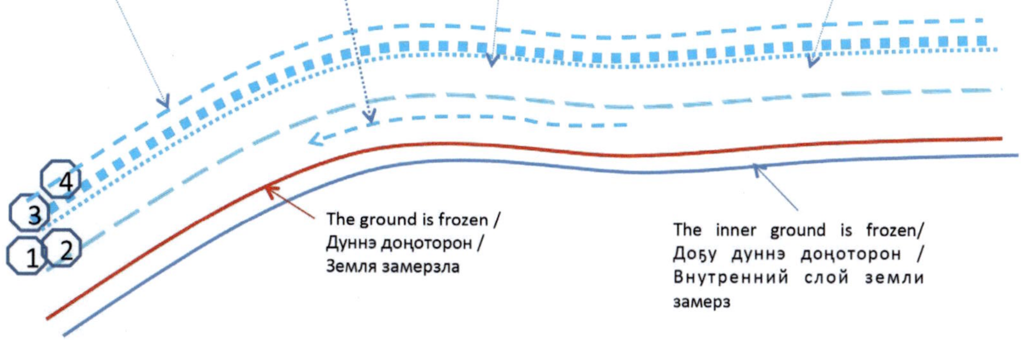

9th step: from (+/-) + 4°C +10°C / 8 этап: от (+/-) + 4°C +10°C: Underneath, water flows out from the melted snow; all the upper snow layers and types are mixed into one layer; the morning and evening frosts form a scab-like layer / Снизу вода вытекает, а сверху смешиваются все виды снега в один слой; при утреннем и вечернем морозах покрывается коркой

The disappearance of the snow cover – Evenki knowledge of norms // <u>from 15 May</u>
Таяние снежного покрова – Норма согласно эвенкийскому знанию // <u>с мая 15 ого</u>

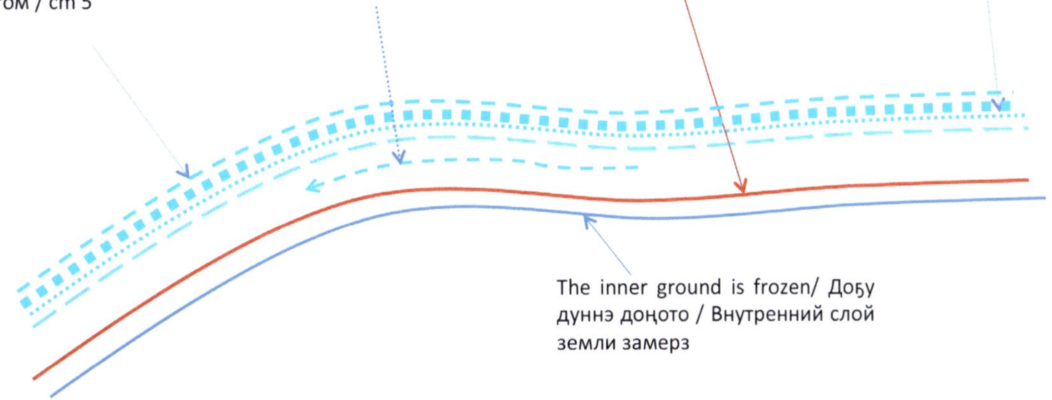

10 th step: from (+/-) + 4°C +10°C / 10 этап: от (+/-) + 4°C +10°C: Underneath, water flows and forces the mixed snow layers to melt: they turn it into water and flow away. The reindeer eat the green plants frozen from last autumn within the snow cover / Снизу вода вытекает и смесь всех видов снега оттаивает, становится водой и все утекает. Олени кушают растения, которые замерзли осенью и сохранились зелеными в снежном покрове.

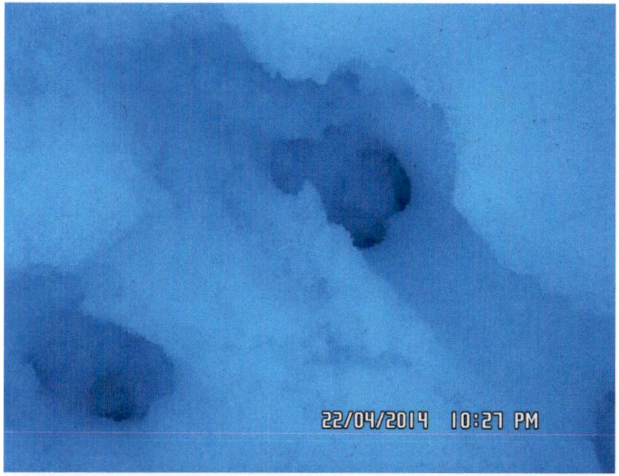

Wild reindeer migrate, a scab-like snow layer forms, water flows under the snow.

Бэйур ӈэнэдйэрэ. Иманна чэӡача, дуннэли мурбуча.

Дикие олени мигрируют, корка снега создалась, вода потекла под снегом.

<u>22 May 2014 / Superposition of several snow types:</u>

=*Cheya* /Чэӡа

= *Buldo* / Булдо

Иманна

= Water / Му

= Ground / Дуннэ

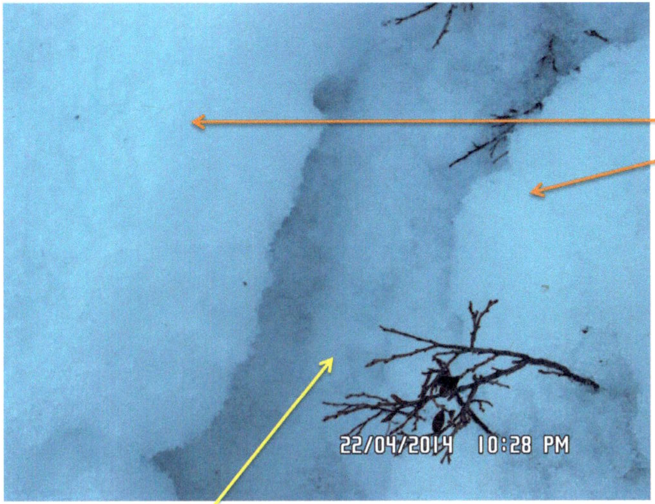

The *buldo* and *cheya* snow types coexist. First *buldo* appears, then *cheya.*

Булдо иманна и *чэҕа* рядом были. Сперва булдо потом *чеҕа* появляется.

The snow melts quicker when the water flows away / Иманна ундйэрэн мурбудйэнэ эйаиндйэрэн / Снег тает быстрее потому, что вода от оттаявшего снега уносит оставшийся снег.

In the evening, a lot of water is generated and flows away.
Кэҕа кэтэ му оча иманна унулэвки мурбулэммэн

Entire small rivers appear from the water created by the melted snow /он биракан одйаран мудук / целые ручья появляются от снега

Water flows under the snow (*murburen*).
Мурбудйэрэн.
Вода течет под снегом.

Wet snow falls (*ulapkun imanna*) 1/3 - Улапкун иманна тыкча – Мокрый снег шел 1/3

Wet snow has just fallen = it melts quickly and the water from soaks the old snow, which in turn also melts quickly.

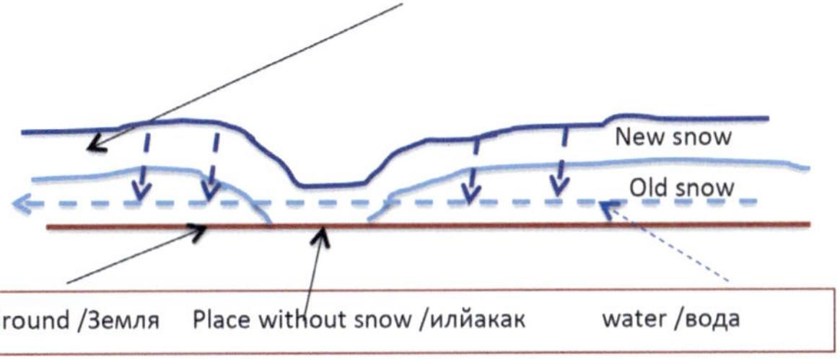

Улапкун иманна тыкча = имакунди ундиӈан. Унэллэкин му элэкэс иманна тыкчадук нонопты иманнавэ унилвканэвки.

Мокрый снег выпал = быстро тает и вода от таяния свежего снега мочит старый снег, который быстро растает.

Wet snow falls (*ulapkun imanna*) 2/3 - Улапкун иманна тыкча – Мокрый снег шел 2/3

Puddles from the melted snow (*tain*) = as the snow starts to melt, puddles are created; these freeze during the morning and evening frosts.

Таин = иманна мубуллэкин, таин овки, элтанэкин доӈотовки.

Лужи *таин* создаются как только снег начинает таять. Замерзают при утреннем и вечернем морозах

Wet snow
Улапкун иманна
Мокрый снег

Wet snow falls (*ulapkun imanna*) 3/3 - Улапкун иманна тыкча – Мокрый снег шел 3/3

Uncomfortable wet snow – even dogs do not like sitting on it / Собакам не сидится на мокром снеге / Нинакир эвкил тэӡэттэ улапкун иманнаду.

The wet snow sticks to trees and bushes
Мокрый снег прилипает к деревьям и кустам

The snow on the branches melts quicker; it drips from the branches to the roots of the tree. The snow on the ground melts quicker as a result, especially on the southern side.

Иманна ундйэрэкин ирйактал гарадуктын намарача иманна имакунди унилэвки; ундйэнэ мурбуӡа чургидйэвки, тар ирйактал тэкэндун имакунди ундйэвки иманна. Северни олдонду иманнакакун.

Как только снег начинает таять, сперва с ветвей тает, оттуда капает на корни деревьев и там на земле быстрее тает, особенно с южной стороны.

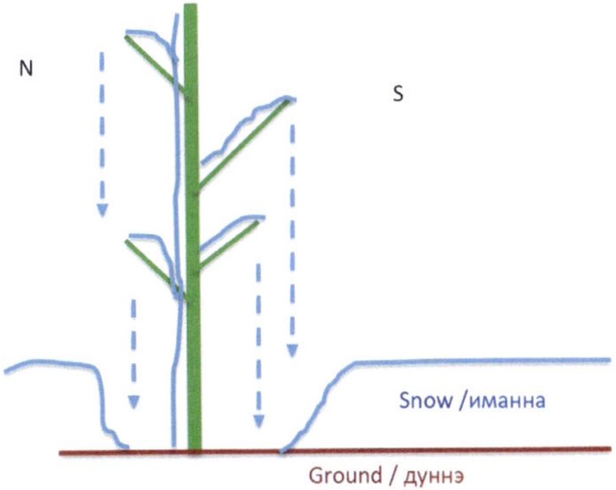

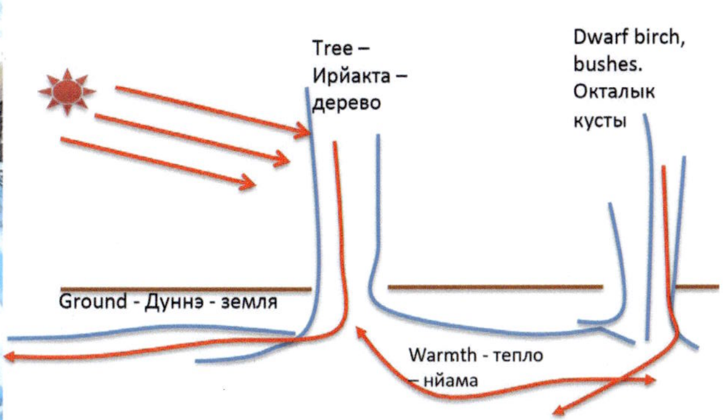

The snow melts faster along the roots of trees and bushes due to the sun: the ground along the root becomes visible. Heat travels through the ground.

Иманна ундйэрэн ирйактал, октакар тэкэрдули (нйамава танэвкил дылачадук) = дуннэ ичэвулэвки. Нйама дуннэдолан ӈэнэрэн.

Снег тает быстрее вдоль корней деревьев и кустов (они втягивают тепло от солнца) = земля виднеется. Тепло идет изнутри земли.

The large river basin thaws at a faster rate because of the snow on the stones: as they absorb the warmth from the sun, the snow there melts faster.

Амннунна дйолол ундйэрэ - амнунна ундйэрэн имакунди.

Широкий бассейн реки быстрее тает из-за камней, на которых снег тает быстрее из-за того что камни теплоту солнца притягивают.

Stony places, taking warmth from the sun, force the snow covering them to melt.

Дйологил ундйэрэ дылачадук нйамалдйэнэ.

Каменистые места тают от теплоты солнца

Places without snow (*iliakak*) / Илйакаӄил / Места где снег растаял

Several *Iliakak* / Илйакаӄил

Reindeer graze in places without snow (*iliakak*). Just as the first spring starts, people nomadise to places without snow so that the reindeer can graze well. Reindeer know where these places are and go there independently .

Орор оӈкодйэвкил илйакаӄилду. Нйэӈнэ элэкэс оча, илйакаӄилтыки нулгивкил бэйэл, орор айат оӈкодотын. Орор мартын илйакаӄилтыки урувкил.

Олени кормуются на местах без снега (*илйакак*). В начале весны, люди кочуют в место илйакак, чтобы олени хорошо кормовались. Олени сами тянутся к *илйакак*.

On the slopes of mountains and hills, there are places without snow – reindeer go to them independently to graze.

Эллэӈэду илйакакча – орор мартын эллэӈэду оӈкодйэвкил.

На склонах полностью стало без снега – олени туда сами ходят кормоваться.

8 May /мая 2014
When snow melts intensively, it creates a small flooding.

Иманна тэтэкунди ундйэрэкин – бира мудакуча.

Когда снег сильно тает, образуются небольшие наводнения.

13 May / мая 2014
The river level declined, the ground dried out, and the snow disappeared: leaves started to apear.

Бира арбача, дуннэ олгочо, иманна ачэн унча, абданна элэкэс йулдйэрэ.

Река спала, земля высохла, снега нету, листья только распускаются.

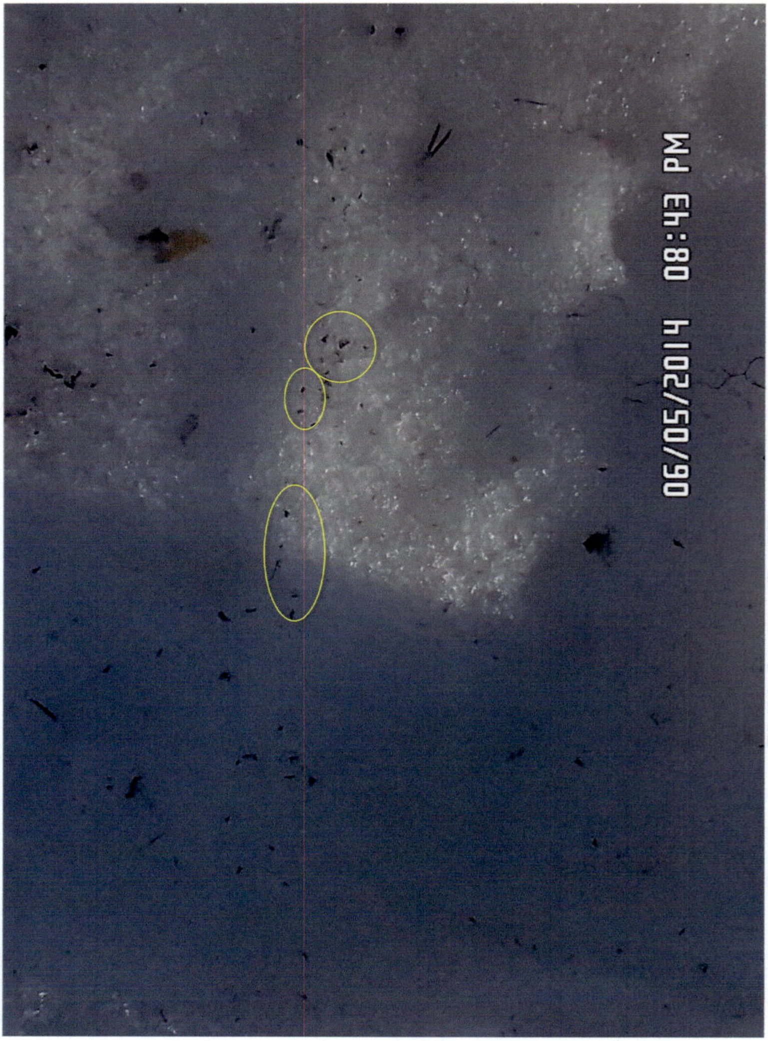

Small larva from various insect species wake up from hibernation and **eat the snow.**
Куликуар мйэлчал иманна тутуичал иманнавэ дйэпчэммэн.
Личинки разных насекомых просыпаются, расползаются по снегу и кушают снег.

***Cheya* snow** (a scab-like layer), formed by the morning and evening frosts, can support a man (70-80 kg): water flows underneath.

ЧЕЂА = кеђа чеђавки элтанэкин бэйэвэ дэбдэвки (70-80 кг) иманна ойолын.

Снег *чэђа* формируется вечером при заморозках и выдерживают вес человека (70-80 кг) по верху снега, пока вода течет под ним.

© S. Gabyshev, A. Lavrillier 2013

Deep snow as a herder / Глубокий снег как пастух

© A. Lavrillier, 2015

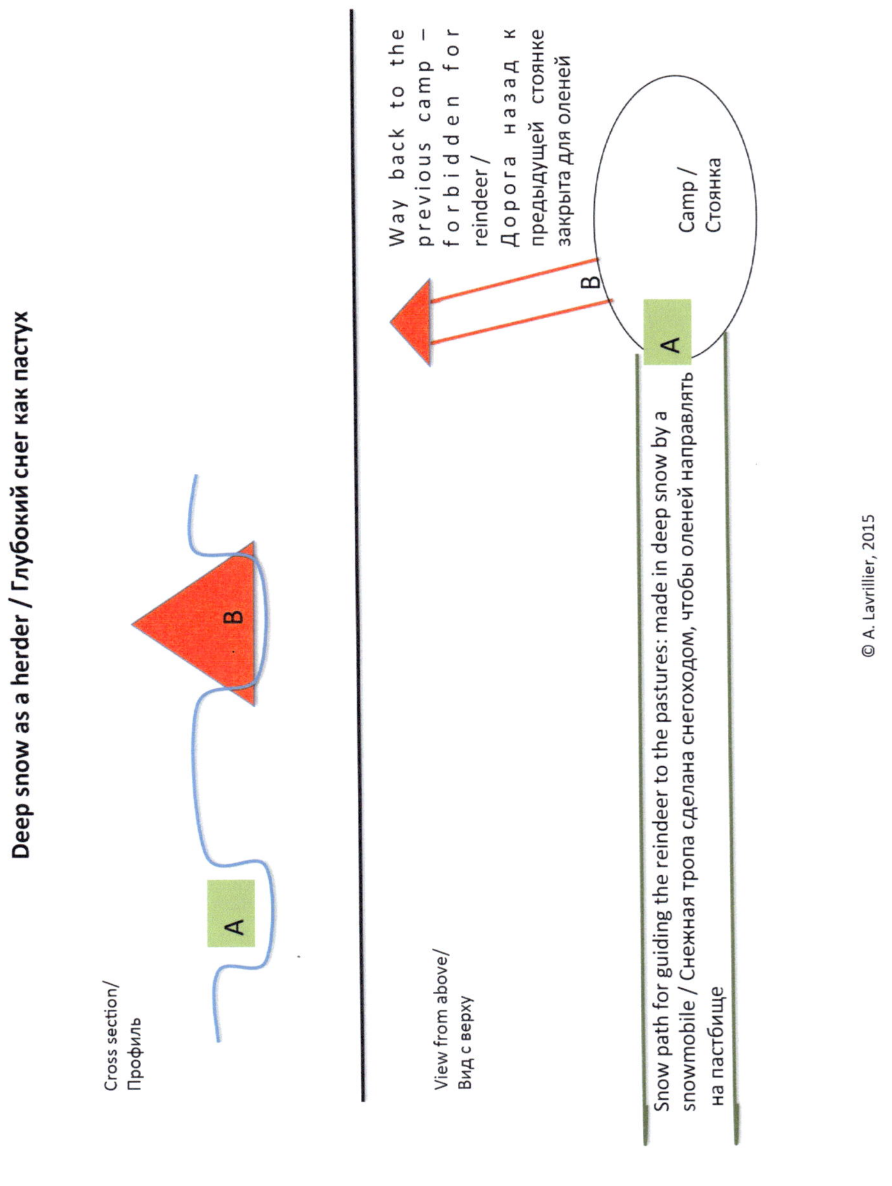
Deep snow as a herder / Глубокий снег как пастух
Cross section/
Профиль
A
B
View from above/
Вид с верху
Way back to the previous camp – forbidden for reindeer / Дорога назад к предыдущей стоянке закрыта для оленей
B
Camp / Стоянка
A
Snow path for guiding the reindeer to the pastures: made in deep snow by a snowmobile / Снежная тропа сделана снегоходом, чтобы оленей направлять на пастбище
© A. Lavrillier, 2015

ULAN (-) – SIRIKTE (Fissure in the ice) – BUKTE (Icing blister)
УЛАН (Наледь) – СИРИКТЭ (Трещина во льду) – БУКТЭ (Наледный бугор пучения)
Norm for the formation of *ulan* on large rivers / Норма формирования *улана* на больших реках

1 *Univun* appear downstream where there is a strong current. On river banks, the deep layer of snow does not allow the ice layer to become thicker or even to freeze at all.

Унивур эйар бивкил. Бира дйапакалын нэмкукар дйукэ бивки. Бира дйапкалын иманна суӊтатмар бивки – эвки доӊкотонирэ иманна дйукэвэ.

Унивун появляются внизу по течении реки, где сильное течение. По берегам глубокий слой снега не дает льду формировать толстый слой или не позволяет вообще замерзать.

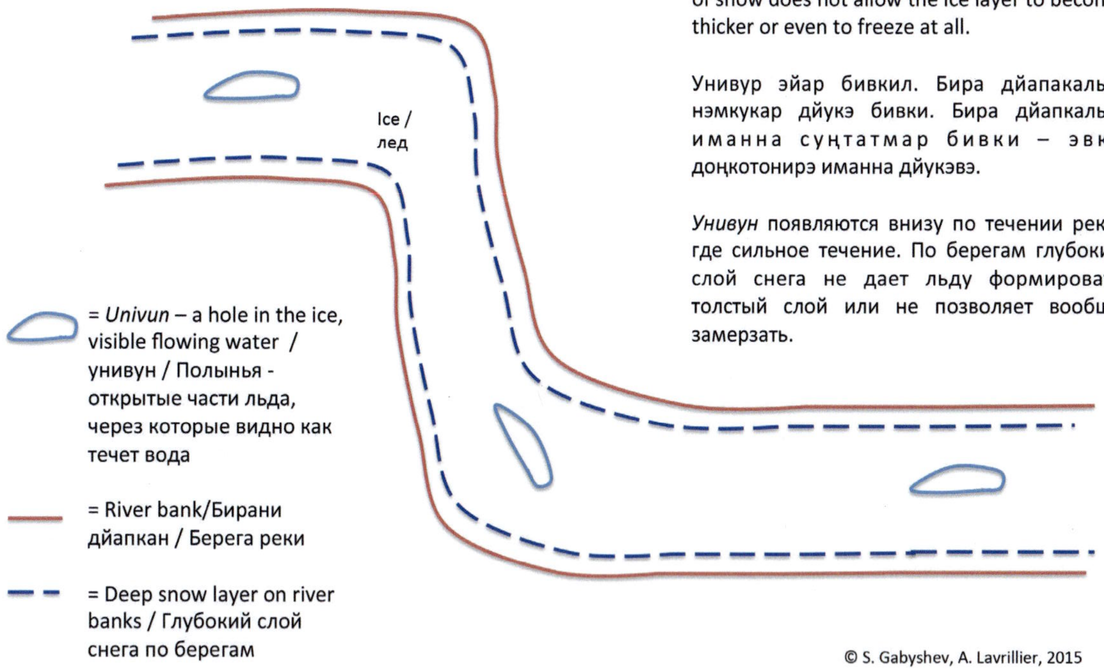

ULAN (-) – SIRIKTE (Fissure in the ice) – BUKTE (Icing blister)

УЛАН (Наледь) – СИРИКТЭ (Трещина во льду) – БУКТЭ (Наледный бугор пучения)

Norm for the formation of *ulan* on large rivers / Норма формирования *улана* на больших реках

= River bank / Бирани дйапкан / Берега реки

= Deep snow layer on river banks / Глубокий слой снега по берегам

= Closed *univun* – a hole in the ice. There is usually visible flowing water, but it is sealed by ice / Унивур самипчал/Полынья - закрылись льдом.

= *Ulan* – water flows out from the ice / *улан* – му йудйэпки дйапкалдук / *Улан* –наледь вода выходит

= Steep river banks *emker*/ эмкэр / Крутой берег реки с обрывом

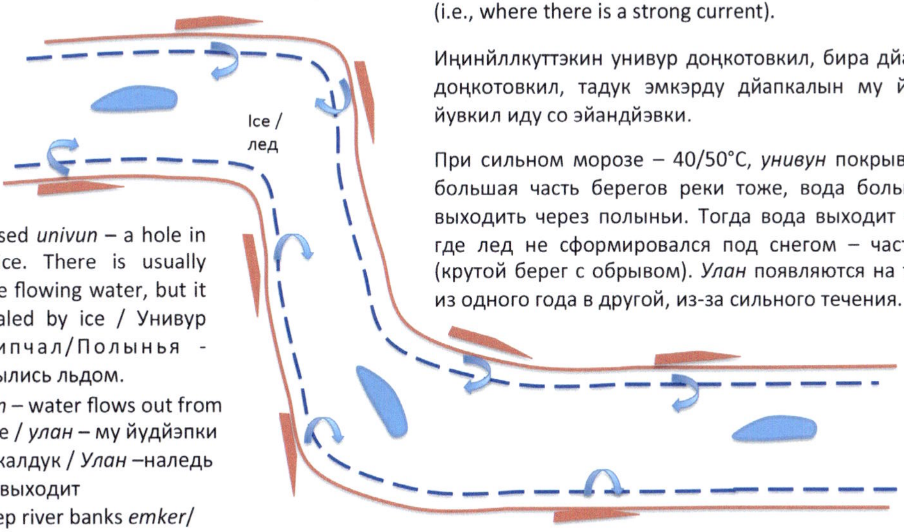

2 During harsh cold spells (-40/50°C), *univun* and most river banks are sealed by ice: the water can no longer flow through these. The water flows through the parts of the river banks where ice did not form because of the deep snow layer (which provides thermal insulation). This happens mostly regularly under steep *emker* river banks. The *ulan* appears each year in the same place (i.e., where there is a strong current).

Иӈинйллкуттэкин унивур доӈкотовкил, бира дйапкалын нйан доӈкотовкил, тадук эмкэрду дйапкалын му йулэвки. Улар йувкил иду со эйандйэвки.

При сильном морозе – 40/50°C, *унивун* покрываются льдом, большая часть берегов реки тоже, вода больше не может выходить через полыньи. Тогда вода выходит через берега, где лед не сформировался под снегом – часто под *эмкэр* (крутой берег с обрывом). *Улан* появляются на тех же местах из одного года в другой, из-за сильного течения.

Ulan – water flows out from the ice / Улан

© A. Lavrillier, 2016

ULAN (-) – SIRIKTE (fissure in the ice) – BUKTE (icing blister)

УЛАН (Наледь) – СИРИКТЭ (Трещина во льду) – БУКТЭ (Наледный бугор пучения)

Norm for the formation of *ulan* on large rivers / Норма формирования *улана* на больших реках

Cross section / Разрез

1

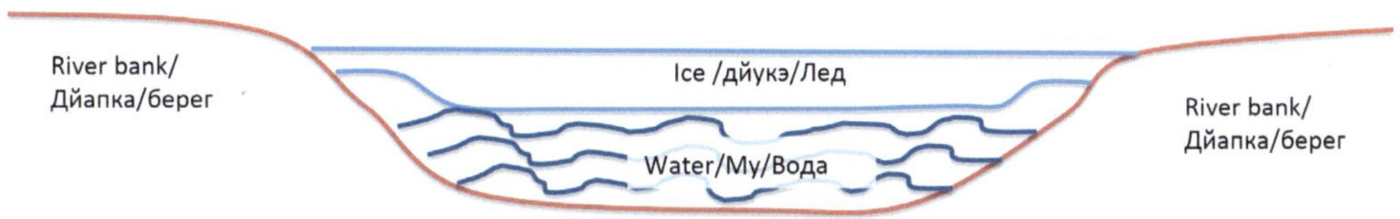

To the end of November-beginning of December: On the river banks, the ice is thin, while in the middle the ice is very thick (cf. Profile view here below).

До конца ноября – начала декабря: Бирани дйапкалын дйукэ нэмкукан бивки. Дулинду дйукэ дэрамкун бивки. По берегам реки лед тонким бывает. Посередине лед толстый. См. Схема с профили здесь ниже.

2

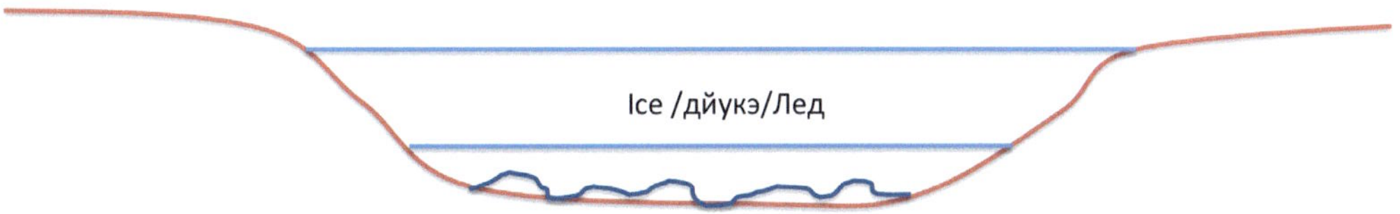

To the end of November-beginning of December: On the river banks, the ice becomes very thick: the water under the ice has significantly decreased (cf. Profile view here below).

До конца ноября – начала декабря: Тадук бирани дйапкалын дйукэ дэрамкун овки иңинйэллэкин, му арбавки / Потом по берегам реки лед толстый становится. Уровень воды падает и мало остается воды (См. Профиль здесь ниже).

ULAN – SIRIKTE – BUKTE / УЛАН – СИРИКТЭ – БУКТЭ
Norm for the formation of *bukte* from *sirikte* on large rivers /
Норма формирования *буктэ* от *сирикт*э на больших реках

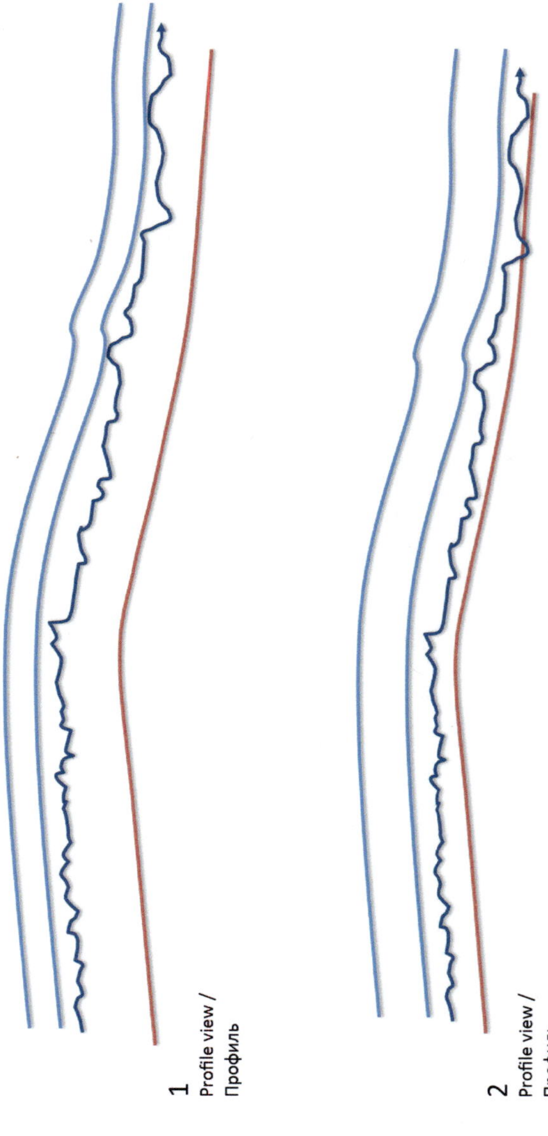

ULAN – SIRIKTE – BUKTE / УЛАН – СИРИКТЭ – БУКТЭ

Norm for the formation of ulan on large rivers / Норма формирования *улана* на больших реках

Cross section / Разрез

3

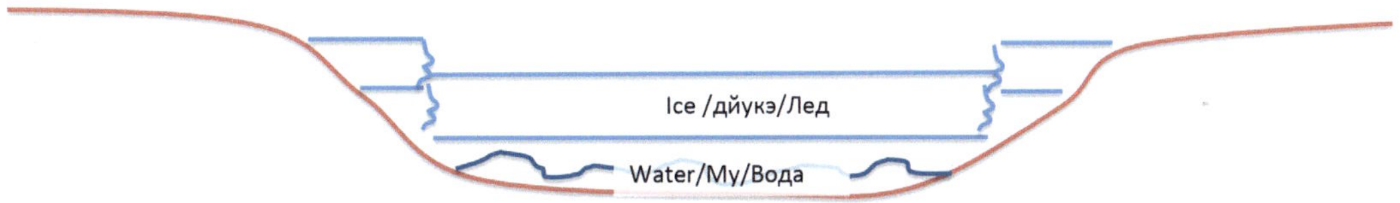

Beginning of December: The ice on the river cracks near the bank and collapses because of the empty spaces within or under the ice layer. This happens because of the decreasing level of the water.

Начало декабря: Дйукэ бираду йолдорговки дйапакалын дулгу дйука тыкивки мудалан. Бира арбавки – дйукэ тыкивки мула.

Лед на реках трескается по берегам и падает из-за пустоледа, оставленного между водой и льдом при замерзании воды.

4

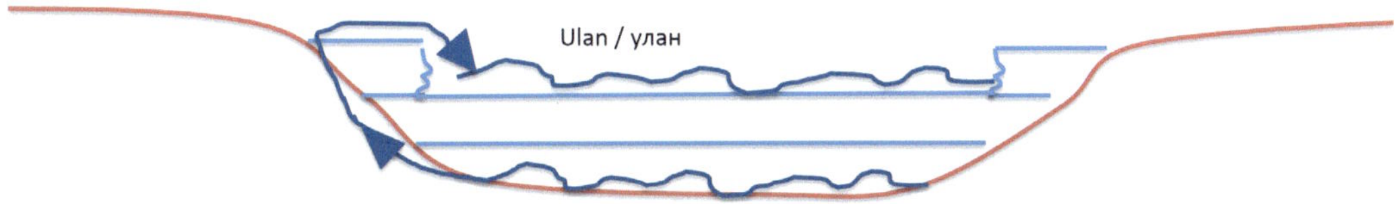

December: Along the river banks, the water flows out and creates *ulan*. *Ulan* appear only during harsh colds (-40 °/-50°C).

Декабря: Бира дйапкалдулы му йулэвки, тадук уланэлэвки. Уланэвки иӊинилкутэкин.

Вода выходит через берега и создается *улан*. *Улан* создается только при сильных морозах - 40 °C -50°C.

ULAN – SIRIKTE – BUKTE / УЛАН – СИРИКТЭ – БУКТЭ

Norm for the formation of ***Sirikte*** on large rivers / Норма формирования *сириктэ* на больших реках

Cross section / Разрез

5

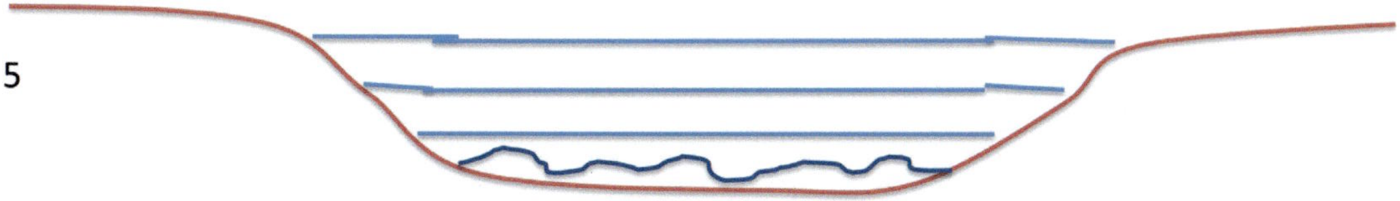

Mid December: All the *ulan* freeze entirely on the river banks and in the middle. The ice becomes very thick, even on the banks of the rivers.

Середина декабря: Улан упкатын доӈкотовки, бира дйапкалын дйукэ дэрамкун овки.

Улан полностью замерзает по берегам и середине реки, лед становится толстым везде и по берегам.

6

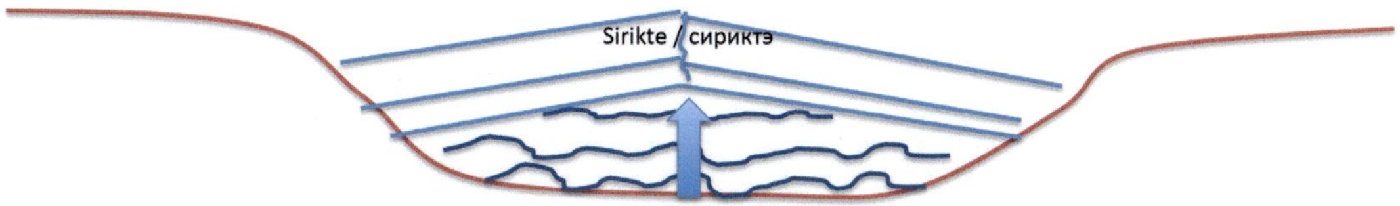

End of December: When the *ulan* freeze near shallow areas of the river, the water and air enclosed under the ice pushes up against the ice, which cracks. It thus becomes a *sirikte*. *Sirikte* appear only during harsh cold.

Конец декабря: Иду эйан бивки, улан доӈкотодйэнэ, му воздухнун самдйэвки дйукэ эрэдун, воздух мунйун анакучэвки дйукэвэ уҕиски– дйукэ йолдорговки - сириктэ овки. Сириктэвки иӈинилкутэкин.

Перед перекатом *улан* замерзает и при этом вода с воздухом, замкнутые подо льдом, заставляют лед трескаться и, таким образом, создается *сириктэ* (трещина в льду). *Сириктэ* создается только при сильном морозе.

ULAN – SIRIKTE – BUKTE / УЛАН – СИРИКТЭ – БУКТЭ

Norm for the formation of ***bukte*** (icing blister) from *sirikte* on large rivers /

Норма формирования *буктэ* (наледный бугор пучения) от *сириктэ* на больших реках

Cross section / Разрез

7

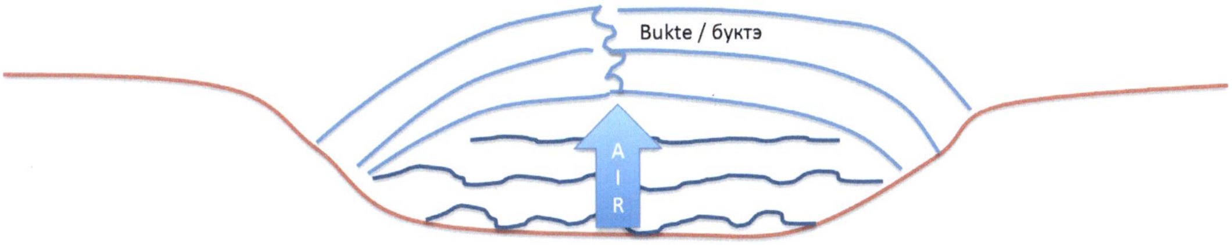

January: Close to the shallow areas of the river, the water and air push up against the ice and create a *sirikte*, from which appears a *bukte*. *Bukte* appear only during harsh cold (Cf. Profile view here below).

Январь: Иду эйан бивки, му воздухнун анакучэвкил дйукэвэ уӻиски, тадук сириктэдук – дйукэ мукчэргэвки – тыка буктэ овки. Буктэдйэвки иӊинилкутэкин.

Перед перекатом вода с воздухом толкают лед вверх и создается *сириктэ* (трещина в льду). От *сириктэ* лед надувается. *Буктэ* формируются при сильных морозах (См. Профиль здесь ниже).

ULAN – SIRIKTE – BUKTE / УЛАН – СИРИКТЭ – БУКТЭ
Norm for the formation of ***bukte*** (icing blister) from *sirikte* on large rivers /
Норма формирования *буктэ* (наледный бугор пучения) от *сириктэ* на больших реках

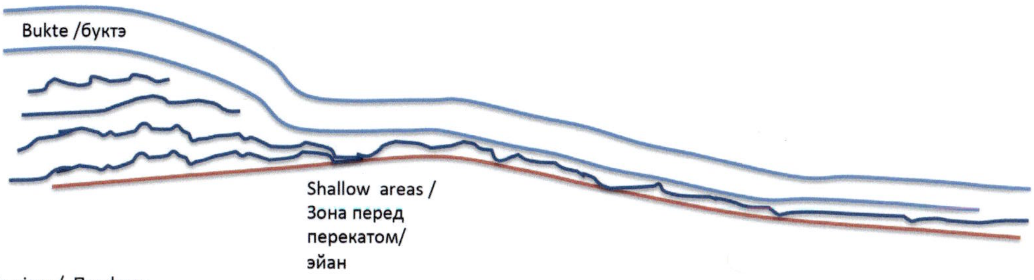

7

Profile view / Профиль

<u>January:</u> Close to shallow areas in the river, the ice narrows the passage used by the flowing water. This creates pressure, which in turn forms a *sirikte*. Afterwards, the ice rounds out and forms a *bukte* (Cf. Cross section here above).

<u>Январь</u>: Иду эйан бивки, дйукэ тырэвки (придавляет) мувэ эрэлан (дно), тадук му воздухнун анакучэвки дйукэвэ – сириктэ овки, тадук нйан анакучэвки - дйукэ мукчэргэвки – буктэ овки.

Перед перекатом лед ужимает воду с воздухом до / почти до/ дна реки. Это создает давление, при которой образуется *сириктэ*, потом лед надувается, создавая *буктэ* (См. Разрез здесь выше)

ULAN – SIRIKTE – BUKTE / УЛАН – СИРИКТЭ – БУКТЭ

Norm for the formation of *bukte* (icing blister) from *sirikte* on large rivers /

Норма формирования *буктэ* (наледный бугор пучения) от *сириктэ* на больших реках

Cross section / Разрез

8

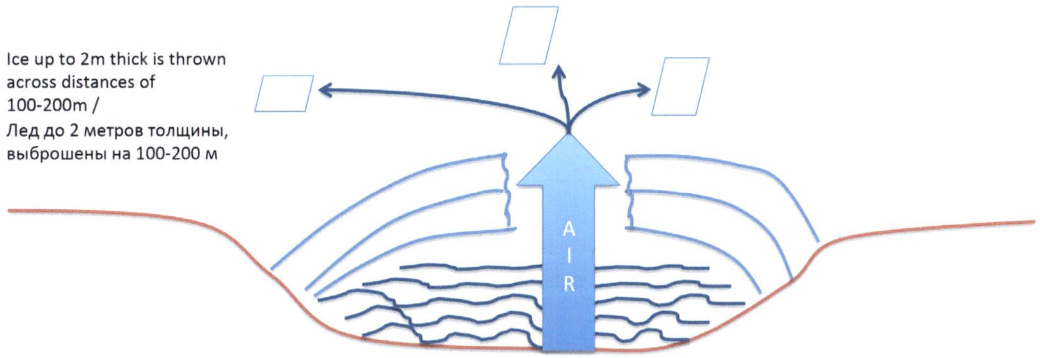

<u>End of January:</u> Near the shallow places of the river, water and air push up against the ice: they explode outwards, throwing large pieces of ice (up to 2m thick) across distances of 100-200 m.

<u>Конец Января:</u> Иду эйан бивки, му воздухнун анакучэвкил дйукэвэ уӈиски, тадук му воздухнун улталэвки дйукэвэ, дйукэкурбэ (2 метровыйкурбэ) на 200 – 300 м. Ноддавки.

Перед перекатом под напором воды с воздухом *буктэ* взрываются и раскидывают большие льдины (до 2 м.) на 100-200 м. *См. схема с профили ниже*.

ULAN – SIRIKTE – BUKTE / УЛАН – СИРИКТЭ – БУКТЭ

Norm for the formation of ***bukte*** (icing blister) from *sirikte* on large rivers /
Норма формирования *буктэ* (наледный бугор пучения) от *сириктэ* на больших реках

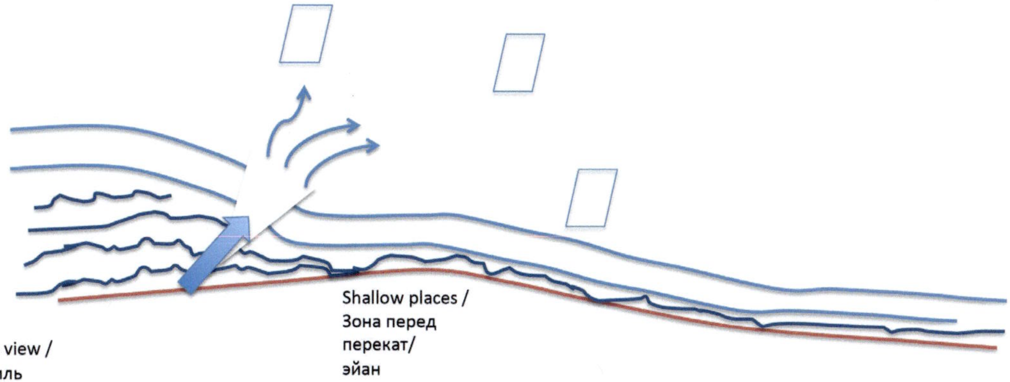

8
Profile view /
Профиль

<u>End of January:</u> Near the shallow places of the river, water and air push up against the ice: they explode outwards, throwing large pieces of ice (up to 2m thick) across distances of 100-200 m.

<u>Конец января</u> = Иду эйан бивки, му воздухнун анакучэвкил дйукэвэ уҕиски, тадук му воздухнун улталэвки дйукэвэ, дйукакурбэ (2 метровыйкурбэ) на 100 – 200 м. Ноддавки.

Перед перекатом под напором воды с воздухом *буктэ* взрываются и раскидают большие льдины (до 2 м.) на 100-200 м (См. Разрез выше)

ULAN – SIRIKTE – BUKTE / УЛАН – СИРИКТЭ – БУКТЭ
Norm for the formation of ***bukte*** (icing blister) from *sirikte* on large rivers /
Норма формирования *буктэ* (наледный бугор пучения) от *сириктэ* на больших реках

A bukte has exploded/Буктэултаргэча / Буктэ взорвался
(Cf. diagram Ice anomalies)
30 January / января 2015 - Умуксэ дапту даҕадун

ULAN – SIRIKTE – BUKTE / УЛАН – СИРИКТЭ – БУКТЭ
Norm for the formation of ***bukte*** from *sirikte* on large rivers /
Норма формирования *буктэ* (наледный бугор пучения) от *сириктэ* на больших реках

The first explosion occurred on 29 January during the day and threw big pieces of ice downstream. The next day, this ice went under water. The second explosion occurred on the night of 29 January or on the morning of 30, throwing new pieces of ice around.

Первый взрыв случился 29-ого января днем, и куски льда были раскиданы вниз. На следующий день они оказались под водой. Второй взрыв был 29-ого ночью или 30-ого утром и раскидал новые льды.

ULAN – SIRIKTE – BUKTE / УЛАН – СИРИКТЭ – БУКТЭ
Norm of formation ***bukte*** from *sirikte* on big rivers /
Норма формирования *буктэ* (наледный бугор пучения) от *сириктэ* на больших реках

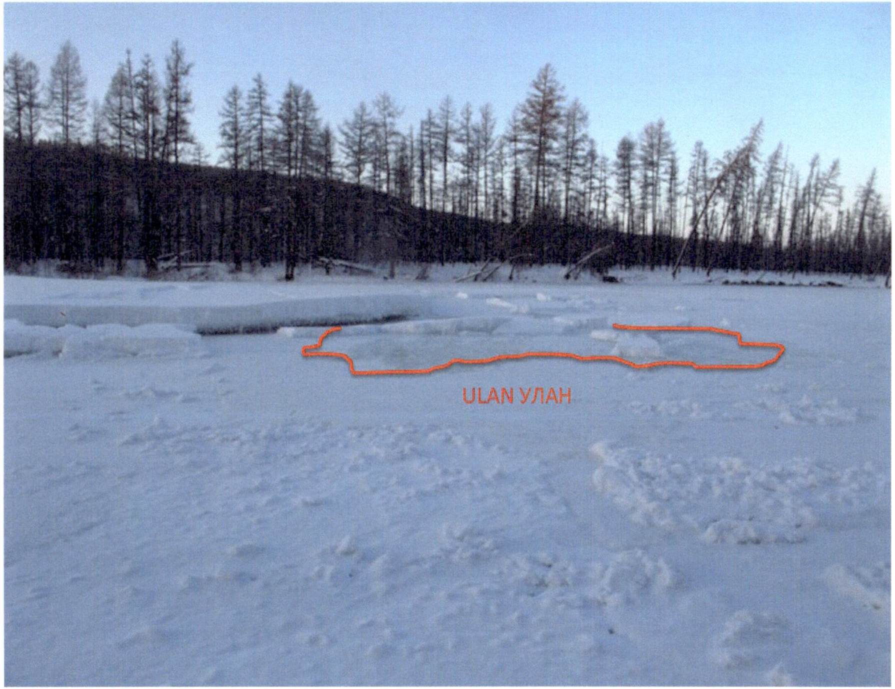

After the explosion, a lot of water flowed out from the under the ice, creating a *ulan* and then freezing / Ултаргэча, мукун йуча, тадук уланылча доңкотоммэн/ После взрыва много воды вышло, новый *улан* образовался и замерз.

ULAN – SIRIKTE – BUKTE / УЛАН – СИРИКТЭ – БУКТЭ
Norm for the formation of ***bukte*** from sirikte on large rivers /
Норма формирования *буктэ* (наледный бугор пучения) от *сирикmэ* на больших реках

Water protrudes from the river and flows according to the current / Бира додукин му йудйэрэн, эйандйэрэн/
Из вне речки выходит вод, течет.

A large **bukte** (icing blister) / һэгдыкун буктэ / Большой буктэ (наледный бугор пучения-

© A. Lavrillier, S. Gabyshev, 2015

Camp Ahikte / Imakta
Ulan / Улан
A large steep river bank (*emker*) / hэгдыкун эмкэр/ Большой крутой берег реки с обрывом (эмкэр)
Camp Tabakerka
Current - Эйан - течение
Cf. См. Part III – Ideal camp
© A. Lavrillier, S. Gabyshev, 2015

***Ulan* – for drinking water** / *Улан* – для питьевой воды

ULAN BUKTE - 2016 Anomaly - A lot of rivers did not produce *bukte* but made too many *ulan*/ Кэтэ бирал эчил буктэрэ, уландйэвкил / Аномалия – Многие реки совсем не создали наледные бугры пучения (*буктэ*), зато много было *улан*.

The Evenki always check the depth of an *ulan*. They verify with a knife or axe the solidity and depth of the surface layers of ice. First, they walk on it for verification and then decide whether they can go along it on their vehicles. By reindeer, it is almost always possible to move along the surface, but it is dangerous to travel by snowmobile because it might sink into the ice and water (cf. next picture).

Эвенки всегда проверяют глубину *улан.* Проверяют ножом или топором крепость льда и глубину под первым слоем льда. После обследования дороги пешком, решают: можно проехать или нет. Оленями почти всегда можно проехать без проблем, а на снегоходе это очень опасно, потому что снегоход может провалиться или утонуть.

Large *ulan* can potentially be deep. After verification, nomads prefer to make a new road with snowmobile on the right-hand river bank. There are always *ulan* in such places.

Большой *улан*, потенциально глубокий. После обследования дороги пешком предпочитают делать новую дорогу снегоходом по левому берегу. На этом месте всегда бывают *улан*.

Consequences of *ulan* – transport problems / Последствия от *улан* – проблема транспорта

© A. Lavrillier, S. Gabyshev, 2015

3 OBSERVING AND PREDICTING NORMS, ANOMALIES, AND TRANSFORMATION

© A. Lavrillier (top), © V. Gabyshev (below)

3.1 Landscape components for an ideal camp

Topographic types needed for a good camp / Топографические типы нужны для хорошей стоянки

- *Ulan:* close for water / Улан даҕадун – мулакит / Улан рядом для воды
- A river for fishing and providing horsetail for grazing / Река для рыбалки и на пастбище есть хвощ
- *Emker* (steep river bank) are suitable places for setting up camp due to the *keteme* (dry and hard ground) / *Эмкэр* хорошо для установки стоянок, там твердая сухая земля *кэтэмэ*
- *Elleŋe* slope – the reindeer graze well there due to the snow conditions / *Эллэӈэ* склон горы – олени кормуются хорошо (мягкий снежный покров)
- *Kerain* – on the top, the snow is shallow and good for grazing; there are also birch trees for sledges and other items / На вершине *кэраина* мелкий снег и хороший свежий ягель; есть крепкие березы для нарт и других изделий (cf. Topographic typology)
- *Nepteke*: Reindeer find this a comfortable place to lie down at night / *Нэптэкэ*: олени хорошо ночуют на этом ровном месте.
- In a landscape like a *kapchan*, one can find *makchika* / Макчикакакун таду / На таком ландшафте есть *макчика*
- *Oio*: a good place for hunting and reindeer herding / *Ойо*: места для охоты на диких оленей

Табакерка урикит айа: Орор оӊкодйэвкил эллэӈэду, капчарду макчикар, кэраин ойодун нэптэкэ бивки иманна арбркукан таду олыһин орор айат оӊковкил. Сивак нйан биһин. Таду упкатын биһин. Таду айа бэйуктэдйэми.

Улан – муликит - Ulan - Drinking water

Camp Tabakerka

3.2 Snow road construction and anomalies

Snow roads – a real construction / Снежная дорога - Настоящая конструкция

Snow road construction: Snow roads with frozen and hardened foundations are crucial for the very survival of tł nomads. They 'construct' it by going along the same path repeatedly from the very first installation of the snow cover the autumn and throughout the snowy period. Thanks to the physics of the snow and the evening and morning fros (*eltan*), which are used as 'tools', the snow layer on the road freezes and hardens every time someone go along i creating a multi-layered snow cover. On the banks of the road, the snow is soft: one can sink deep.

Конструкция снежной дороги: Снежные дороги с замерзшей и твердой основой являются важнейшими д/ выживания кочевников. Поэтому эвенки «строят» эти дороги, проезжая всегда по одним и тем же тропам г несколькy раз с начала установления снежного покрова осенью до конца снежного периода. Благодаря физи снега и вечерним и утренним морозам (*элтан*), которые применяются как «инструмент» строения, слои снега н дороге твердеют и замерзают при каждом проезде кочевников – этим способом создается многослойнь снежный покров на месте дороги. Но по бокам дорог снег мягкий и глубокий и можно легко провалиться н нартах или снегоходах.

© A. Lavrillier, 2016

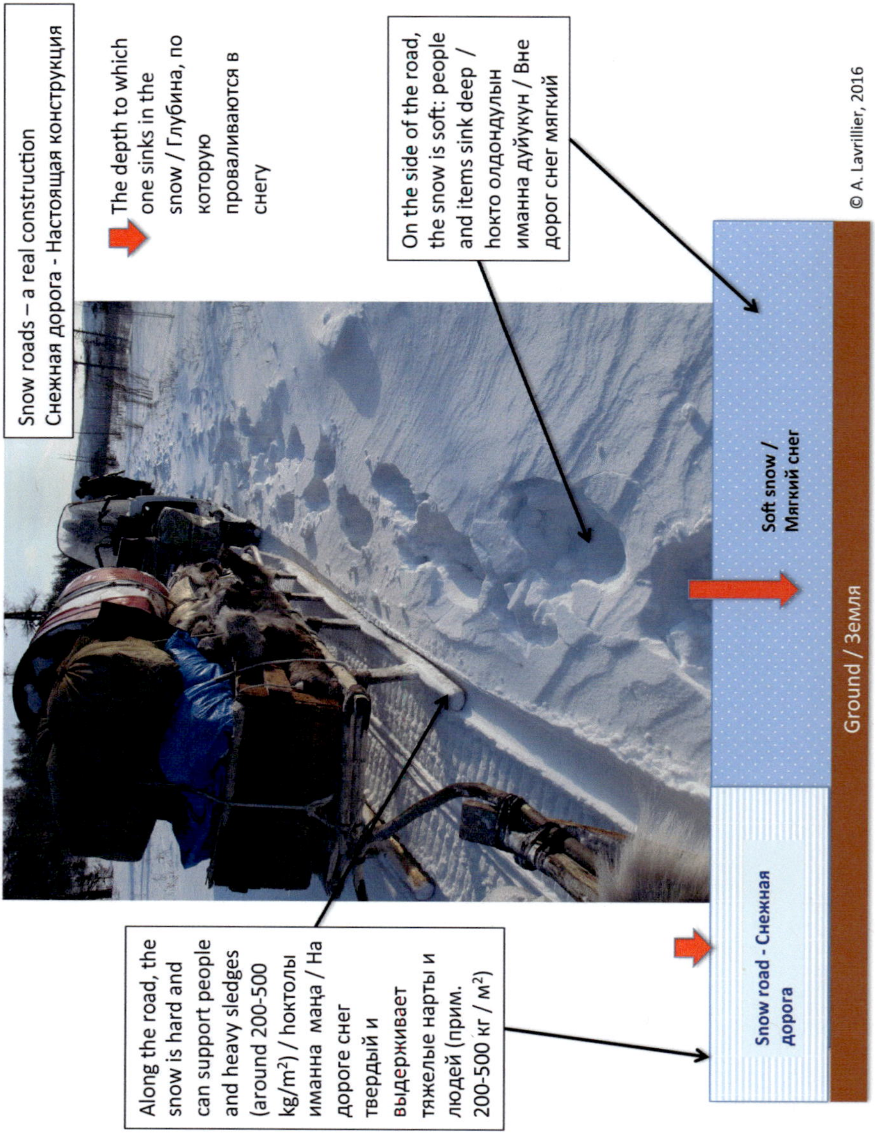
Snow roads – a real construction
Снежная дорога - Настоящая конструкция
The depth to which one sinks in the snow / Глубина, по которую проваливаются в снегу
On the side of the road, the snow is soft: people and items sink deep / нокто олдондулын иманна дуйукун / Вне дорог снег мягкий
Along the road, the snow is hard and can support people and heavy sledges (around 200-500 kg/m²) / ноктолы иманна маңа / На дороге снег твердый и выдерживает тяжелые нарты и людей (прим. 200-500 кг / м²)
Snow road - Снежная дорога
Soft snow / Мягкий снег
Ground / Земля
© A. Lavrillier, 2016

Snow path – **dealing with deep and soft snow** /
Снежная тропинка – разобраться с глубоким и мягким снегом

© A. Lavrillier, 2016

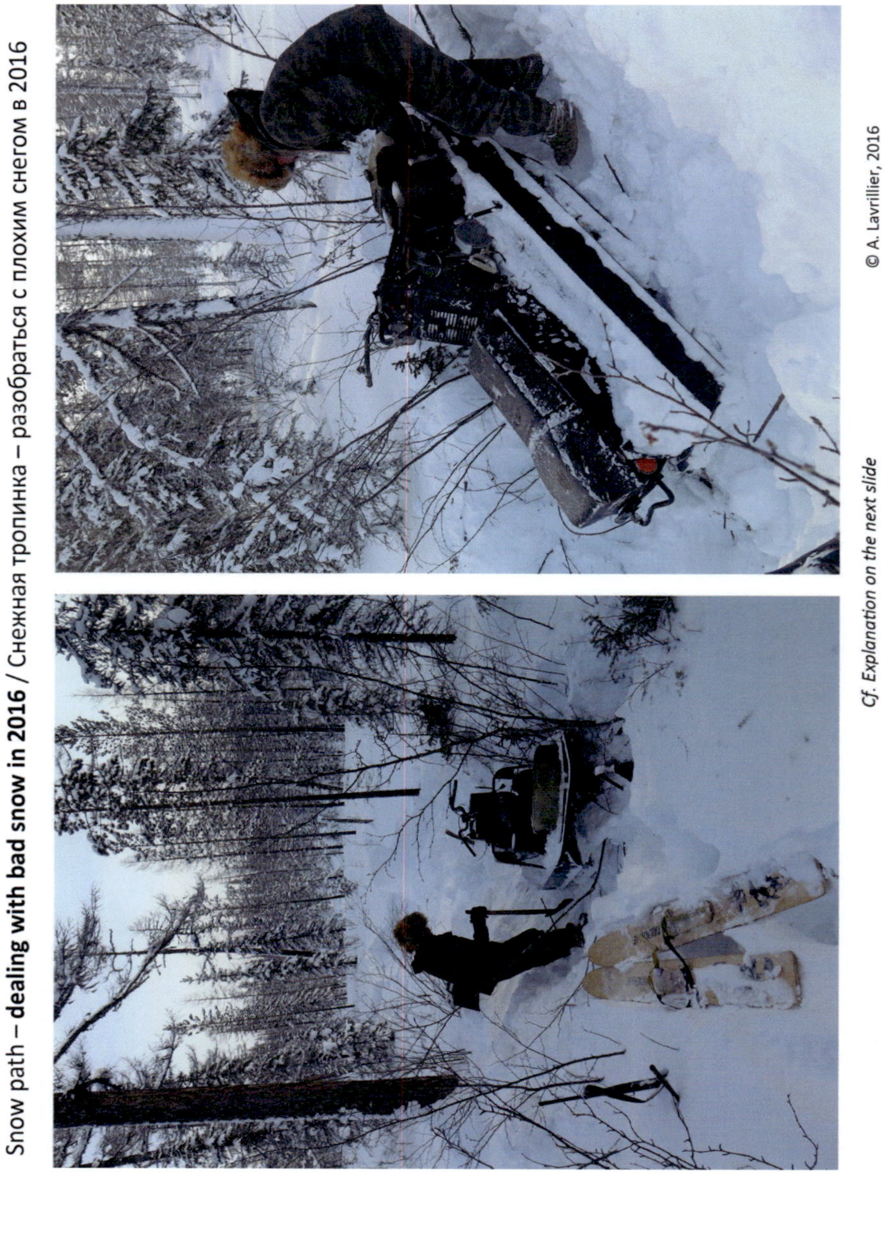
Snow path – **dealing with bad snow in 2016** / Снежная тропинка – разобраться с плохим снегом в 2016
Cf. Explanation on the next slide
© A. Lavrillier, 2016

Snow path – **dealing with bad snow in 2016** / Снежная тропинка – разобраться с плохим снегом в 2016

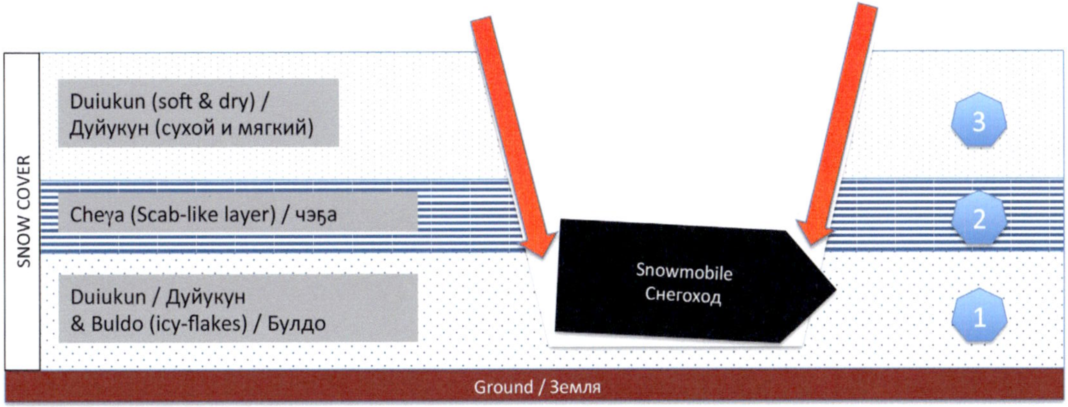

Because of the bad snow conditions, the snow cover was too crumbly and collapsed under sledges and snowmobiles, which sank. 1) The first snow layer fell all at once and was too deep (the *duiukun* partly turned into *buldo* snow); 2) Successive phases of warming and freezing created a scab-like layer (*cheγa*) of snow that (because of thermic isolation) did not allow the lower layer to become compact (as is normally the case) and did not reduce the depth of first snow layer; 3) A very deep layer of soft snow (*duiukun*) was deposited in the most recent snowfall. As a result, the entire snow cover was too fragile and collapsed under the weight of reindeer sledges or snowmobiles.

Из-за плохого снежного покрова (снег слишком мягкий и порождает проблемы продвижения) часто нарты и снегоходы проваливаются в снегу. И так: 1. первый снежный слой выпал сразу глубоким и слишком мягким (*дуйукун* частично превратился в *булдо*); 2. Потепление, потом замерзание создали корку снега *чэҕа*, которая из-за изоляции, которую она создает, не дала снежному слою стать компактнее и оседать (уменьшая при этом толщину снежного покрова); 3. Очень глубокий слой мягкого снега (*дуйукун*) выпал на старую поверхность снежного покрова. В итоге весь снежный покров слишком нестабильный и в нем регулярно проваливаются нарты или снегоходы.

Cf. Illustration on the previous slide © A. Lavrillier, 2016

3.3 Analysis of grazing pastures

(View from below) The reindeer have eaten horsetail. Horsetail always grows on river banks (between the river bank and *kerain iɣak* a small stony hill with a flat, long top. Snow depth from 54-66 cm (80 in the middle).

Орор сивакчал

(вид снизу) Олени кушали хвощ: Хвощ (*сивак*) всегда бывает на берегу реки (между рекой и каменистой маленькой горкой с плоской и длинной вершиной (*кэраин иҕак*) / Глубина снега здесь от 54 см до 66 см (а в середине по 80 см).

(View from the top/ Вид сверху):
From one grazing pasture to another / от одного пастбища к другому.

©A. Lavrillier, S. Gabyshev 2014

Reindeer droppings and pasture types - Ориктэ оӈкодук – Помет оленей и пастбище

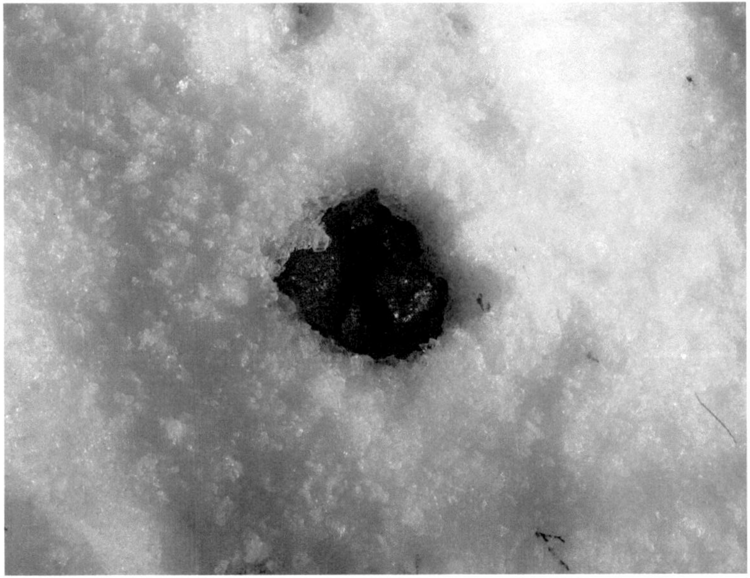

Reindeer droppings after eating horsetail. This plant is good for the reindeer, as they gain weight (similar to when they eat mushrooms).

Помет оленей (*ориктэ*) зелёный после употребления хвоща (*сивак*), как летом. Хороший (наедаются, как после грибов).

Reindeer grazing zones, or how to adapt to bad snow conditions (or bad access conditions) by using the diversity of the topographic landscape types /
Зоны корма оленей или как адаптироваться к аномалиям снежного покрова (и доступ к корму), используя разновидность топографии
Zone 2: On the slope of a small hill – lichen (*liavukte*) 2014
Зона 2: на склоне маленькой горы – ягель (*лйавуктэ*)
©A. Lavrillier, S. Gabyshev 2014

On the slope of a *kerain* (a small hill with a flat and long top), the snow is not deep and the reindeer graze well. Sometimes, reindeer prefer to graze in places with shallow snow, i.e. on slopes. On such slopes, lichen can always be found.

Кэраилду орор айат оӈкодйовкил, таргачинду эрэӡэр лйавуктэ бивки.

На склоне *кэраин* (маленькой горки с плоской и длинной вершиной) -мелкий снег. Олени иногда предпочитают кормоваться там где снег мелкий, т.е. на склонах. На таких склонах, всегда есть ягель.

©A. Lavrillier, S. Gabyshev 2014

On slopes, the snow is not deep but it can have various qualities: there might be a soft (*duiukun*) or a hard layer (*chuiur*), but rarely icy flakes (*buldo*). Snow depth: from 23-31cm down hill to 28-45cm up hill.

На склонах, снег мелкий, но разного качества: мягки (*дуйукун*) или наст (*чуйур*), но редко зернообразное (*булдо*). Глубина снега от 23 до 31 см внизу и от 28 см до 45 см на верху.

©A. Lavrillier, S. Gabyshev 2014

Grazing pastures on the slope under the trees / *Лйавуктэ*, кэраинду ирйактэ эрэдун / Корм на склоне под деревом.

©A. Lavrillier, S. Gabyshev 2014

Upper part of the slope (45cm of snow) – species of lichen / Верхняя часть склона (45 см снега) и виды ягеля.

©A. Lavrillier, S. Gabyshev 2014

Reindeer droppings and pastures types - Ориктэ оӊкодук – Помет оленей и пастбище

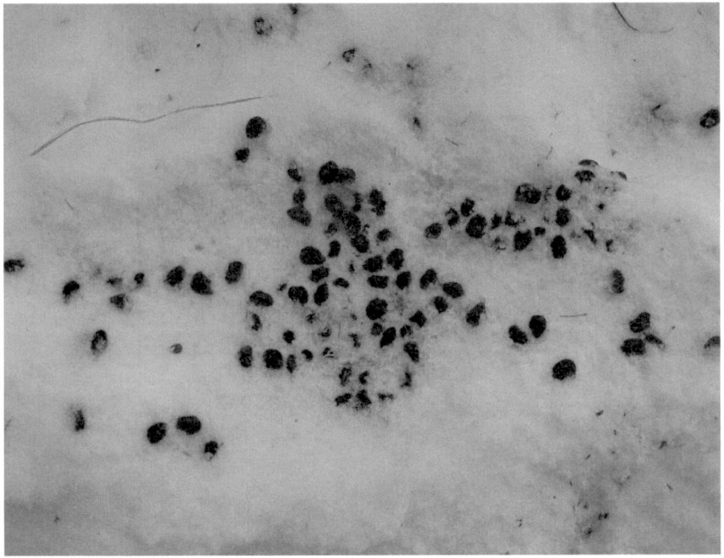

Reindeer droppings (*orikte*) after eating lichen (*liavukte*) – small and dry.

Помет оленей (*ориктэ*) после поедания ягеля (*лйавуктэ*) - Маленькие и сухие.

Close to *pinus pumila*, soft snow (*duiukun*) and grazing pastures can be found. Reindeer graze in such places.

Болгиктэл даҕадун бибкил дуйукун иманна. Болгиктэл даҕадун эрэҕэр оҥко бивки. Орор таду оҥкодйоксовкил.

Рядом со стлаником всегда мягкий снег (*дуйукун*), там всегда хороший корм для оленей, олени там кормуются.

©A. Lavrillier, S. Gabyshev 2014

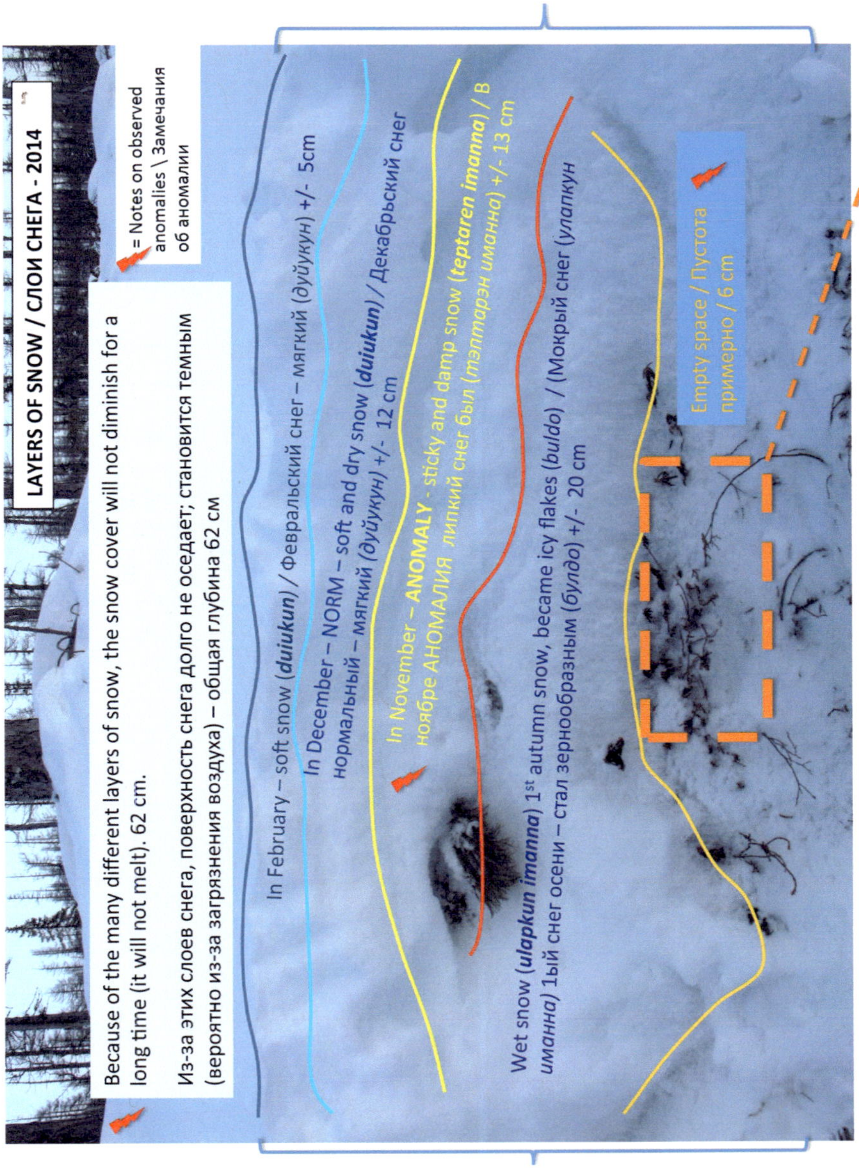

©A. Lavrillier, S. Gabyshev 2014

Lichen from under the snow /
Корм из-под снега

Snow depth along the reindeer path (almost at the top of the hill) / Глубина снега по оленьей тропинке почти на вершине холма / 120 cm

Between the small places (*ir*) where reindeer dig in the snow to graze, they walk around a lot in order to find good snow conditions and the best grazing pastures.

Между точками кормления (*up*) олени много ходят в поиске мест с не слишком глубоким снегом и с хорошим кормом.

The grazing pasture is good: it is dry and free from snow / Оҥко айакикин олгокин / Ягель хороший, сухой, свободный от снега.

Reindeer grazing zones, or how to adapt to bad snow conditions (or bad access conditions) by using the diversity of the topographic landscape types /
Зоны корма оленей или как адаптироваться к аномалиям снежного покрова (и доступ к корму), используя разновидность топографии
Zone 3: In a flat small river basin – lichen (*liavukte*) 2014
Зона 3: в ровном бассейне речки – ягель (*лйавуктэ*)
©A. Lavrillier, S. Gabyshev 2014

©A. Lavrillier, S. Gabyshev 2014

In *ialbuka* moss, the reindeer graze on a particular lichen made of small tubes: they are the same colour as the lichen itself. Reindeer also eat the leaves of bog blueberries.

Во мху *йалбука* олени кушают особенный ягель в форме трубочек такого цвета, как ягель. Олени также кушают листья голубики.

Winter 2016 - LICHEN GRAZING QUALITIES / КАЧЕСТВА ПАСТБИЩ 2016
Grazing *liavukte* / оҥко / корм *лйавуктэ*
A – The reindeer do not eat this / Олени не кушают
A
A
A
Grazing *lulgikte*, *liuvukte* / оҥко / корм *лулгиктэ*, *лйувуктэ*
©A. Lavrillier, S. Gabyshev 2016

Grazing near River Cheenchelir / Пастбище р. Чээнчэлир 2016

Grazing *liavukte* / оңко / корм *лйавуктэ*

©A. Lavrillier, S. Gabyshev 2016

Grazing near River Cheenchelir / Пастбище р. Чээнчэлир 2016

Hard snow / твердый снег / Маӈа иманна

Grazing *liavukte* embedded in ice / оӈко дйукачи / корм *лйавуктэ* со льдом

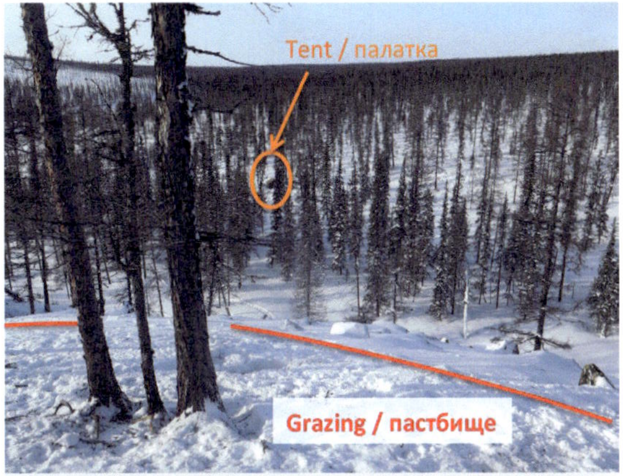

The hard snow (*maŋa imanna*) is transformed into icy flakes (*buldo*) / Твердый снег (*маӈа иманна*) превратился в зернообразный снег (*булдо*) / Маӈа иманна булдо овки

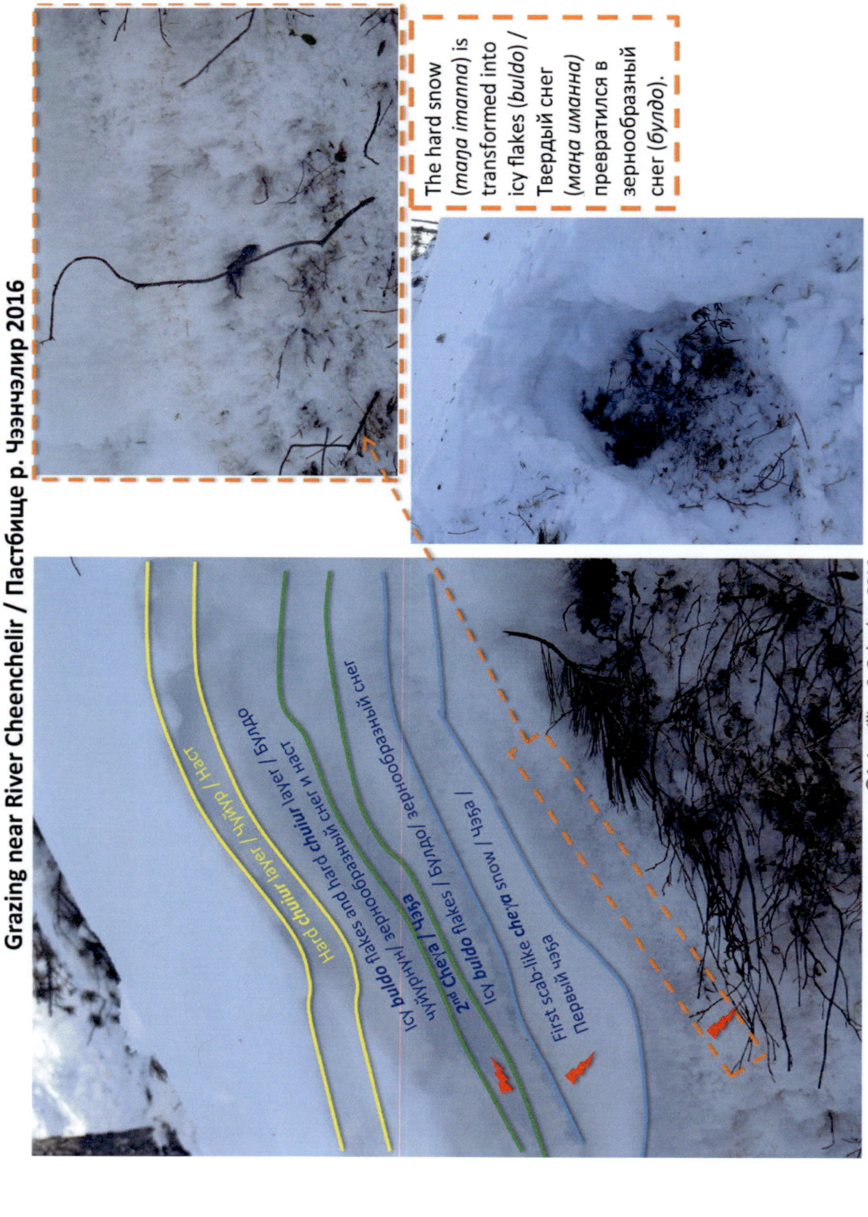
Grazing near River Cheenchelir / Пастбище р. Чээнчэлир 2016
Hard chuiur layer / Чүйүр / Наст
Icy buldo flakes and hard chuiur layer / Булдо
чүйүрнүн/ зернообразный снег и наст
2nd Cheɣa / Чэҕа
Icy buldo flakes / Булдо/ зернообразный снег
First scab-like cheɣa snow / Чэҕа /
Первый чэҕа
The hard snow (maŋa imanna) is transformed into icy flakes (buldo) / Твердый снег (маӊа иманна) превратился в зернообразный снег (булдо).
©A. Lavrillier, S. Gabyshev 2016

Grazing near River Cheenchelir / Пастбище р. Чээнчэлир 2016

Grazing on the larch bark / Оӊко ирйакталду / Корм на коре деревьев

3.4 Anomalies in the snow cover 2013–2014

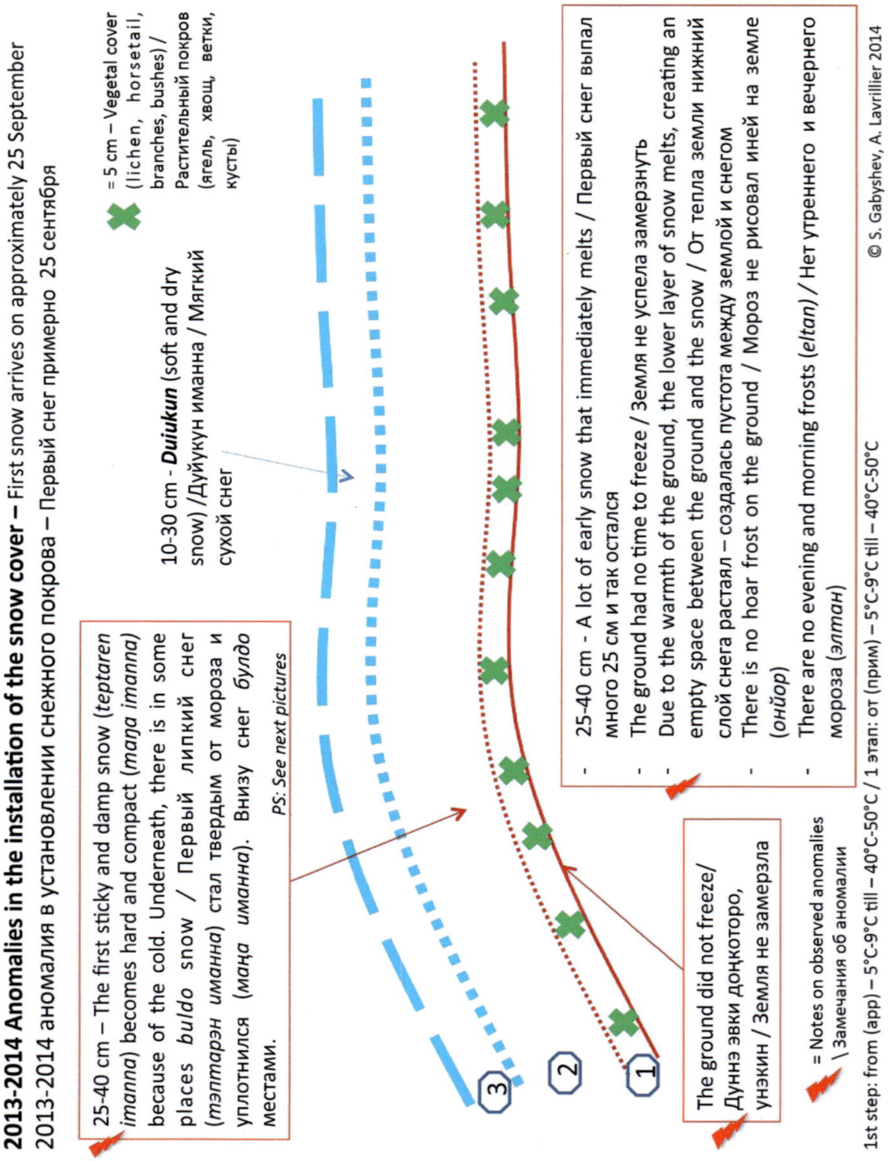

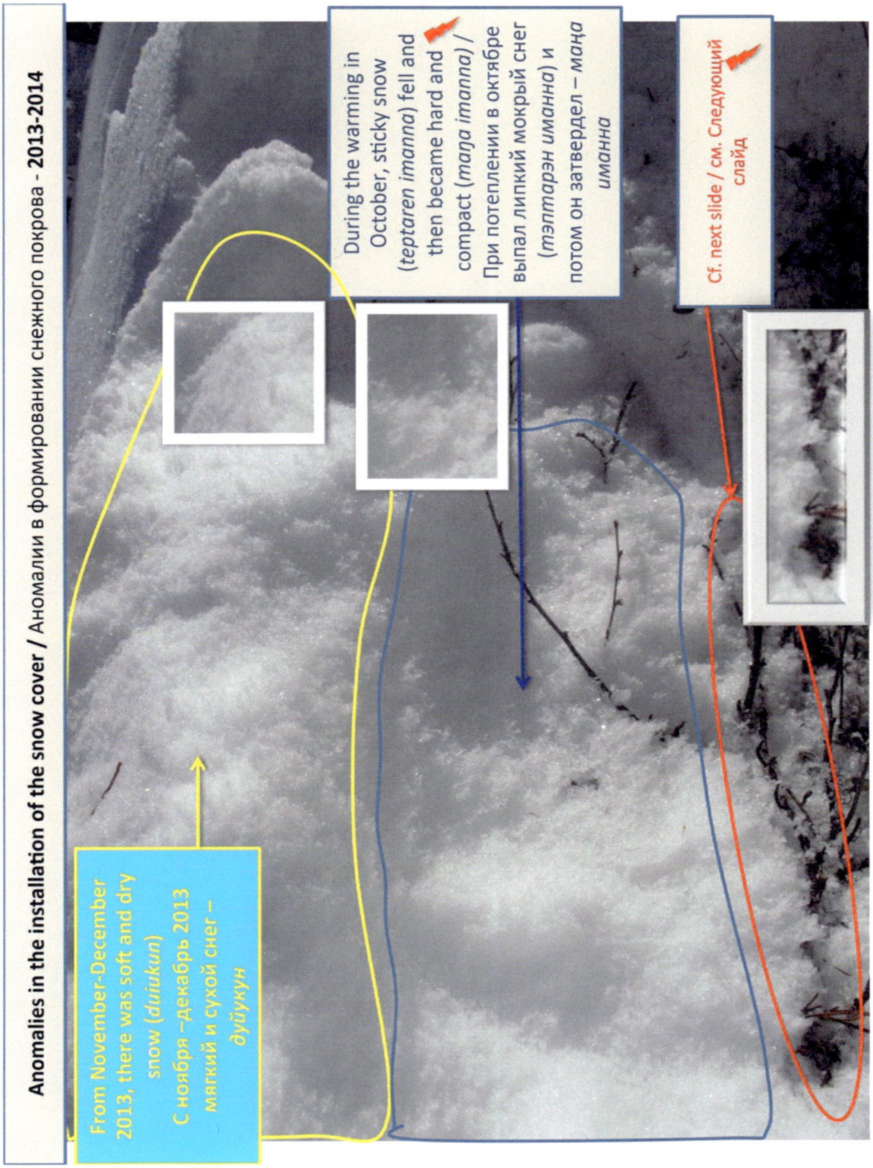

© A. Lavrillier, S. Gabyshev, 2014

PS: See previous diagram

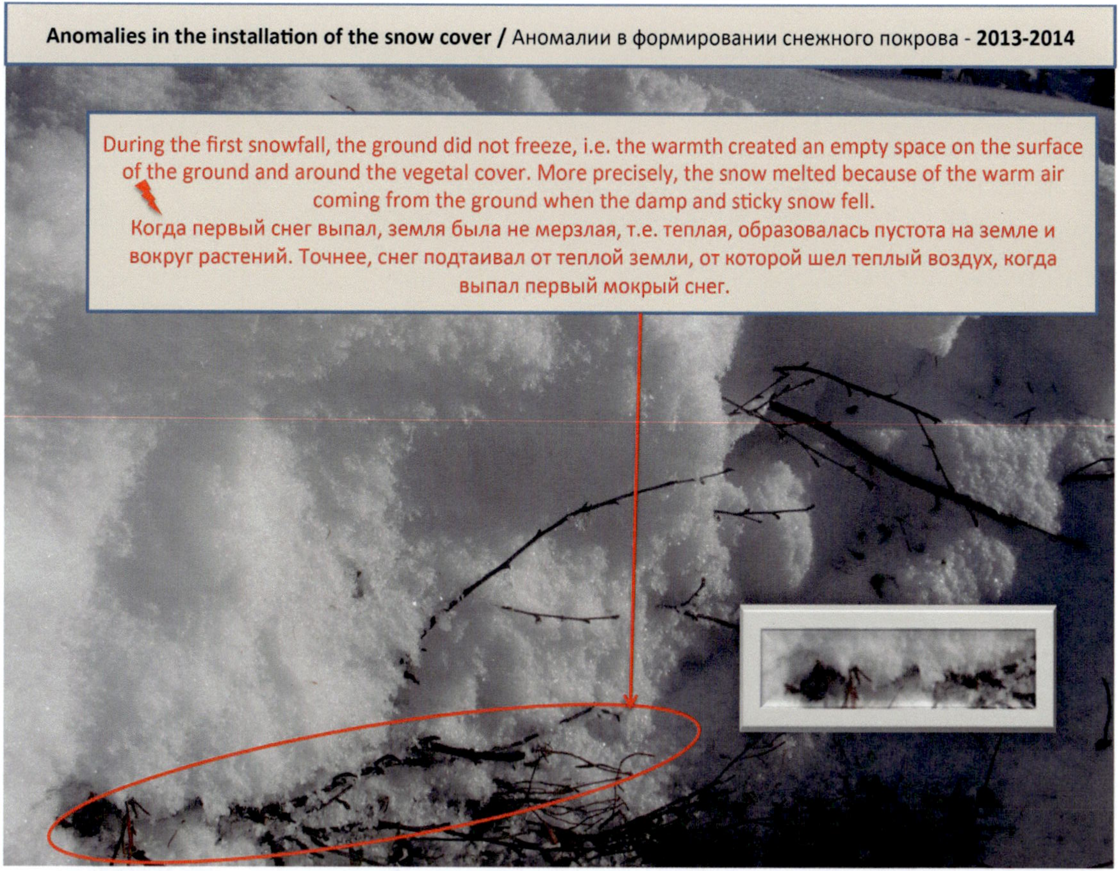

PS: See previous diagram

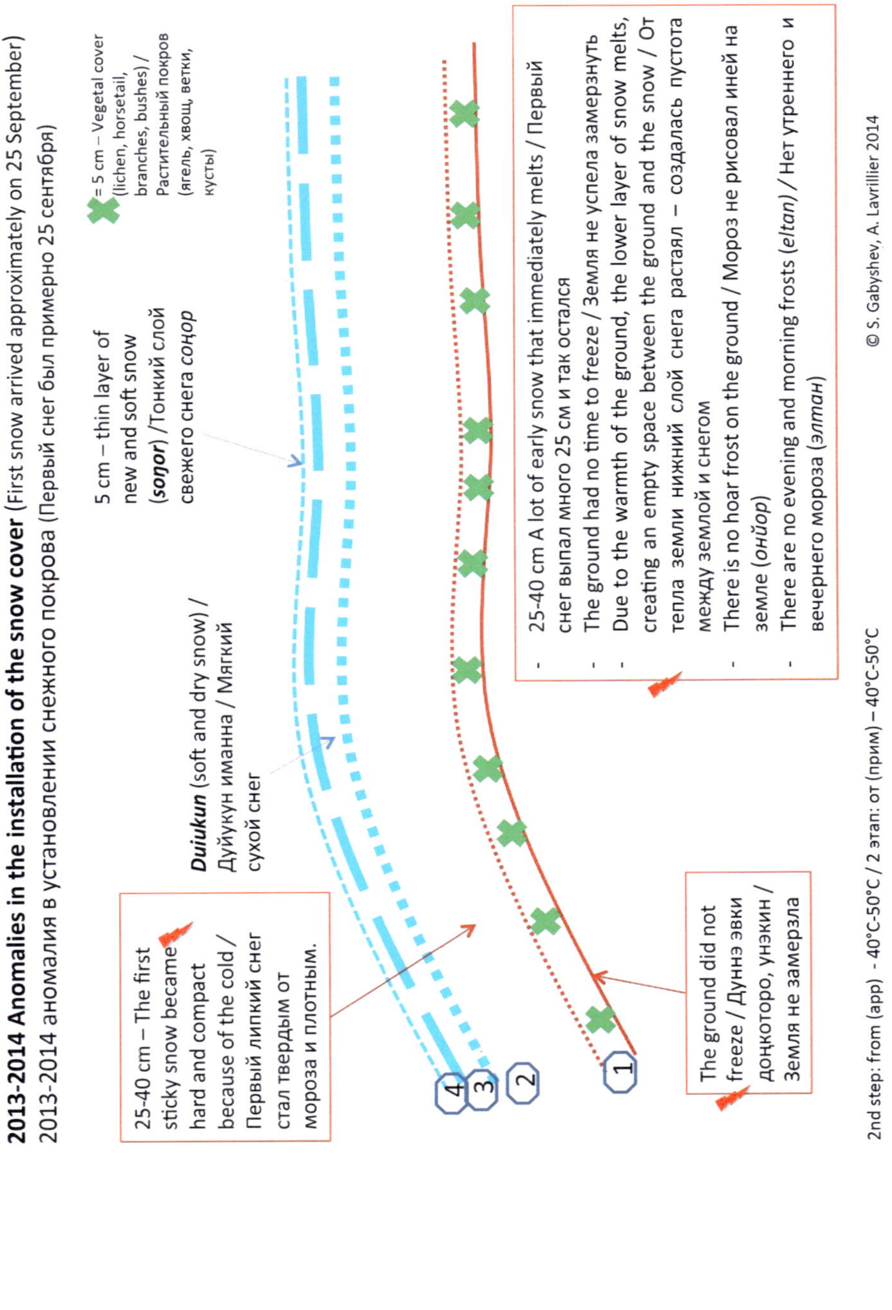
2013-2014 Anomalies in the installation of the snow cover (First snow arrived approximately on 25 September)
2013-2014 аномалия в установлении снежного покрова (Первый снег был примерно 25 сентября)
= 5 cm – Vegetal cover (lichen, horsetail, branches, bushes) / Растительный покров (ягель, хвощ, ветки, кусты)
5 cm – thin layer of new and soft snow (**soŋor**) /Тонкий слой свежего снега *соҥор*
Duiukun (soft and dry snow) / Дуйукун иманна / Мягкий сухой снег
25-40 cm – The first sticky snow became hard and compact because of the cold / Первый липкий снег стал твердым от мороза и плотным.
4
3
2
1
- 25-40 cm A lot of early snow that immediately melts / Первый снег выпал много 25 см и так остался
- The ground had no time to freeze / Земля не успела замерзнуть
- Due to the warmth of the ground, the lower layer of snow melts, creating an empty space between the ground and the snow / От тепла земли нижний слой снега растаял – создалась пустота между землой и снегом
- There is no hoar frost on the ground / Мороз не рисовал инеей на земле (*онйор*)
- There are no evening and morning frosts (*eltan*) / Нет утреннего и вечернего мороза (*элтан*)
The ground did not freeze / Дуннэ эвки доҥкоторо, унэкин / Земля не замерзла
2nd step: from (app) - 40°C-50°C / 2 этап: от (прим) – 40°C-50°C
© S. Gabyshev, A. Lavrillier 2014

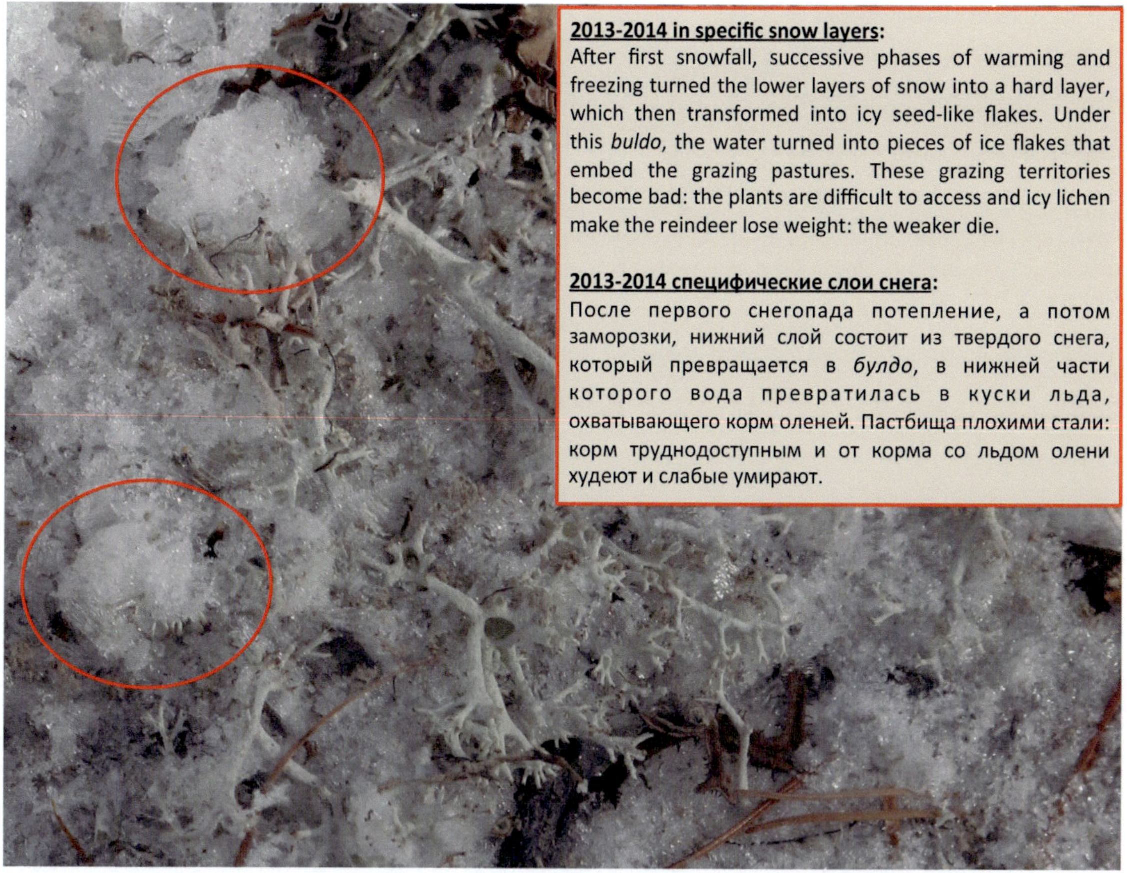

© A. Lavrillier, S. Gabyshev, 2014

© A. Lavrillier, S. Gabyshev, 2014

2013-2014 Anomalies in the installation of the snow cover

2013-2014 аномалия в установлении снежного покрова

Consequences

For hunting: dogs get tired from walking in such snow, as they sink into it: the sable can run away from the dogs on the surface of the snow (in October, November). It is also difficult to hunt wild reindeer either by foot or while riding a reindeer (the animal gets tired very quickly). Wild reindeer do not like to walk in the hard snow: they tend to go to places with shallow snow or high mountains. In the mountains, there is always soft and dry snow (*duiukun*). In contrast, red deer and elk walk in all snow types, so it is possible to hunt them: however, humans and domestic reindeer become very exhausted.

Reindeer herding: Herders tend to nomadise to places with a relatively shallow snow cover composed of soft snow types (*duiukun* or *buldo,* for instance). Indeed, domestic reindeer do not like to graze in hard snow and go to places with shallow and soft snow (like rivers known by the herders). They may also go to mountain rivers with steep banks: as the snow is not compact here because of the slopes, reindeer can dig easily. (cf. diagram Analysis of grazing pastures and next slides)

Последствия

Для охоты: Тяжело собакам бегать по такому снегу, они проваливаются, а соболь наоборот по поверхности обгоняет собак (октябрь-ноябрь). За дикими оленями гнаться тяжело пешком или на верховом олене, который быстро устает. Дикие олени вообще не любят ходить по твердому глубокому снегу и стараются уходить туда, где есть или мелкий снег, или высокие горы. Там снег рыхлый всегда бывает (*дуйукун*). Наоборот, благородный олень, лось по любому снегу ходят, можно на них охотиться, но только человек и олень устают.

Для оленеводства: Оленеводы стараются кочевать туда, где мелкий снег и где снег по мягче (*дуйукун*, *булдо*). Ведь олени не очень любят кормоваться по такому снегу и уходят туда, где снег мелкий или помягче (например на определенных реках). На горных речках с крутыми берегами снег не компактный и выгребается легко, потому что падает на склон. (см. слайд дальше, diagram Analysis of grazing pastures and next slides)

Grazing on small or large slopes is easier for reindeer because when they dig the snow with their hooves, the snow falls away and frees the pastures up. Reindeer prefer to graze in soft and dry snow (*duiukun*). The holes made in the snow by reindeer for grazing are called *ir*.

На маленьком или большом склоне оленям легче кормоваться потому, что когда копытят снег, то снег вниз сам падает. Олени предпочитают кормоваться по мягкому снегу *дйуйкун*. Ямы в снегу сделанные кормующимися оленями называют *ир*.

© A. Lavrillier, S. Gabyshev, 2014

2013-2014 Anomalies at the end of February – Consequences for the thawing of the snow cover 2013-2014 аномалия конца февраля – Последствия таяния снежного покрова

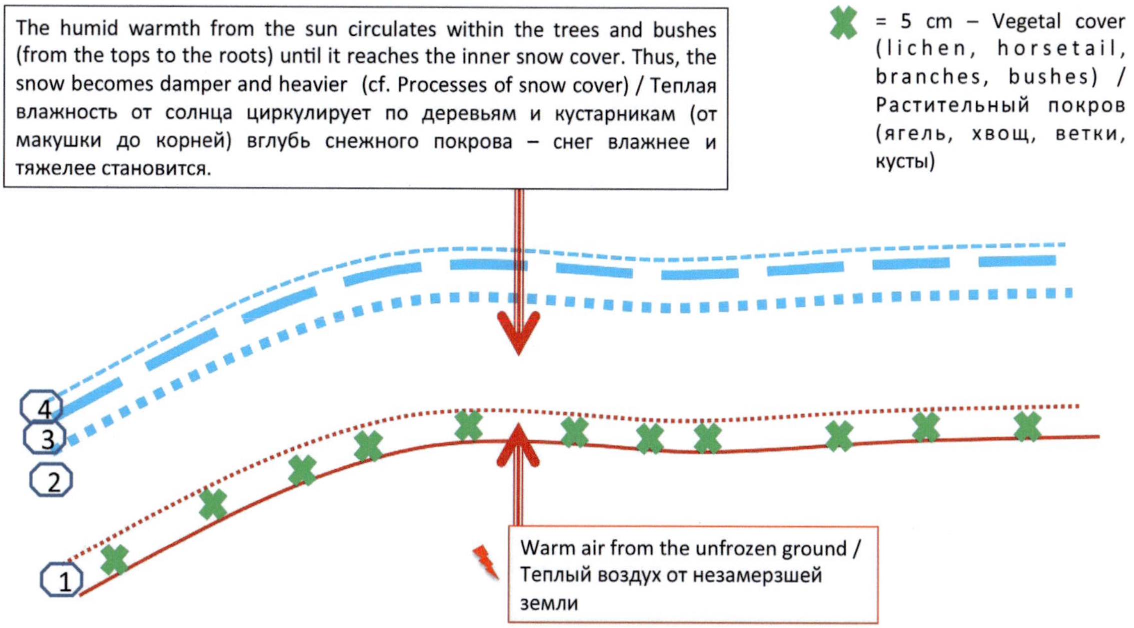

FORECAST: This is a prediction by Semen Gabyshev, which was made according to the Evenki TEK performed in February 2014. Having observed and analysed the melting process, this prediction proved to be right, although it was a week out.

ПРОГНОЗ: С этого слайда этот прогноз сделан Семеном Габышевым согласно эвенкийскому традиционному знанию в феврале 2014г. Данный прогноз оказался точным в разницу на одну неделю.

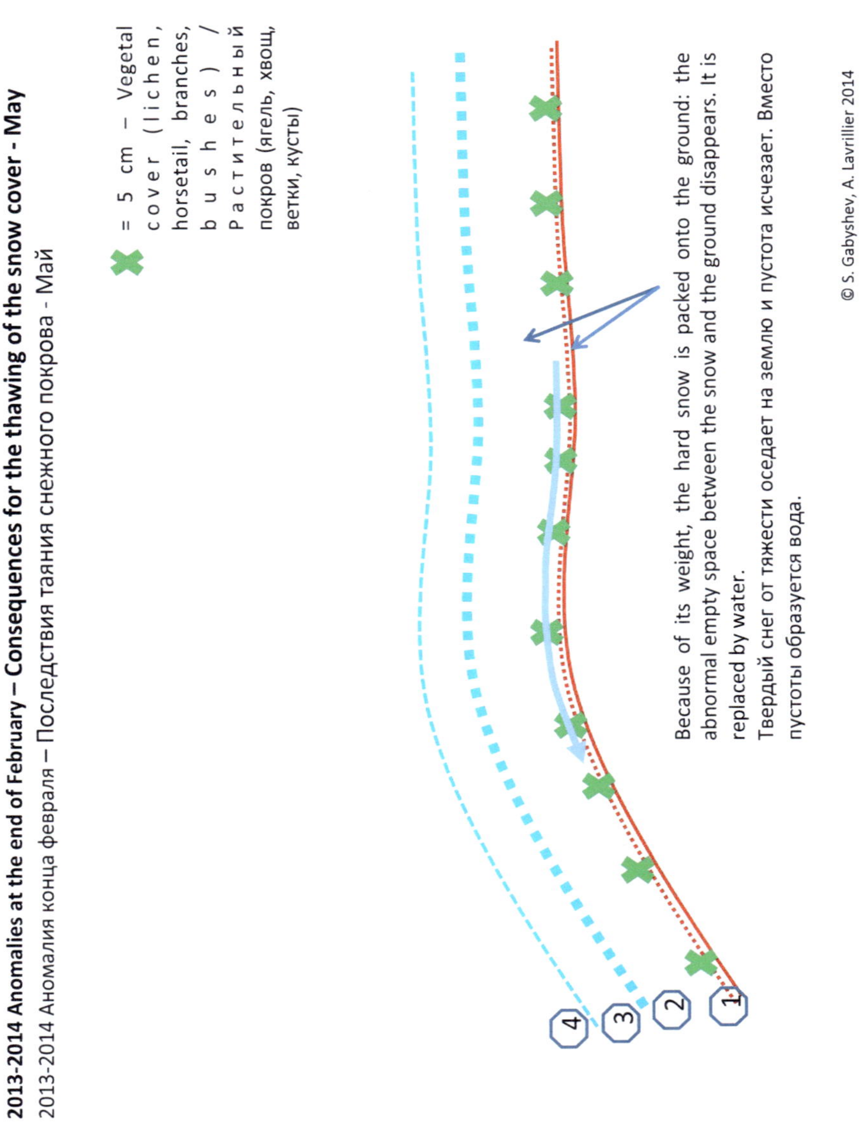
2013-2014 Anomalies at the end of February – Consequences for the thawing of the snow cover - May
2013-2014 Аномалия конца февраля – Последствия таяния снежного покрова - Май
= 5 cm – Vegetal cover (lichen, horsetail, branches, bushes) / Растительный покров (ягель, хвощ, ветки, кусты)
Because of its weight, the hard snow is packed onto the ground: the abnormal empty space between the snow and the ground disappears. It is replaced by water.
Твердый снег от тяжести оседает на землю и пустота исчезает. Вместо пустоты образуется вода.
1
2
3
4
© S. Gabyshev, A. Lavrillier 2014

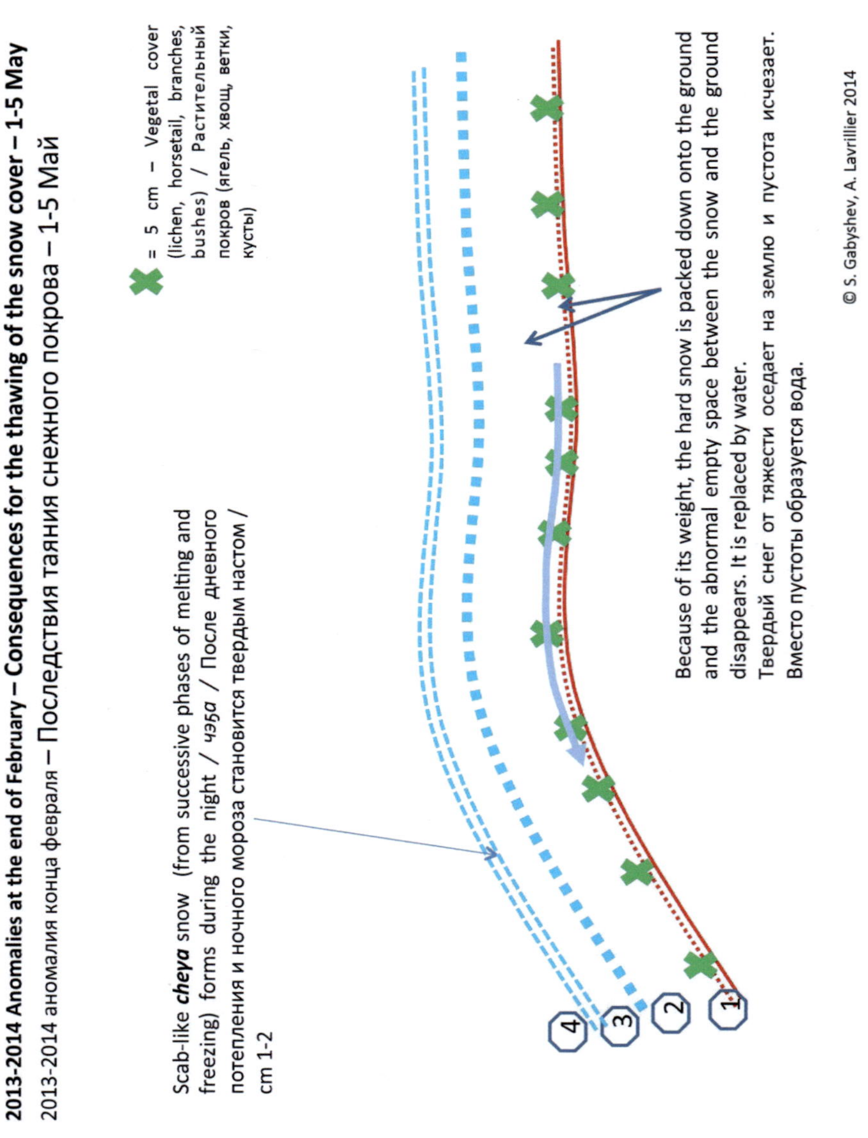

2013-2014 Anomalies at the end of February – Consequences for the thawing of the snow cover – 1-5 May
2013-2014 аномалия конца февраля – Последствия таяния снежного покрова – 1-5 Май
= 5 cm – Vegetal cover (lichen, horsetail, branches, bushes) / Растительный покров (ягель, хвощ, ветки, кусты)
Scab-like *cheya* snow (from successive phases of melting and freezing) forms during the night / чэҕа / После дневного потепления и ночного мороза становится твердым настом / cm 1-2
Because of its weight, the hard snow is packed down onto the ground and the abnormal empty space between the snow and the ground disappears. It is replaced by water.
Твердый снег от тяжести оседает на землю и пустота исчезает. Вместо пустоты образуется вода.
1
2
3
4
© S. Gabyshev, A. Lavrillier 2014

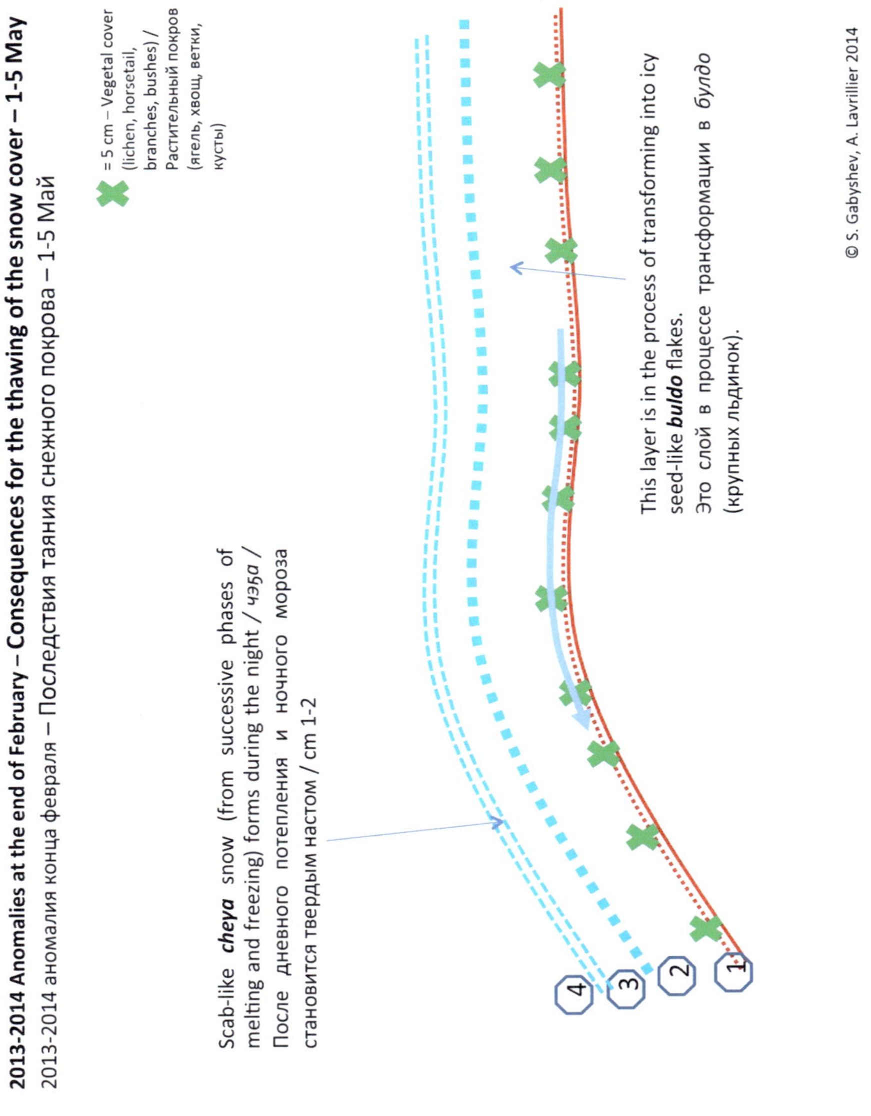

2013-2014 Anomalies at the end of February – Consequences for the thawing of the snow cover – 1-5 May
2013-2014 аномалия конца февраля – Последствия таяния снежного покрова – 1-5 Май
= 5 cm – Vegetal cover (lichen, horsetail, branches, bushes) / Растительный покров (ягель, хвощ, ветки, кусты)
Scab-like ***cheya*** snow (from successive phases of melting and freezing) forms during the night / *чэҕа* / После дневного потепления и ночного мороза становится твердым настом / cm 1-2
This layer is in the process of transforming into icy seed-like ***buldo*** flakes.
Это слой в процессе трансформации в *булдо* (крупных льдинок).
1
2
3
4
© S. Gabyshev, A. Lavrillier 2014

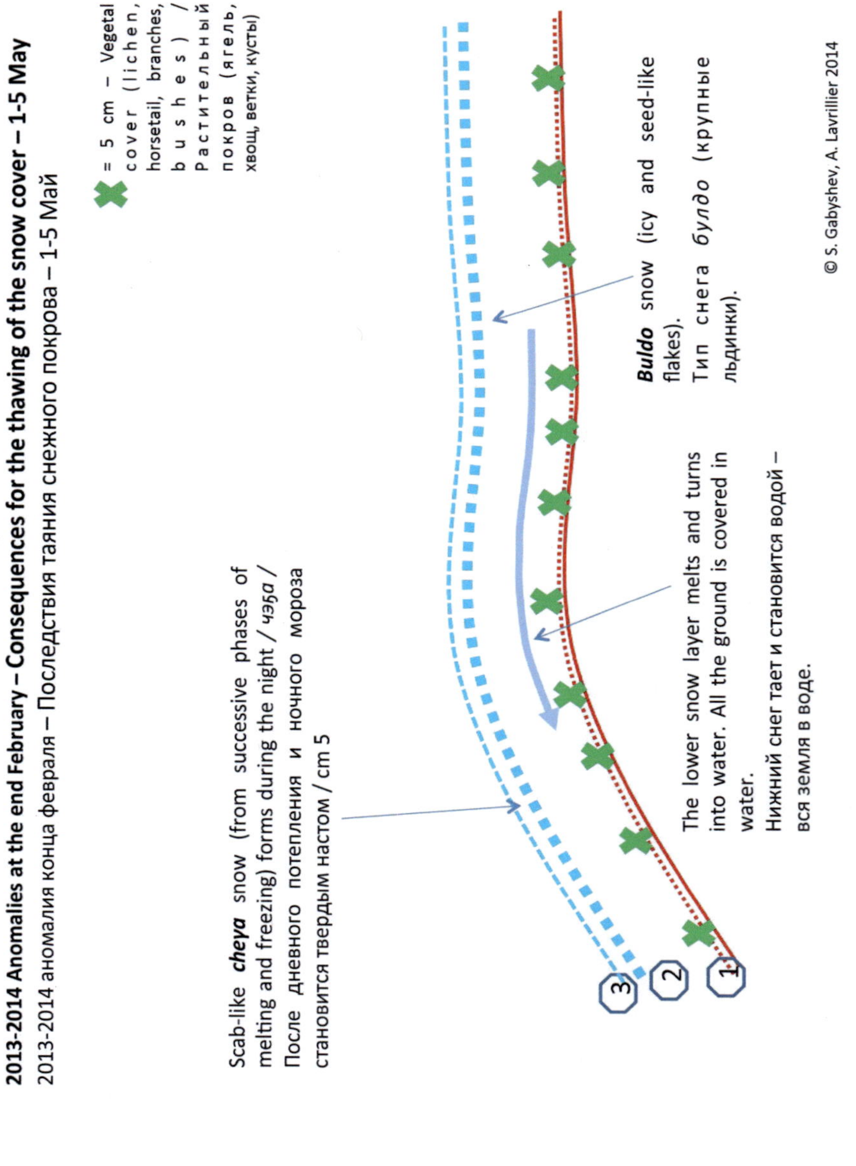
2013-2014 Anomalies at the end February – Consequences for the thawing of the snow cover – 1-5 May
2013-2014 аномалия конца февраля – Последствия таяния снежного покрова – 1-5 Май
= 5 cm – Vegetal cover (lichen, horsetail, branches, bushes) / Растительный покров (ягель, хвощ, ветки, кусты)
Scab-like cheya snow (from successive phases of melting and freezing) forms during the night / чэҕа / После дневного потепления и ночного мороза становится твердым настом / cm 5
Buldo snow (icy and seed-like flakes).
Тип снега булдо (крупные льдинки).
The lower snow layer melts and turns into water. All the ground is covered in water.
Нижний снег тает и становится водой – вся земля в воде.
1
2
3
© S. Gabyshev, A. Lavrillier 2014

2013-2014 Anomalies at the end February – Consequences for the thawing of the snow cover

Consequences

Hunting: people hunt for their own use only (not for selling) because the meat will dry and rot in *salgyn* air (cf. Evenki climatology: warm and cold airs). The sable and wild reindeer have their calving seasons at this time, so the Evenki prohibit hunting wild reindeer. In any case, tracks disappear quickly on the melting snow. In addition, the seeds of various species of berries freeze and die from the frost (*beɣivki*) (cf. Vegetal cover typology). This means that there will be no berries to harvest in the autumn and that sables will very seldom, if at all, pass through territories affected by this kind of snow cover anomaly.

For reindeer herding: The seeds of tussock cotton-grass (*nirgakte*) freeze and die (*beɣivki*). This means that there will be no harvest in the spring and reindeer will not be able to graze on this grass (which is beneficial for them after the long winter) (cf. Vegetal cover typology). In addition, because the snow melts quickly, newly-born calves do not leave tracks, which means one can lose the herd. Shallow snow makes hunting easy for predators, so they become a danger to the reindeer. Reindeer disperse over great distances independently on this snow cover (*erekeltavkil*). People have difficulty keeping the herd gathered and the percentage of living calves at the end of the spring is low.

Последствия

Для охоты: Охота только для себя, не на продажу, ведь мясо высохнет или сгниёт от воздуха *салгын* (см. Evenki climatology: warm and cold airs). У соболей и диких оленей отел, эвенки себе запрещают охотиться на них. Следы на тающем снегу быстрее исчезают. Кроме того, семена разных ягод погибли от мороза (*бэӻивки*) (см. Vegetal cover typology). Это значит, что на этих местах, тронутых такими аномалиями, урожай ягоды не будет осенью и соболя не будут ходить.

Для оленеводства = Семена пушицы (*ниргактэ*) погибли от мороза (*бэӻивки*), и значит весной урожая не будет и олени не смогут кушать это очень нужное растение для их организма после зимы (см. Vegetal cover typology).

Кроме этого, из-за того что снег тает быстро, следов не видно. Только родившихся телят трудно находить, ведь снега нету. Так можно оленей терять. По мелкому снегу хищникам легче охотиться чем по глубокому снегу и они являются угрозой для телят. Олени сами по себе далеко ходят (*эрэкэлтавкил*). Человеку трудно держать стадо собранным, и сохранять олений приплод.

Snow turned upside down by reindeer. It has become hard and compact.

Использованный и перевернутый оленями снег, затвердел и склеился.

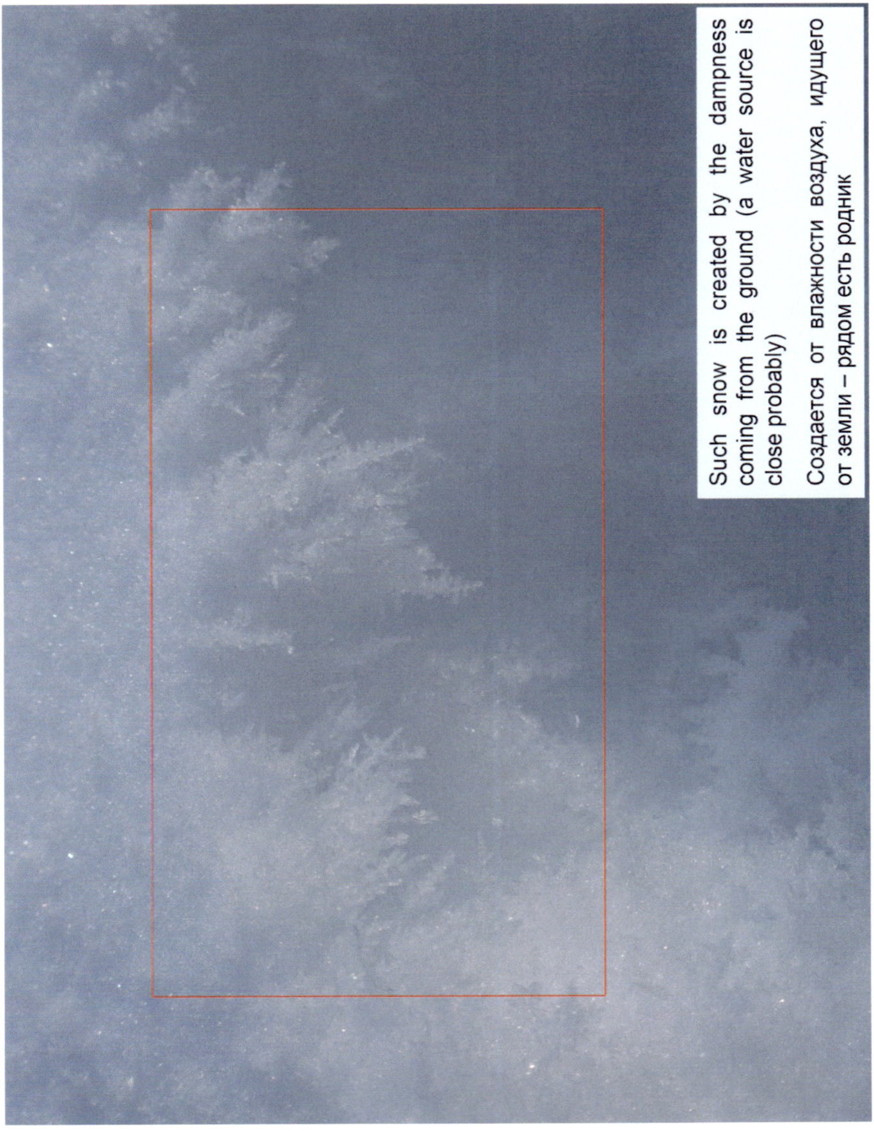

© A. Lavrillier, S. Gabyshev, 2014

3.5 Anomalies in the snow cover 2014–2015

2014-2015 Anomalies in the installation of the snow cover – (similarities with 1974)
2014-2015 Аномалия в установлении снежного покрова – (похож на 1974)

First snow / Первый снег

1 – 1st snow – 23 September = - 2 or -3 °C = Anomaly like in 1974 г.– a lot of snow falls at once and that will not melt
1 - 1ый снег – 23 сентября = - 2°C or -3 °C = Аномалия как в 1974г – много снега выпало сразу, который уже не растаял

= Notes on observed anomalies \ Замечания об аномалии

2014-2015 Anomalies in the installation of the snow cover – (similarities with 1974)
2014-2015 аномалия в установлении снежного покрова – (похож на 1974)

2 End September – beginning of October: The first *duiukun* snow fell at a temperature of -8 °C at a depth of around 10 cm. With the frost, it turned into *buldo* snow; on 25-26 September, the surface snow melted when a positive temperature (0 or +1°C) occurred. It quickly froze again, causing a layer of *cheya* to form.
Nomads noticed that during the autumns of 2013, 2014, and 2015, the snow type *cheya,* specific to the spring, appeared in autumn. It caused many grazing problems for reindeer. (cf. next 2 slides for detailed study)
2 Конец сентября - начало октября: иҥиниктэрэн -8 °C у, иманна тыкчан глубинан см 10, иҥиинниктэдйэрэкин булдо иманна овки; 25-26 сентября 0 или + температуры – ойгу иманна унйачан, тадук иҥинэктэдйэрэкин чэӡа иманна овки.
2 Конец сентября - начало октября: сначала при холоде -8 °C, снег *дуйукун* еще выпал см 10, и из-за морозов превратился в *булдо*; а в 25-26 сентября 0 и плюсовые температуры и поверхность снега растаяла и сразу из-за морозов превратилась в корку снега *чэӡа.*
Кочевники замечают, что 2013г., 2014г. и 2015г. снег *чэӡа* (типичный для весны), устанавливался осенью. Этот снег создает большие трудности для доступа оленей к пастбищам (см. Более углубленный анализ в следующих 2 слайдах)

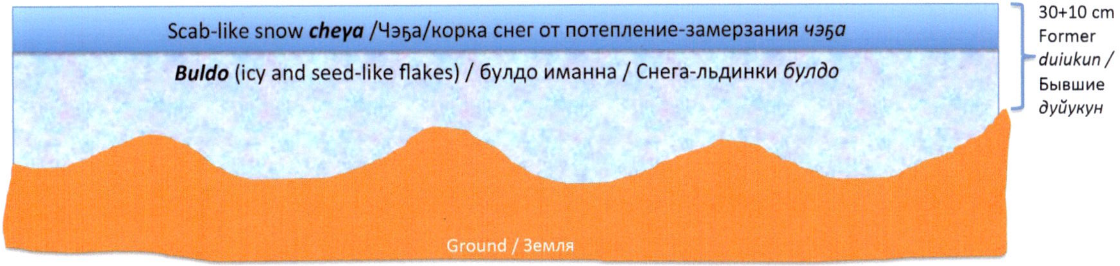

2014-2015 Anomalies in the evolution of the snow cover – (similarities with 1974) – **end Septembre – 10 October**
2014-2015 Аномалия эволюция снежного покрова – (похож на 1974) - конец сентября – 10 ого октября

ANOMALY – Deep snow cover and scab-like layer *cheya* / Суӈта чэҕанун /
АНОМАЛИЯ – глубокий снег и корка снега *чэҕа*

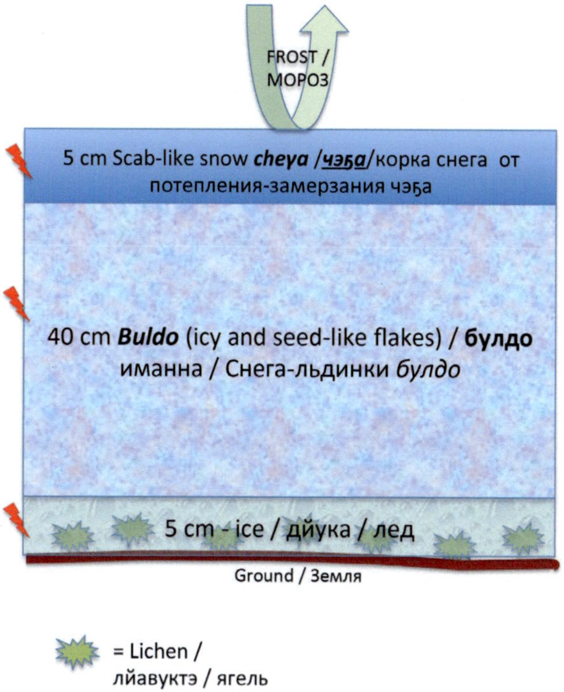

2 bis The deep layer of *buldo* snow, when covered by a *cheya* layer, does not allow the frost to go through the snow layer to the ice in which the grazing pastures are embedded. This ice was created by humidity from the snow thawing during abnormally warm conditions. The frost from the air normally disintegrates the ice and frees the vegetation. However, in this case, the grazing pastures will remain embedded in ice all winter. If wild or domestic reindeer graze on icy lichen, they loose weight: the weaker may die.

Толстый слой снега *булдо* покрытый коркой *чэҕа* не дает морозу проникнуть до льда (лед образовался от влаги после снега), который охватывает олений корм, а мороз мог бы раскрошить лед и освободить корм. Так корм со льдом останется на всю зиму: домашние и дикие олени не смогут кормоваться. Если они кормуются ягелем со льдом, то они сильно худеют, и самые слабые пропадают.

PS: see the norm in next slide /
см. Норма в следующем слайде

2014-2015 Anomalies in the evolution of the snow cover – (similarities with 1974) – **end Septembre – 10 October**
2014-2015 Аномалия эволюции снежного покрова – (похож на 1974) - Конец сентября – 10 ого октября

NORM – shallow snow /
НОРМА – мелкий снег

The thin layers of *buldo* and *cheya* snow are not too deep, which means that the frost can reach the ground. It cracks the ice in which the lichen is embedded, allowing for wild or domestic reindeer to graze freely and normally.

Неглубокий слой *булдо* и тонкий *чэɓа* дают морозу доходить до земли. От этого мороза трескается лед и освобождается ягель ото льда. Домашние и дикие олени могут хорошо кормоваться.

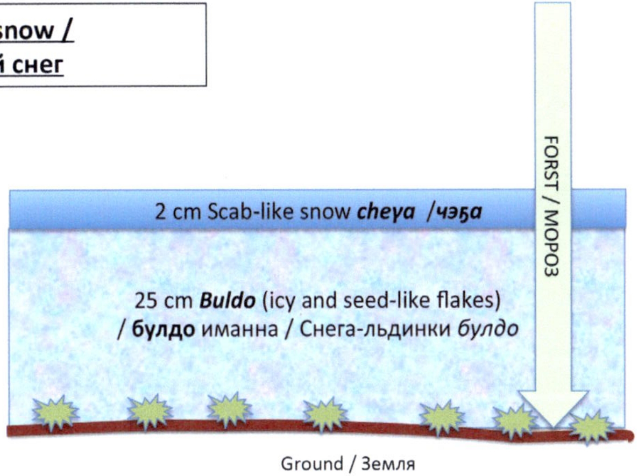

= Lichen / лйавуктэ / ягель

2014-2015 Anomalies in the installation of the snow cover – (similarities with 1974)
2014-2015 аномалия в установлении снежного покрова – (похож на 1974)

3 – <u>From 5 – 10 October</u>: It was warm (+ 2°C): the snow was thawing. With the evening frost (*eltan*), the water froze and the *cheya* snow turned into *sy* snow. This is a hard snow type: it was 5-7 cm deep. This was a surprisingly thick layer of *sy:* it could support a person. While grazing, the reindeer were soaked by such snow and froze. Thus, the reindeer could not graze – it was an extreme weather event.

3 – <u>С 5 по 10 октября:</u> Нйама бичан + 2°C, иманна ундйэрэкин, кэҕа элтанэвки мул донготовкил, иманна чэҕа сыкун овки. Маӈна иманна сынйун – 5-7 см. Дэрамкун бичан, бэйэл ойолы ӈэнэктэчатын. Оӈко улапиҕа, му орэн, тадук доӈкотовки.

3 – <u>С 5 по 10 октября:</u> Было тепло + 2°C снег таял, вечером от мороза воды от снега замерзли, и снег *чэҕа* стал большим слоем снега *сы*. Это смесь твердого снега с *сы* – 5-7 см. Удивительно толстый слой *сы*, что люди по поверхности ходят. Олений корм намок, а потом весь замерз вместе с *сы*. Олени не могли кормоваться – и это стало таежной экстремальной ситуацией.

Cf. see the rest of text in the next slide /
см. Продолжение текста на след слайд

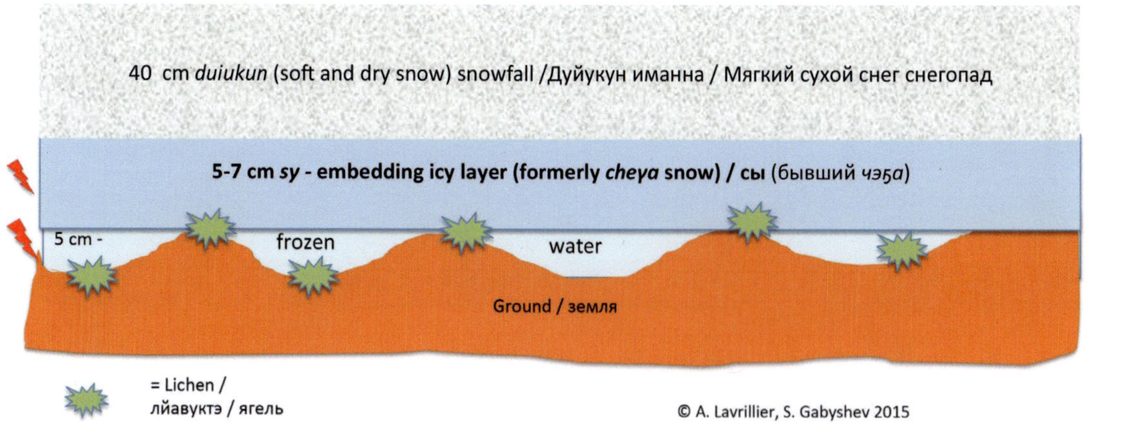

© A. Lavrillier, S. Gabyshev 2015

2014-2015 Anomalies in the installation of the snow cover – (similarities with 1974)
2014-2015 аномалия в установлении снежного покрова – (похож на 1974)

3 bis - Consequences – the grazing pastures are embedded in frozen water, so the reindeer have to travel far to find food. They go to tussock fields (*kever*) downstream in the pingo or to the slopes of mountains, since the snow melts in such locations.

For this extreme event, the Evenki term-concept *sydiŋan* is often used. The first explanation nomads gave of this term was: 'the domestic reindeer and game will move away from our usual areas of nomadisation in order to reach places where the snow is shallow. People need to move away from the usual areas to other places before it is too late'. Later, they explained this term in more detail, saying that it derives from the name of the snow type *sy*: they explained all the processes expressed here in the diagrams. In some places, there was a double extreme event: the grazing pastures were embedded within the *sy* snow type while very deep snow was also present.

3 bis - Тадук – оңко му овки, орор ңэнэмарил, кэвэрилбэ урувкил эйакуки муркикарду иманна унивки тарилду оңкодйэвкил, эллэңилду иду иманна унча тала урувкил.

3 bis - Последствия – олений корм в воде замерз, олени постоянно ходят голодными: на кочкарник (*кэвэр*) вниз по течению - там снег тает на бугре *мурки*, или на склонах гор. Олени уходят туда, где снег растаял.

Об этой экстремальной климатической обстановке люди сперва говорили одним эвенкийским термином *сыдиңан*. Сперва объяснили этот термин, говоря, что «домашние олени и дичь будут уходить от обычных мест кочевания туда, где есть места с мелким снегом. Людям тоже надо откочевать срочно от мест, где глубокий снег, пока не поздно». Только потом, кочевники объяснили более детально, что этот термин идет от названия снега *сы* и всего процесса, который тут показан диаграммами. В некоторых местах двойная проблема была: тип снега сы и очень глубокий снег.

2014-2015 Anomalies in the installation of the snow cover – (similarities with 1974)
2014-2015 аномалия в установлении снежного покрова – (похож на 1974)

4 – <u>From 15 – 25 October</u>: As soon as the frost began (at around -22°C), the *sy* snow type turned into the snow type *buldo*. This did not solve the problem, since the pastures were still embedded in ice.

4 – С <u>15 по 25 октября</u>: Иңиниктэллэкин -22 °С, сы булдо иманна овки. Оңко эрэӡэр дйукэду бивки, орор ңэнэмарил.

4 – <u>С 15 по 25 октября</u>: От похолодания прим -22 °С, тип снега *сы* превращается в тип снега *булдо*. Проблема не решена, потому что олений корм до сих пор охвачен льдом.

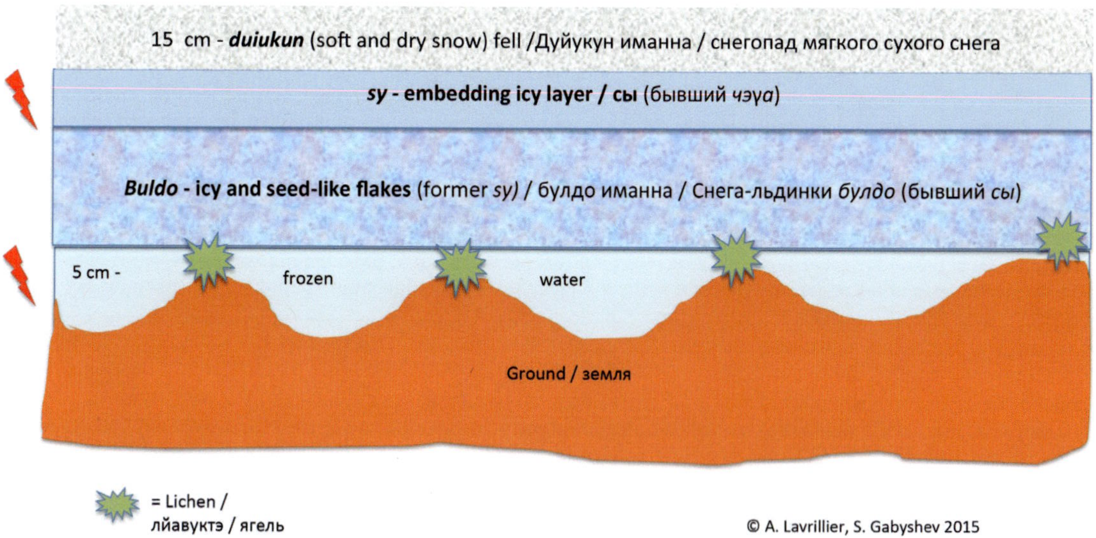

© A. Lavrillier, S. Gabyshev 2015

2014-2015 Anomalies in the installation of the snow cover – Appearance of new snow type?

2014-2015 аномалия в установлении снежного покрова – (похож на 1974) – Новый тип снега?

At the ends of February and December 2014, a new snow type appeared during the abnormal warm periods (-5°C or -8°C) which interspersed long periods of -30-40°C. A few nomads recognised this snow type as 'April snow', which is specific to the warm days of spring: however, the majority of the nomads (including the elders) had never seen such snow before. It seems that this was a new snow type for this region.

В конце февраля и в декабре 2014: появился новый тип снега в период потепления (с -5°C или - 8°C) во время периода с -30°C-40°C. Только редкие таежники идентифицировали этот снег как «апрельский снег», специфический для теплых дней весны, хотя большинство эвенков (особенно пожилые люди) говорили, что ни разу в жизни такого снега не видели. Во всяком случае, этот тип снега кажется новым.

A thin layer of *chuiur* in a place where winds are moderate which is covered by 'April snow'/ Слои тонкого *чуйура,* где средний ветер дует, покрытый «апрельским снегом»

2014-2015 Anomalies in the installation of the snow cover – January - February 2015 – in some places
2014-2015 аномалия в установлении снежного покрова – Январь - Февраль 2015 - местами

ANOMALY – A deep *duiukun*, and *buldo* snow cover cannot form: the pastures are embedded within *sy*/
АНОМАЛИЯ - Суӊта дуйукун, булдо эвки орэ /
Глубокий слой *дуйукун* и *булдо* не может сформироваться и *сы* охватывает олений корм

The thick layer of *duiukun* snow type does not allow the frost through to transform the *duiukun* and *sy* snow types into *buldo*. *Sy* was created by humidity from the snowfall and embedded the grazing pasture. The frost from the air should normally disintegrate the *sy* and transform *duiukun* into *buldo* (cf. diagram *buldo*; *sy*), thus freeing the vegetation. Here, however, the pastures remained embedded in the ice all winter. If wild or domestic reindeer graze upon icy lichen, they loose weight: the weaker may die.

Толстый слой снега *дуйукун* не дает морозу проникнуть внутрь и превратить снег *дуйукун* и снег *сы* в тип снега *булдо*. *Сы* создался от влаги после снегопада и охватывает корм. Мороз от воздуха обычно должен трескать *сы* и превратить *сы и* тип снега *дуйукун* в *булдо* (см. Диаграммы *булдо; сы*) и освободить корм. Но здесь корм со льдом останется на всю зиму: домашние и дикие олени не смогут кормоваться. Если они кормуются ягелем со льдом, то они сильно худеют и самые слабые пропадают.

PS: see the picture of the anomaly and norm in next slides / см. Норма в следующем слайде

ANOMALY – A deep *duiukun*, and *buldo* snow cover cannot form /
АНОМАЛИЯ - Суӈта дуйукун, булдо эвки орэ / Глубокий слой *дуйукун* и *булдо* не может сформироваться

ANOMALY / АНОМАЛИЯ:

Too much *duiukyn* snow does not allow the frost to transform the *duiukun* and *sy* snow into *buldo*.

Слишком много снега *дуйукун* и он не дает морозу проникнуть внутрь и трансформировать *дуйукун* и *сы* в *булдо,* чтобы освободить растительность.

© A. Lavrillier, S. Gabyshev 2015

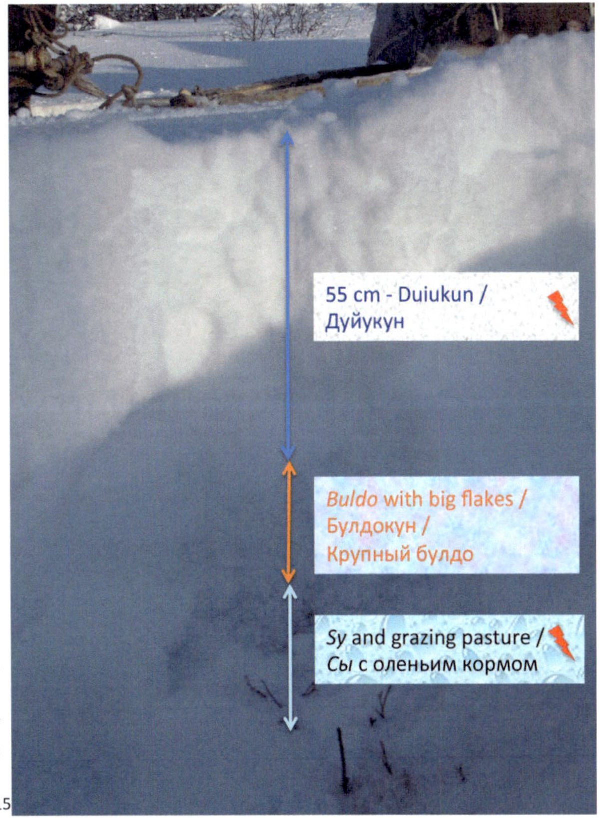

2014-2015 Anomalies in the installation of the snow cover – January - February 2015 – in some places
2014-2015 аномалия в установлении снежного покрова – Январь - Февраль 2015 - местами

NORM – The depth of the *duiukun* snow is 'up to the knee': the frost transforms the *duiukun* and *sy* into *buldo,* freeing the grazing pastures.
НОРМА - Тип снега *дуйукун* «до колено», мороз превращает *дуйукун* и *сы* в *булдо* – освобождает олений корм.

The layer of *duiukun* snow is not too deep, which means that the frost can reach the ground. It 1) transforms the *duiukun* into *buldo* (cf. diagram buldo) and 2) cracks the *sy* within which the lichen is embedded, allowing domestic and wild reindeer to graze freely and normally.

Слой *булдо* снега небольшой, значит мороз может доходить до земли. Этот мороз превращает *дуйукун* в *булдо* (cf. diagram buldo) и освобождает ягель ото льда. Домашние и дикие олени могут хорошо кормоваться

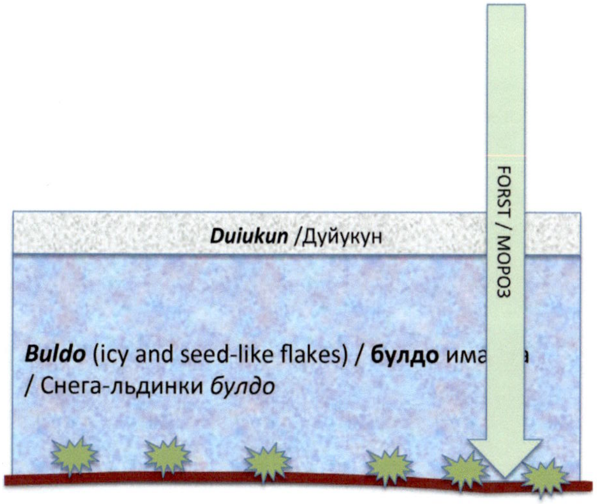

= Lichen /
лйавуктэ / ягель

2014-2015 Anomalies in the installation of the snow cover – mid January 2015 – Moving troubles

2014-2015 аномалия в установлении снежного покрова – середина января 2015 – Проблемы передвижения

ANOMALY – Abnormal warming: a deep *debdeme* layer (fluffy snow) abruptly appeared and was covered by a hard *chuiur* layer created by strong winds. Thus, the frost could not transform the *debdeme* into *buldo*

АНОМАЛИЯ – Глубокий слой пышного снега (*дэбдэмэ*) выпавший за раз и верхний наст, созданный пургой, не дают морозу проникнуть сквозь снежный покров и превратить дэбдэмэ в снег типа *булдо*

During an abnormally warm storm, a deep layer of *debdeme* (fluffy snow) was suddenly deposited and covered by a hard *chuiur* layer created by the storm's strong winds; the frost thus could not transform the *debdeme* into *buldo*, which normally packs down the snow and reduces the depth of the snow cover.

Consequences: for one-to-two weeks, this anomaly created a lot of transportation problems: the nomads got stuck in the middle of their trips and had to spend hours digging their reindeer sledges or snowmobiles out of the porridge-like snow. In addition, the reindeer could not feel the road under the thick snow and hurt their breasts by rubbing them against the hard *chuiur* layer when pulling sledges or grazing. It also created grazing problems for the reindeer because of the snow depth.

Аномальная теплая пурга положила толстый слой пышного и влажноватого снег типа *дэбдэмэ*. Одновременно, из-за пурги создался твердый наст от ветра *чуйур*. Тот не дает морозу проникнуть внутрь и превратить снег *дэбдэмэ* в *булдо*. Это позволило бы уменьшить толщину снега (слои снежного покрова уменьшаются при трансформации снега в снег *булдо*).

Последствия: в течение одного-двух недель многие транспортные проблемы появились: кочевники застряли в середине пути и потратили много часов, выкапывая оленьи упряжки или снегоходы от этого снега, похожего на кашу. Мало того, олени не чуяли дороги под глубоким снегом и ранили себе грудь об твердый надув снега (*чуйур*), когда в упряжке или кормуются. Это также осложнило кормление оленей из-за глубины.

3.6 Ice anomalies (Ulan bukte)

2013-2014 - Anomaly – *ulan*, *bukte* in small rivers and *amnunna*

Аномалия – *улан*, *буктэ* - Маленькие реки и *амнунна*

First, a considerable snowfall deposited deep and wet snow on unfrozen ground. As a result, there was no *ulan* on the *amnunna:* the river had frozen into thin layers of ice between which empty spaces appeared. From between these layers, the air escaped outside. The melted water flowed deep under the ice and could not escape. When a larger amount of snow fell in September, the rivers (and ground) did not freeze. Falling on the flowing water, the snow melted. The small amount of snow that fell in October remained on the surface ice layer that appeared during the first frost of 1 October. After 1 October, the lever of water in the rivers decreased: on 10 October, a new layer of ice formed during the frost. The same thing occurred on 20 October. Between these ice layers, empty spaces were created. The freezing of the ground is an essential factor in the normal formation of ice on rivers.

Сперва выпало много мокрого снега, а земля была не замерзшая. В результате на *амнунна* не было *улан,* потому что сама речка, замерзая, образовала разные слои льда с пустотами между ними. Через эти слои воздух выходит. Талая вода течет глубже по дну, и не может выйти наружу. Когда большая масса снега выпала в сентябре, река (и земля) не были замерзшими. Падая на воду, снег таял. Небольшой снег, выпавший в октябре, остался на тонком слое льда, который появился при первых заморозках 1 октября. После 1-ого октября уровень самой реки падал, а в 10-ых числах октября при заморозках образовался еще один тонкий слой льда, и позже в 20-ых числах октября. И поэтому между ними появился пустолед. Замерзание земли является важнейшим фактором для формирования льда рек.

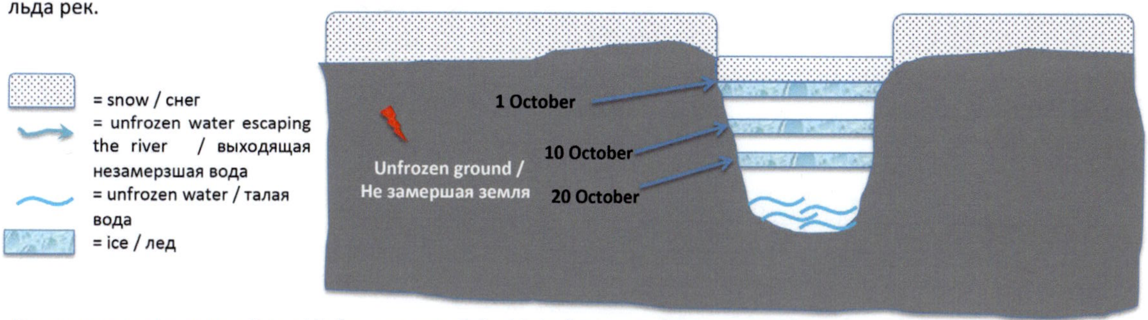

Consequences: because of the thin ice, one can sink: this is dangerous for transportation.
Последствия = для транспорта опасно, из-за тонкого льда можно провалиться.

Norm – *ulan*, *bukte* – in mall rivers, *amnunna*

Норма – *улан*, *буктэ* - Маленькие реки и *амнунна*

First, the ground becomes completely frozen: the upper layer of ice freezes in around two weeks. Afterwards, *duiukun* snow (soft and dry) snow falls. The air contained within the water escapes (together with water) and creates an *ulan* above the ground.

Сперва земля и верхний слой реки полностью замерзают в течение 2х недель. Потом выпадает хороший сухой снег. А после этого воздух, задержанный в воде, выходит наружу вместе с водой и создает *улан* над землей и на льду.

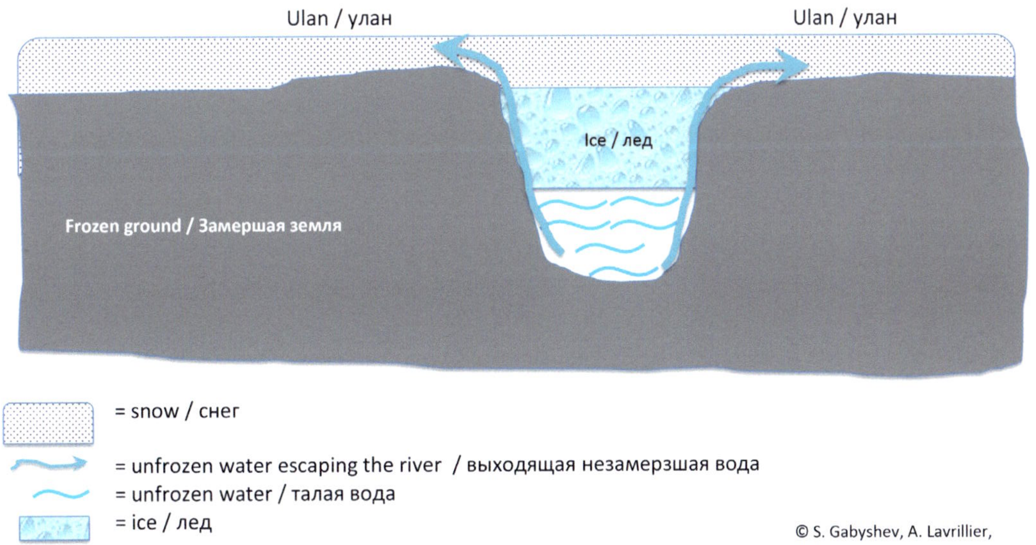

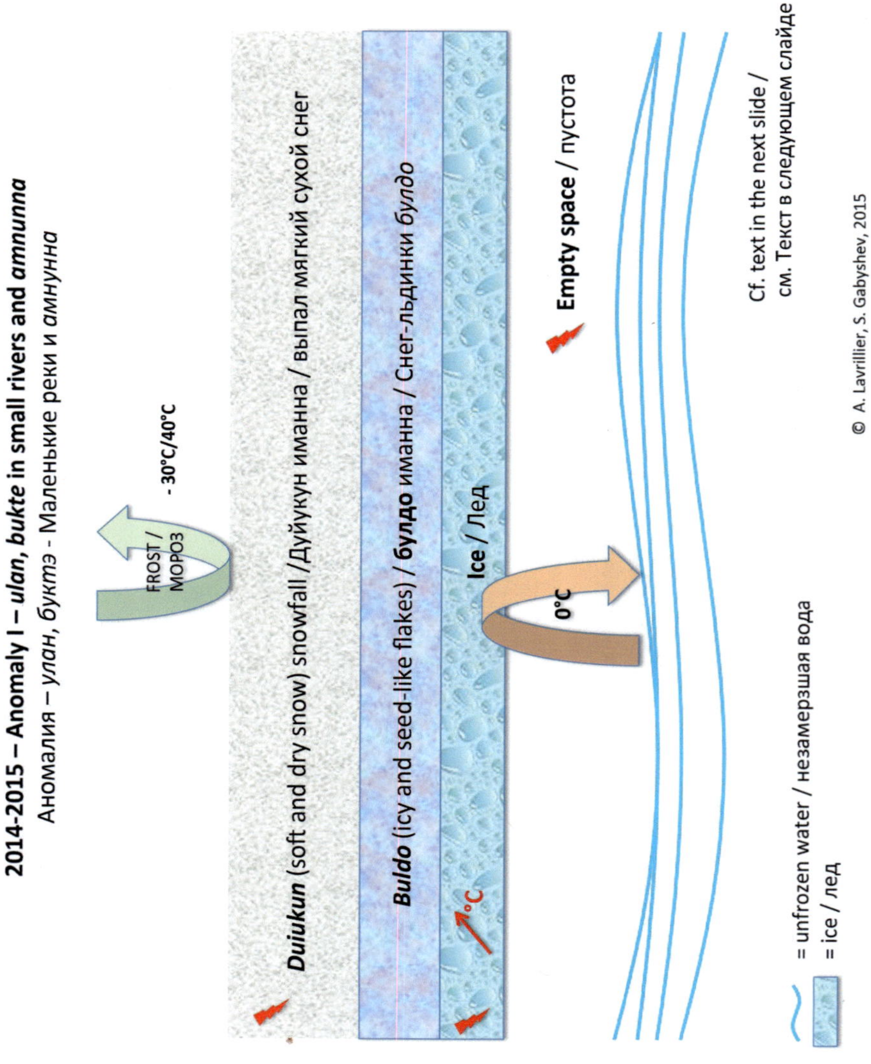
2014-2015 – Anomaly I – ulan, bukte in small rivers and amnunna
Аномалия – улан, буктэ - Маленькие реки и амнунна
FROST / МОРОЗ
- 30°C/40°C
Duiukun (soft and dry snow) snowfall /Дуйукун иманна / выпал мягкий сухой снег
Buldo (icy and seed-like flakes) / булдо иманна / Снег-льдинки булдо
Ice / Лед
°C
0°C
Empty space / пустота
= unfrozen water / незамерзшая вода
= ice / лед
Cf. text in the next slide /
см. Текст в следующем слайде
© A. Lavrillier, S. Gabyshev, 2015

2014-2015 – Anomaly I – *ulan, bukte* in small rivers and *amnunna*

Аномалия – *улан, буктэ* - Маленькие реки и *амнунна*

In 2014-15, the ice was abnormally thin (50% thinner than the norm). In January 2015, the ice was already starting to melt under the snow: many nomads were sinking into it when travelling by snowmobile and reindeer sledge.

The ice was thin because the deep layer of snow did not allow the frost in the outside air to go through the snow cover to the ice layer: equally, the water below did not freeze, so the ice did not become thicker. The ice melted because of the insulation role played by the overly deep snow cover: the ice was affected only by warm water (0°C), so it started to melt (*See the next diagram for comparison with the norm*).

В 2014-2015 гг. лед был ненормально тонким, наполовину меньше нормы. В январе 2015 лед уже начинает таять под снегом и многие кочевники проваливаются на снегоходах и на оленях.

Лед тонкий из-за толстого слоя снега, который не дает морозу проникнуть до льда. Из-за этого лед не становится толще (не замерзает с низу). Дополнительно, толстый слой снега создает термальную изоляцию и лед находится под влиянием теплой температуры воды (0°C) и начинает таять. (*см. Следующая диаграмма для сравнения с нормой*)

2014-2015 - Anomaly – *ulan*, *bukte* in small rivers and *amnunna*
Аномалия – *улан*, *буктэ* - Маленькие реки и *амнунна*

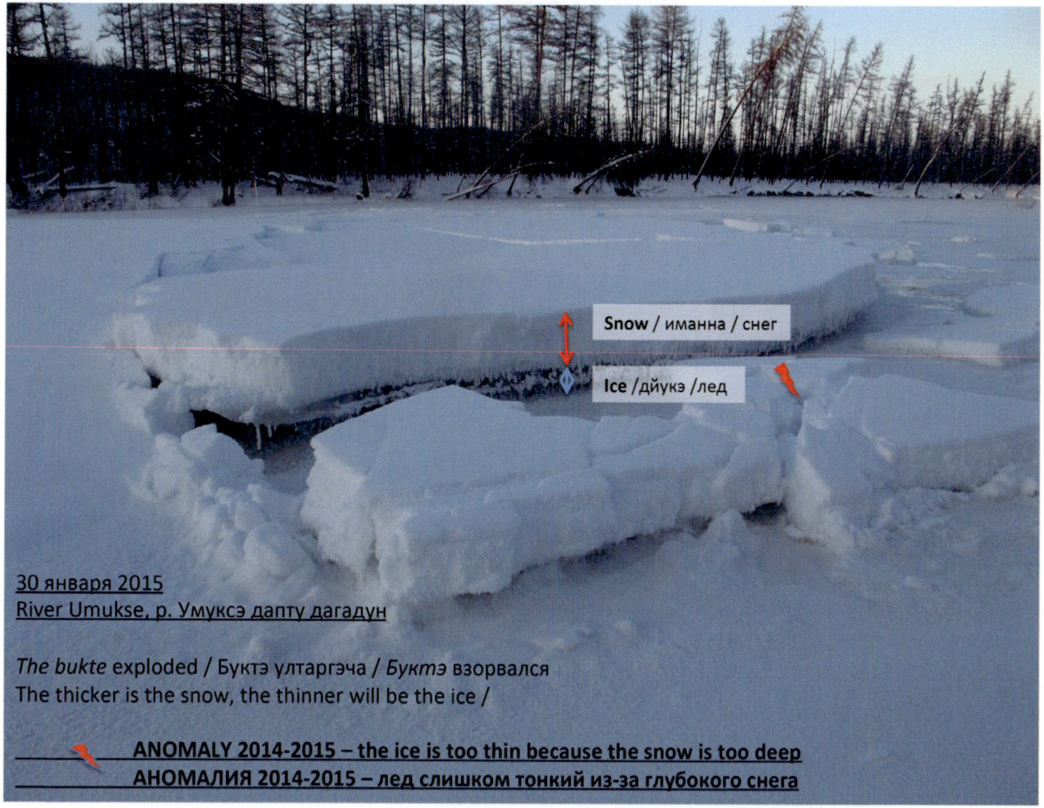

© A. Lavrillier, 2015

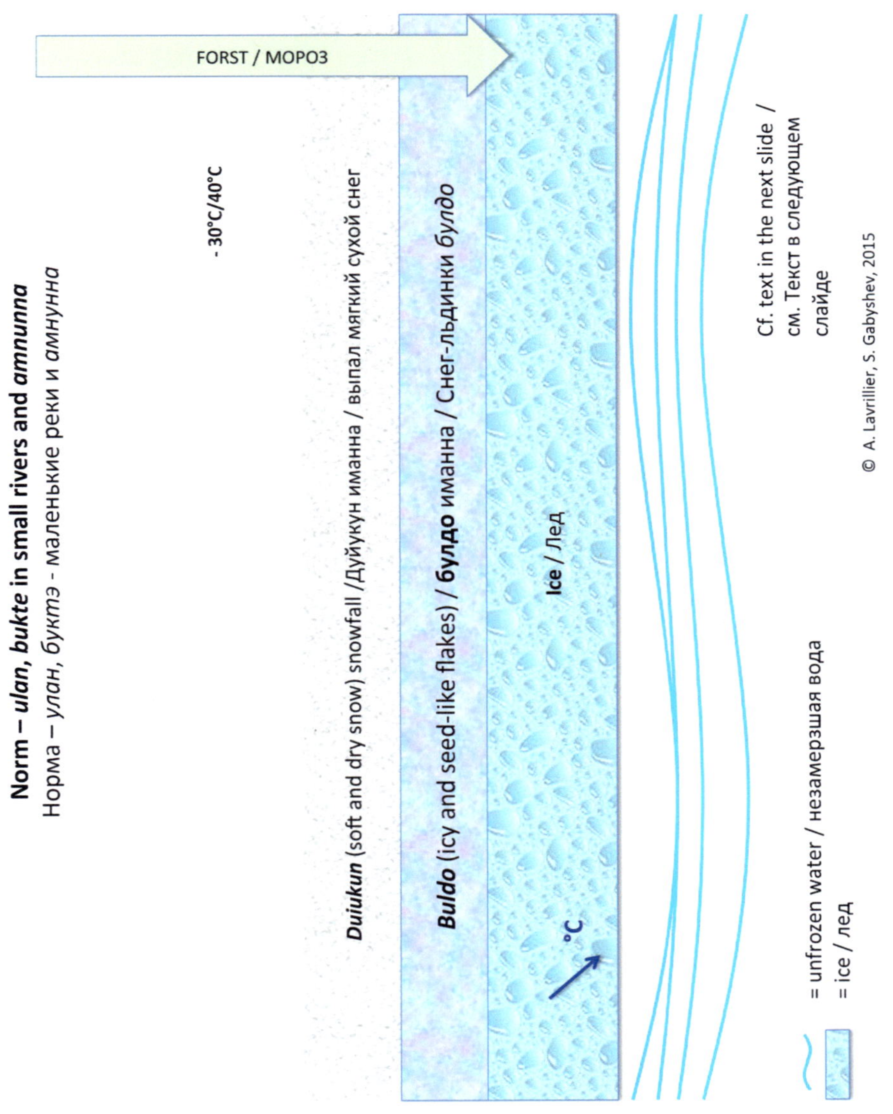
Norm – *ulan, bukte* in small rivers and *amnunna*
Норма – *улан, буктэ* - маленькие реки и *амнунна*
- 30°C/40°C
FORST / МОРОЗ
Duiukun (soft and dry snow) snowfall /Дуйукун иманна / выпал мягкий сухой снег
Buldo (icy and seed-like flakes) / **булдо** иманна / Снег-льдинки *булдо*
Ice / Лед
°C
= unfrozen water / незамерзшая вода
= ice / лед
Cf. text in the next slide / см. Текст в следующем слайде
© A. Lavrillier, S. Gabyshev, 2015

Norm – *ulan*, *bukte* in small rivers and *amnunna*

Норма – *улан, буктэ* - Маленькие реки и *амнунна*

The ice is deep because the thin layer of snow allows the frost from the outside air to move through the snow cover to the ice layer. This makes the ice thicker by freezing the water below. In addition, the ice does not melt because the snow does not insulate it and there is no influence of warm water (0°C). (*see the next diagram for comparison with the norm*)

Лед толстый из-за тонкого слоя снега, который дает морозу проникнуть до льда. Из-за этого лед становится толще (замерзает снизу). Дополнительно, тонкий слой снега не создает термальную изоляцию и лед не находится под влиянием теплой температуры воды (0°C) и не тает. (см. *Следующая диаграмма для сравнения с нормой*)

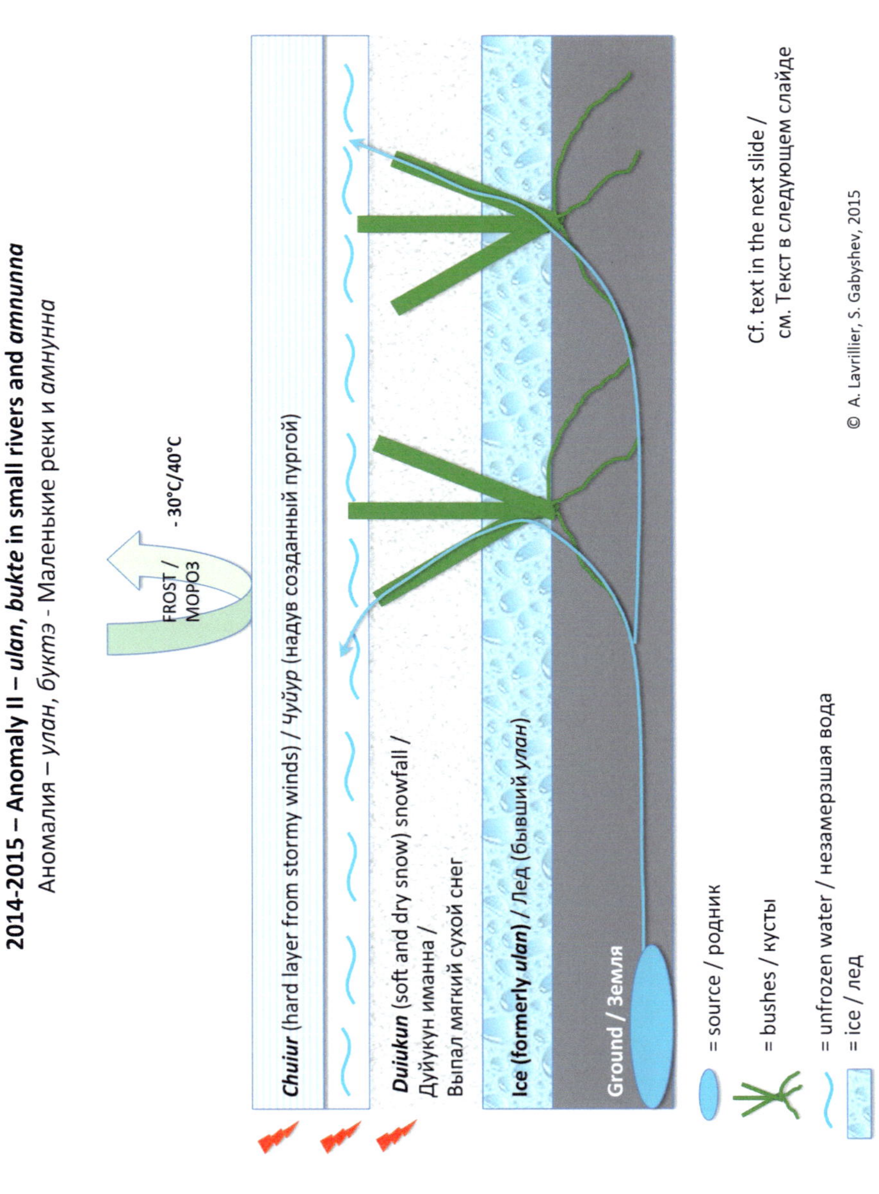

2014-2015 – Anomaly II – *ulan, bukte* in small rivers and *amnunna*
Аномалия – *улан, буктэ* - Маленькие реки и *амнунна*
FROST / МОРОЗ
- 30°C/40°C
Chuiur (hard layer from stormy winds) / *Чуйур* (надув созданный пургой)
Duiukun (soft and dry snow) snowfall /
Дуйукун иманна /
Выпал мягкий сухой снег
Ice (formerly *ulan*) / Лед (бывший *улан*)
Ground / Земля
= source / родник
= bushes / кусты
= unfrozen water / незамерзшая вода
= ice / лед
Cf. text in the next slide /
см. Текст в следующем слайде
© A. Lavrillier, S. Gabyshev, 2015

2014-2015 – Anomaly II – *ulan, bukte* in small rivers and *amnunna*

Аномалия – *улан, буктэ* - Маленькие реки и *амнунна*

An ulan appeared and the water escaped through passages used by the air (bushes and their roots, from under the ice on river banks, and from underground sources). This became a mixture of water and snow that spread everywhere (sometimes it was very deep). Water flowed under the hard surface layer (*chuiur*). As a result of the insulation role played by the *chuiur* layer, the water of the *ulan* could not freeze. The snow cover remained a deep porridge-like layer, which was very bad for transportation.

Улан появились и вода вышла наружу через все, что пропускает воздух: из кустов, по корням или из-подо льда по берегам рек, из родников и в результате вода со снегом распространилась всюду (смесь вода со снегом иногда очень глубокая). Вода потекла под коркой снега *чуйура*. В итоге из-за изоляции, созданной снегом *чуйур*, воды *улан* не замерзают. Снег становится как каша, которая создает много проблем для передвижения.

Norm – *ulan*, *bukte* in small rivers and *amnunna*

Норма – *улан*, *буктэ* - Маленькие реки и *амнунна*

Water escaped through the same passages used by the air: bushes and their roots, from under the ice on river banks, and from underground sources. The frost in the air outside froze the water and created a new ice layer.

Вода вышла наружу через все, что пропускает воздух: из кустов, по корням, из-подо льда по берегам рек, из родников. От мороза с воздуха сразу создается новый слой льда.

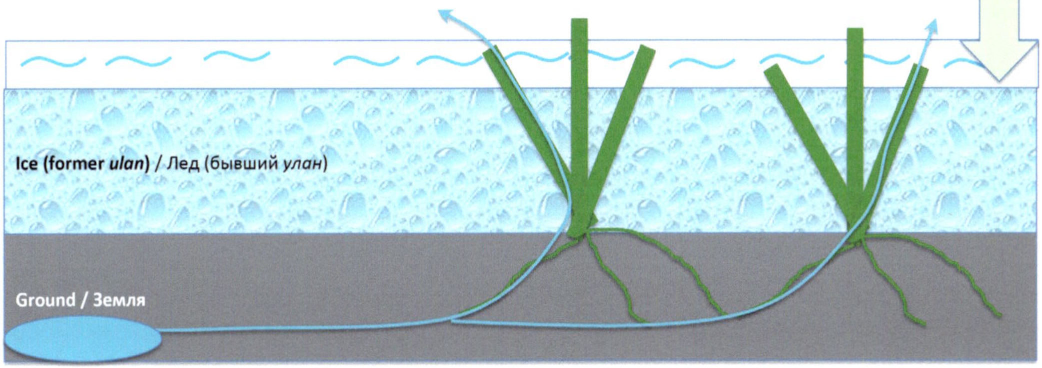

3.7 Landscape transformation due to climate change

Transformation of the landscape due to climate change - Изменение ландшафта с изменением климата
How ice creates a new large river basin (*Amnunna*) - Как лед формирует новый бассейн реки *(амнунна)*
Омакта амнунна одйэрэн дйукэдук.

3-5 year-old larches and birches are still in this place. This means no *bukte* has formed here for 3-5 years / Ирйактэкачар, чалбакар йулчал 3-5 аӈнанни чаҕаду – эрэкин буктэрэ / Молодые деревья (3 – 5 лет). Их присутствие говорит о том, что уже 3-5 лет на этом месте не формировался *буктэ.*

Former *bukte* (icing blister) /Туҕэптэ буктэ / Бывший наледный бугор пучения.

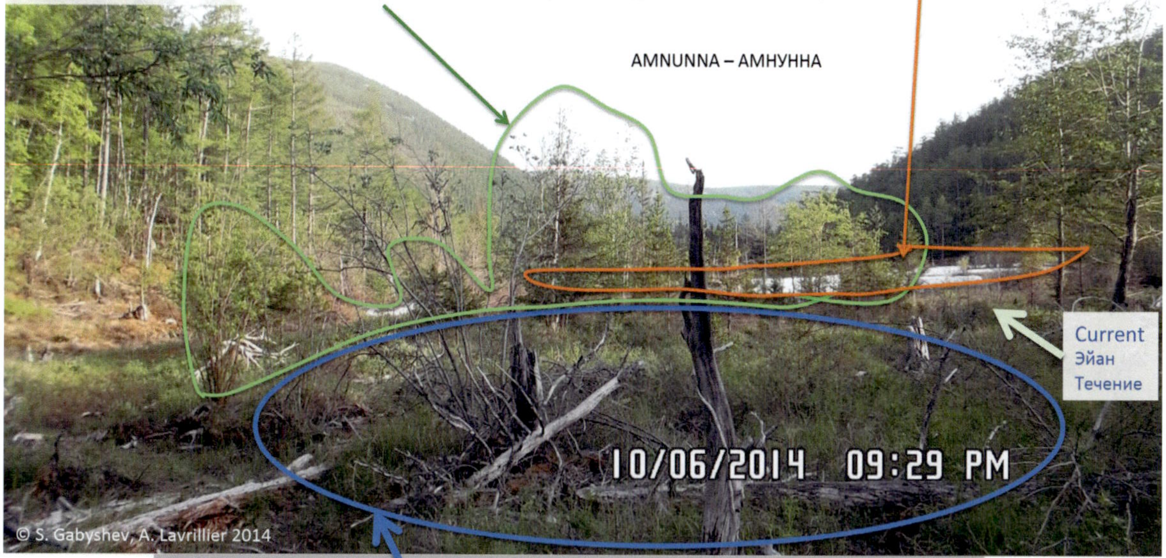

1st step: 1980 – Dead trees crushed by the thick ice from the *bukte* /
Олгокир ирйактэл тэрэпчан нонон дйукэкурдук, тадук олгочол /
1 этап: амнунна 1980 – мертвые деревья раздавлены толстыми льдами *буктэ*

Transformation of the landscape due to climate change - Изменение ландшафта с изменением климата

How ice creates a new large river basin (*Amnunna*) - Как лед формирует новый бассейн реки *(амнунна)*

The *amnunna* is at the centre of Evenki life (cf. Topographic typology). For instance, reindeer can graze on horsetail there and enjoy the freshness of the left-over ice when it is very hot in the summer: there also are far fewer mosquitoes. In the winter, one can take water from the *ulan* (polynia). Domestic and wild reindeer come to the *amnunna* to lick the ice.
Because of climate change, there are fewer *bukte*. If there are **no *bukte***, there will be **no large *amnunna* river basins** because the area will be entirely covered by trees: in other words, it will be a *teŋke* landscape (cf. Vegetal cover typology).
In the 1980s and 1990s, it was very cold and, consequently, there were a lot of *bukte* along the large river basins. Huge amounts of water flowed out from the *ulan* and turned into ice. In the spring, during the thaw, enormous quantities of ice crashed down and crushed the trees, which then dried out. This is why many *amnunna* appeared.
The picture and diagrams were completed on the initiative of S. Gabyshev, who expressed a hypothesis about past and future transformations in the landscape: he derived this hypothesis from both collective Evenki knowledge and his own individual thoughts.

Амнунна со эвэнкилду надо: орор сивакдйэвкил, дйукэлду соңундукин тэӡэчэвкил, мармактал эвкил таду дйэптэ, туӡэ мулидйэми айа уландйэвки, уланду орор имадйэвкил, бэйур нйан.
Нйамалдйэрэкин климат, буктэ адыкулэвкил. Эрэкин буктэрэ амуннаду ирйактэл балдылэвкил. Тадук амнунна ачэн одиңан – ирйактэл балдыдиӡарэ – тэңкэкун одиңан. Нонон (1980-1990 аңнанилду) талы иңикакун бичан, буктэ упкат амуннава уланэнкин, ирйактэл ачэр бичатын.

***Амнунна* находится в центре кочевой жизни эвенков** (cf. Topographic typology). Например, там олени кушают хвощ, лежат на прохладных льдах, когда жарко, и там комары их не кусают. Зимой можно ходить туда за водой. И на наледи *амнунны* домашние и дикие олени спускаются с горы, чтобы лизать соль.
Из-за потепления климата, *буктэ* меньше становится. Если **не будет *буктэ*** – то ***амнунна* скорее всего не будет** – *амнунна* зарастет деревьями, *тэңкэ* станет (cf. Vegetal cover typology). Раньше (в 1980-1990х годах) сильно холодно было и поэтому были очень большие *буктэ* по всему *амнунна* и наледи были, большие массы воды выливались, которые становились льдом. Весной, при таянии льда большая масса льда рухнет и разбивает молодые деревья, которые потом высыхают, и поэтому было много широких бассейнов реки.
Фотография и диаграммы сделаны по инициативе С. Габышева для того, чтобы выразить его гипотезы о прошлом формировании и трансформации в будущем ландшафта, согласно не только традиционному эвенкийскому знанию, но и также согласно собственных размышлений.

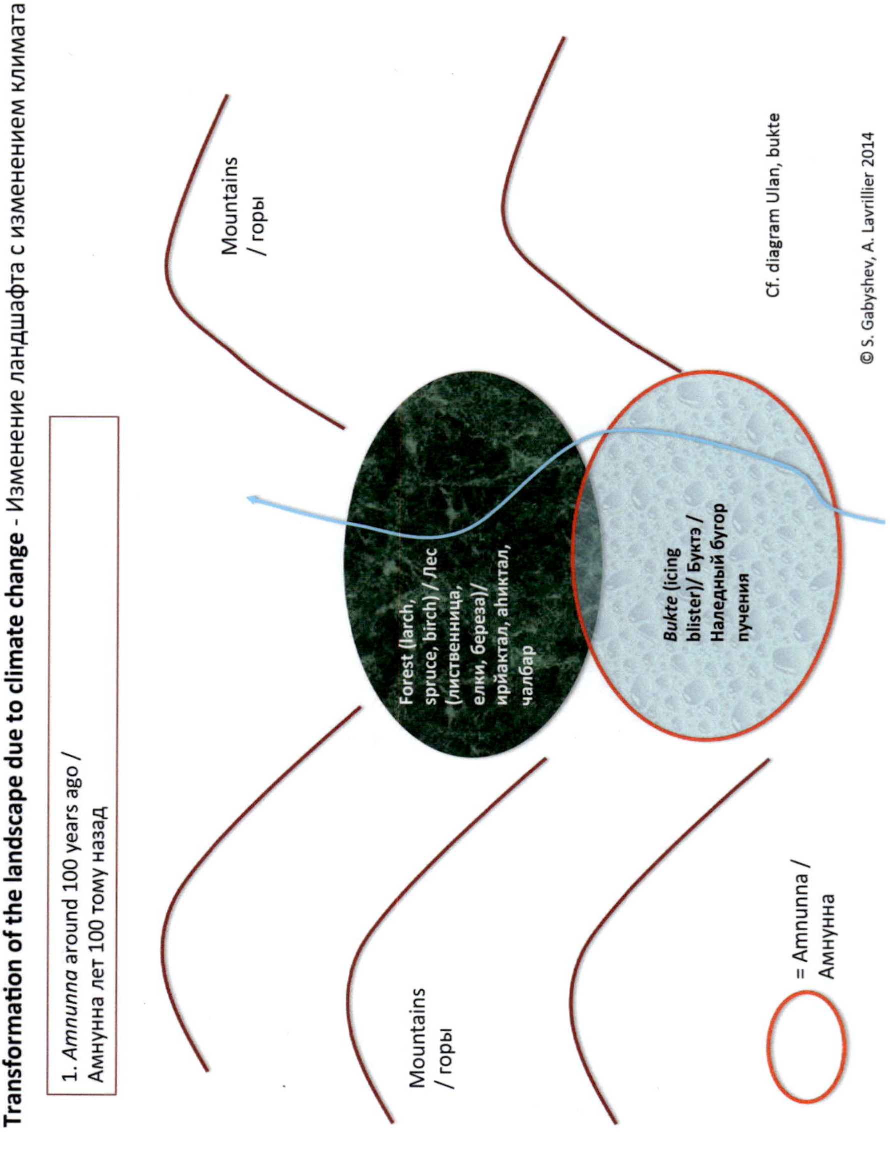
Transformation of the landscape due to climate change - Изменение ландшафта с изменением климата
1. Amnunna around 100 years ago / Амнунна лет 100 тому назад
Mountains / горы
Mountains / горы
Forest (larch, spruce, birch) / Лес (лиственница, елки, береза)/ ирйактал, аhиктал, чалбар
Bukte (icing blister)/ Буктэ / Наледный бугор пучения
Cf. diagram Ulan, bukte
= Amnunna / Амнунна
© S. Gabyshev, A. Lavrillier 2014

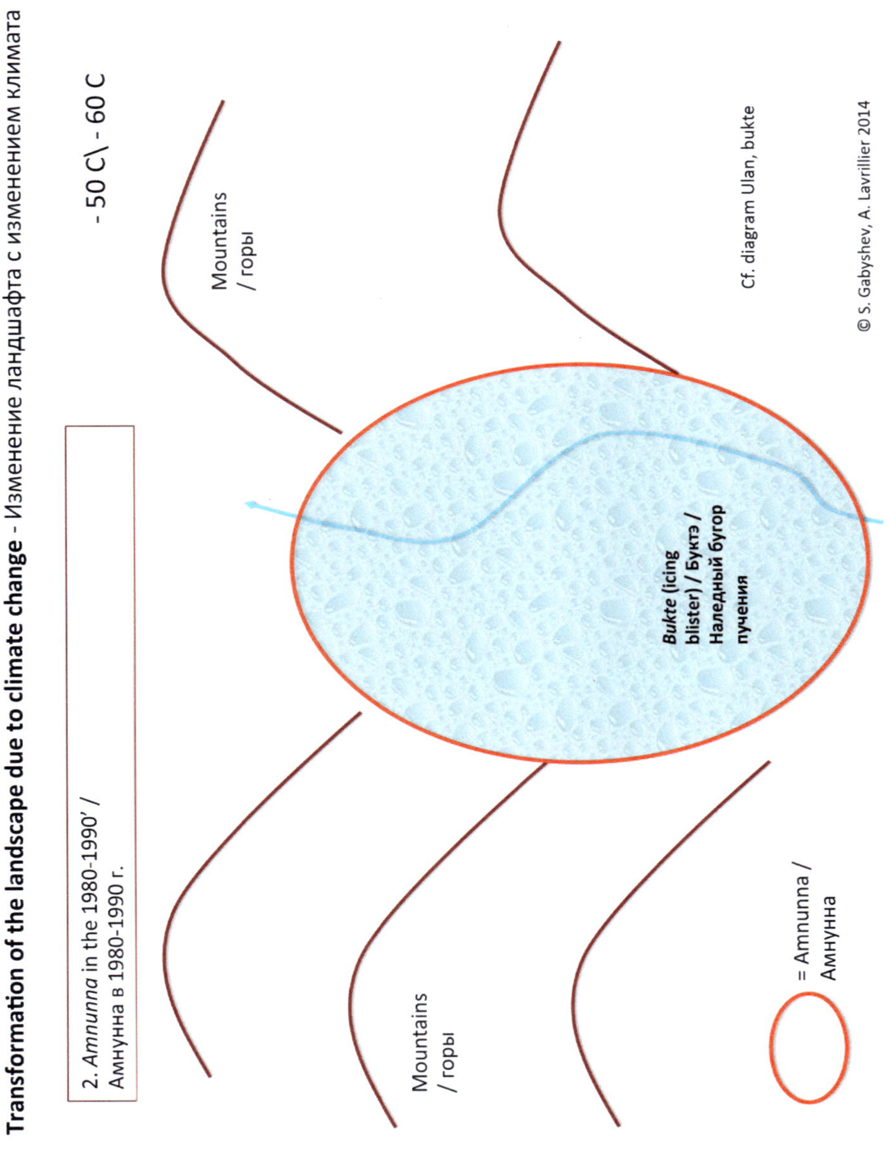
Transformation of the landscape due to climate change - Изменение ландшафта с изменением климата
2. *Amnunna* in the 1980-1990' /
Амнунна в 1980-1990 г.
- 50 C\ - 60 C
Mountains
/ горы
Mountains
/ горы
Bukte (icing blister) / Буктэ / Наледный бугор пучения
Cf. diagram Ulan, bukte
= Amnunna /
Амнунна
© S. Gabyshev, A. Lavrillier 2014

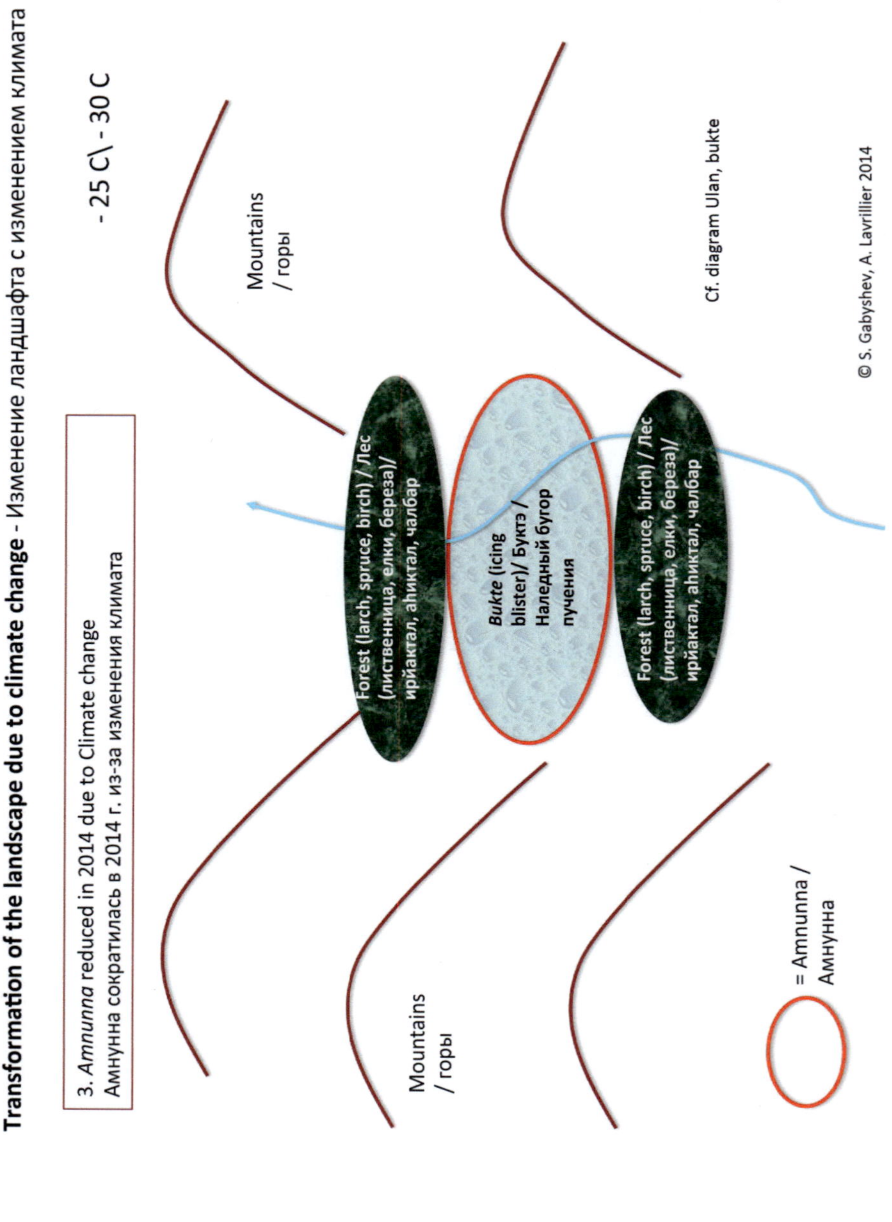
Transformation of the landscape due to climate change - Изменение ландшафта с изменением климата
3. Amnunna reduced in 2014 due to Climate change
Амнунна сократилась в 2014 г. из-за изменения климата
- 25 C\ - 30 C
Mountains / горы
Forest (larch, spruce, birch) / Лес (лиственница, елки, береза)/ ирйактал, аhиктал, чалбар
Bukte (icing blister)/ Буктэ / Наледный бугор пучения
Forest (larch, spruce, birch) / Лес (лиственница, елки, береза)/ ирйактал, аhиктал, чалбар
Mountains / горы
Cf. diagram Ulan, bukte
© S. Gabyshev, A. Lavrillier 2014
= Amnunna / Амнунна

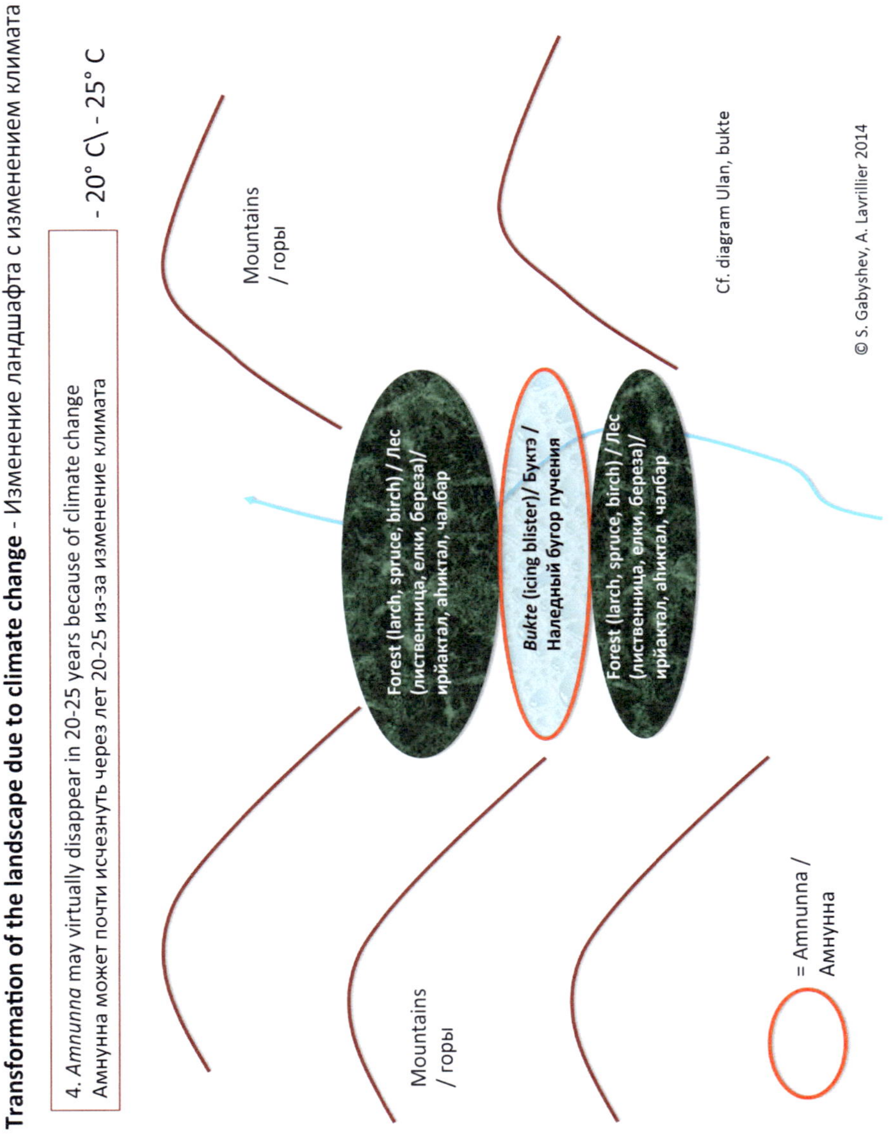
Transformation of the landscape due to climate change - Изменение ландшафта с изменением климата
4. *Amnunna* may virtually disappear in 20-25 years because of climate change
Амнунна может почти исчезнуть через лет 20-25 из-за изменение климата
- 20° C\ - 25° C
Mountains / горы
Forest (larch, spruce, birch) / Лес (лиственница, елки, береза)/ ирйактал, аһиктал, чалбар
Bukte (icing blister)/ Буктэ / Наледный бугор пучения
Forest (larch, spruce, birch) / Лес (лиственница, елки, береза)/ ирйактал, аһиктал, чалбар
Mountains / горы
Cf. diagram Ulan, bukte
= Amnunna / Амнунна
© S. Gabyshev, A. Lavrillier 2014

3.8 Inter-annual comparisons and modelling (norms, anomalies)

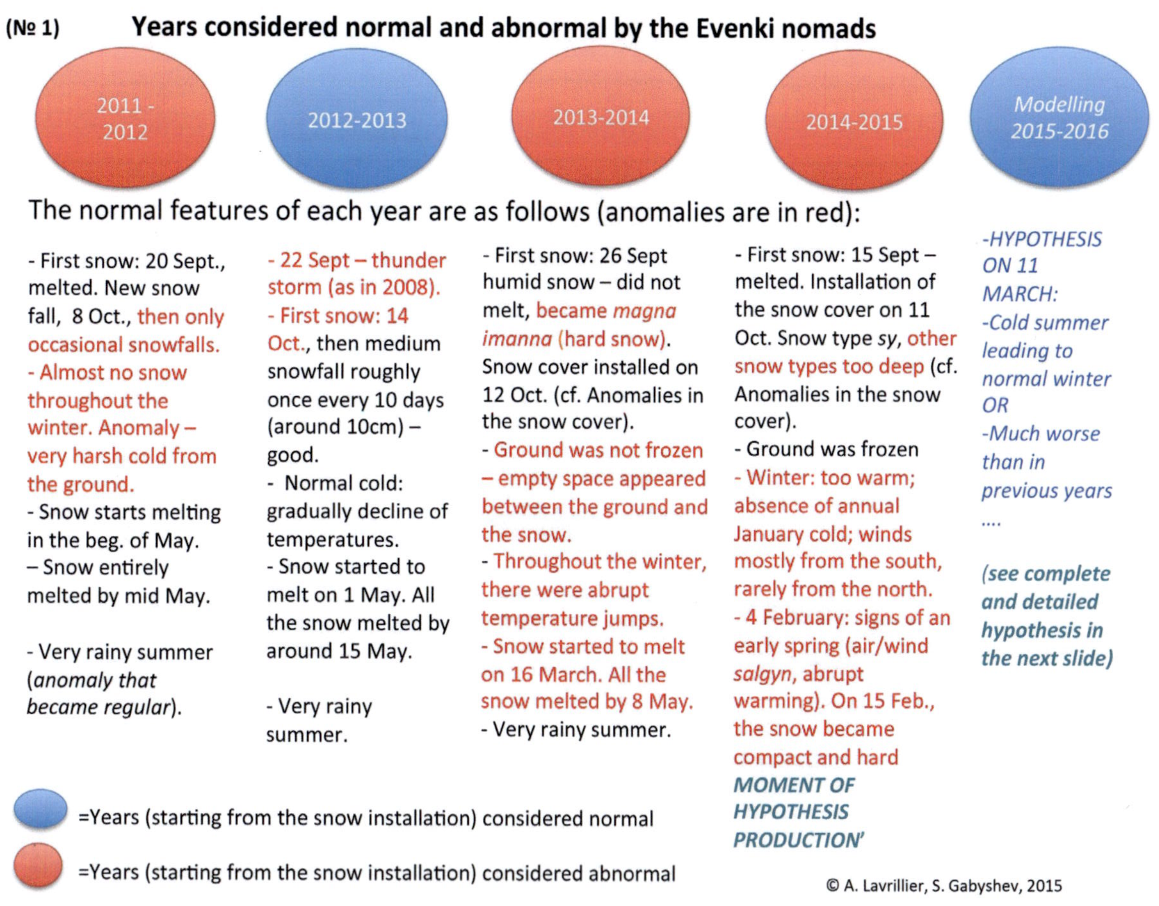

(№ 1) Годы считаются нормальными или аномальными оленеводами…

А по фактам следующие характеристики (в красном – аномалии)…

2011 - 2012	2012-2013	2013-2014	2014-2015	*Прогноз 2015-2016*
- Первый снег – 20/09 растаял, выпал 8 октября, потом редко выпадал. - Снега не было почти всю зиму- Аномальный холод от земли. - Снег нач. таять в нач. мая – снег полностью растаял в середине мая. - Очень дождливое лето (*регулярная аномалия*).	- 22/09 – гроза (//2008). -Первый снег- 14 октября, потом выпадал поэтапно раз в неделю по 10 см примерно – хорошо. Нормальный холод – постепенное похолодание. - Снег начал таять 1 мая – снег полностью растаял 15 мая. - Очень дождливое лето.	- Первый снег-26/09 влажный снег – не растаял, стал *маңа иманна*. Установился 12 окт (cf. Anomalies in the snow cover). - Земля талая была – пустота между землей и снегом. - Всю зиму резкие перепады температуры. - Снег начал таять 16 марта – снег полностью растаял 8 мая. - Очень дождливое лето.	- Первый снег-15/09 – растаял. Установился 11 окт. Местами снег *сы*, слишком глубокий снег (cf. Anomalies in the snow cover). - Земля мерзлая. - Зима: слишком теплая; крещенских морозов не было; частые южные ветра; редко северный ветер. - 4 февраля: Признаки ранней весны (*салгын*, и очень тепло резко стало); 15 фев.: Снег компактный (*марны*). ***..МОМЕНТ ПРОГНОЗА***	*-ПРОГНОЗ НА 11 МАРТА:* *-Лето холодное будет, значит нормальная зима будет ИЛИ еще хуже, чем в прошлые годы* ….. -***(см. гипотезы в след. диаграмме)***

=год (с момента установления снежного покрова) с нормальным

=год (с момента установления снежного покрова) с аномальным снежным периодом

(№ 2) **Evenki modelling of the climate and environment /**
Эвенкийское моделирование климата и природы

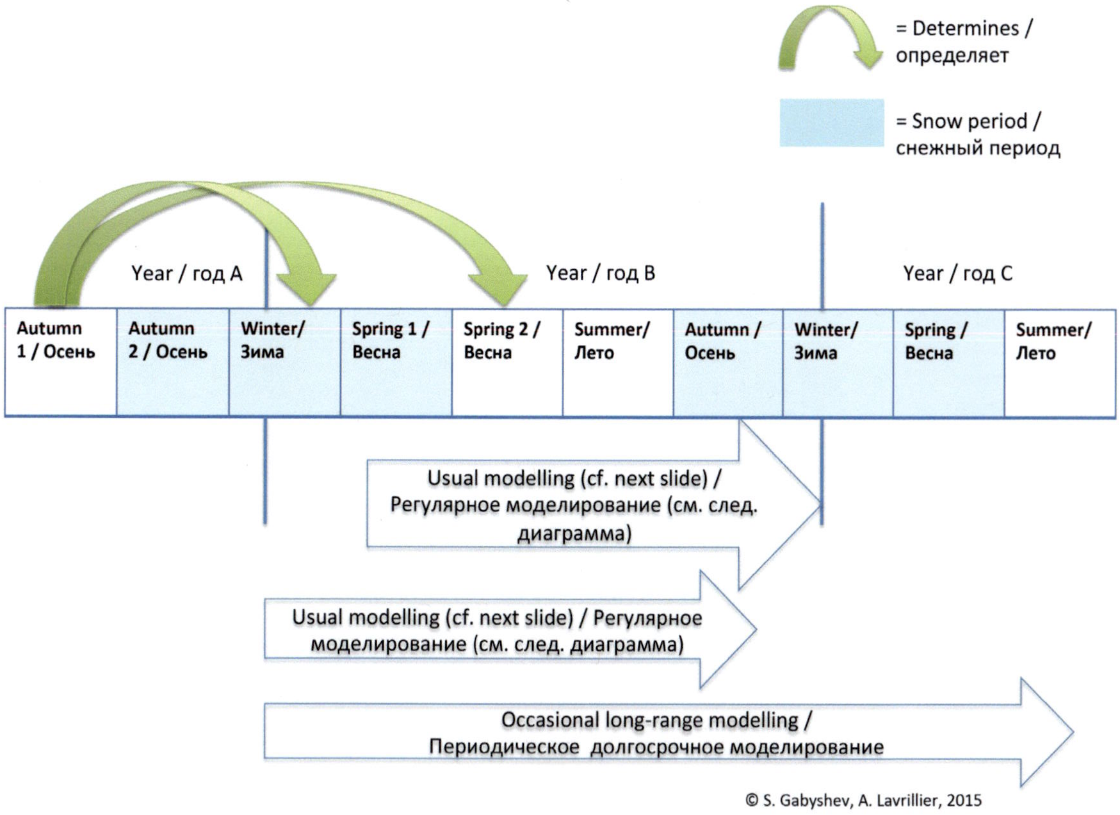

Evenki modelling of the climate and environment / эвенкийское моделирование климата и природы

The Evenki regularly, if not constantly, 'model' (cf. Lavrillier, Gabyshev, Rojo 2016) at least 8-10 months ahead (and sometimes several years ahead). Indeed, several times throughout the year (especially when they observe anomalies), they produce hypotheses for events expected in the climate and the natural environment. They immediately produce several hypotheses based on observed phenomenon or received knowledge, engaging the typologies we presented. For instance, from each assumption they advance several hypotheses for possible developments. Each hypothesis is related not only to the climate, but also to the vegetal and animal realms and to the consequences for the Evenki. Sometimes, they produce hypotheses for three or more years ahead; this can be compared to 'modelling' the inter-influences between climate, landscape, and humans. (cf. Introduction, Evenki climatology, Conclusion)

Slides 1 and 3 were completed on Lavrillier's initiative in order to compare the anomalies for each year and to express Gabyshev's oral hypothesis in writing. The second slide is a diagram created on the initiative of Gabyshev, which was first drawn in a notebook.

Несколько раз в год, особенно если наблюдается аномалия, эвенки делают прогноз на несколько месяцев вперед о всевозможных изменениях в природе и климате. Выдвигают сразу несколько гипотез в зависимости от проведенных наблюдений. Например, от каждого предположения выдвигают несколько прогнозов возможных развитий. Каждый прогноз касается не только климата, но и растительного и животного мира, а также как и последствия этих изменений для эвенков. Иногда кочевники делают выводы на 3-4 года вперед, тогда это можно было бы сравнить с моделированием взаимосвязи климата, природы, и человека.

Слайды 1 и 3 сделаны по инициативе Лаврилье, с целю сравнить аномалии между годами и также представить в письменном виде прогнозы Габышева. Слайд 2 был сделан по инициативе Габышева, сначала нарисованный на блокноте.

= Consecutive normal events / подследственные нормальные факты

= Consecutive anomaly / подследственная аномалия

= Observed facts наблюдаемые факты

= Modelled changes / Прогнозируемые изменения

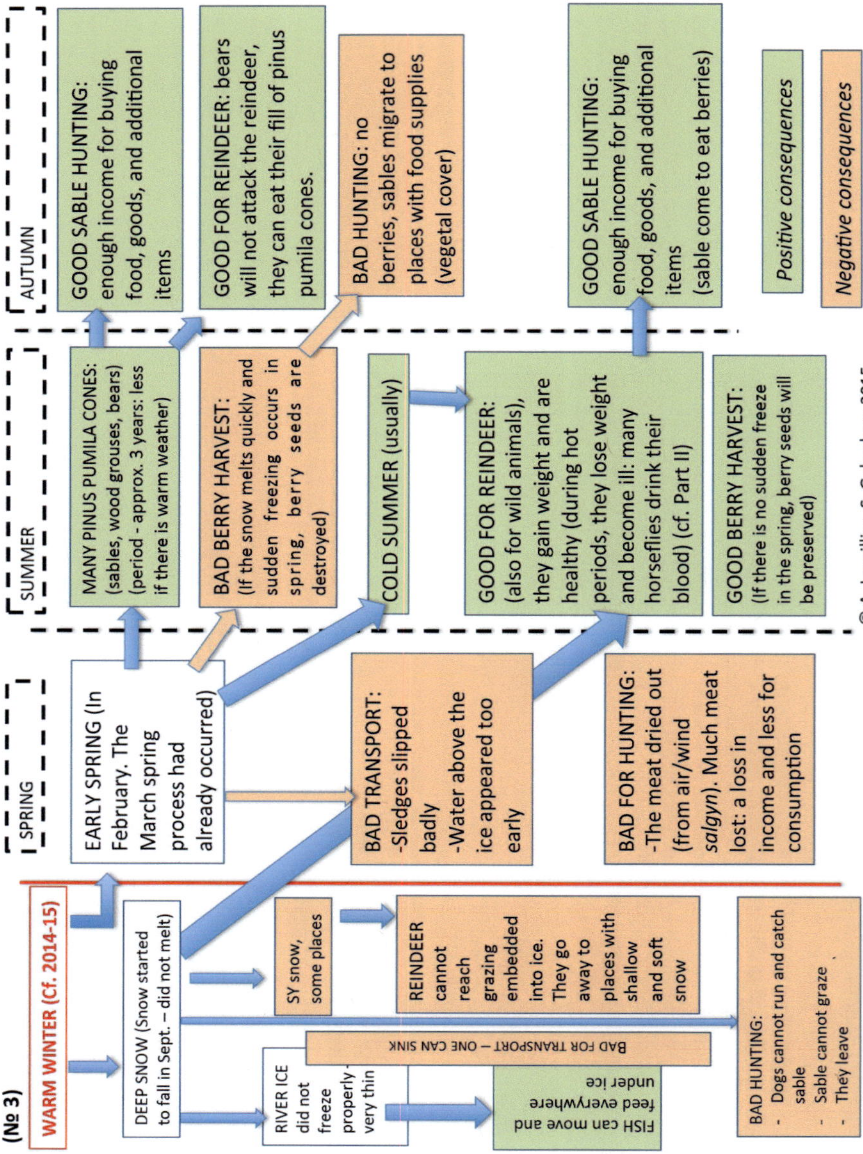
(№ 3)
WARM WINTER (Cf. 2014-15)
DEEP SNOW (Snow started to fall in Sept. – did not melt)
RIVER ICE did not freeze properly - very thin
FISH can move and feed everywhere under ice
BAD FOR TRANSPORT – ONE CAN SINK
SY snow, some places
REINDEER cannot reach grazing embedded into ice. They go away to places with shallow and soft snow
BAD HUNTING: - Dogs cannot run and catch sable - Sable cannot graze - They leave
SPRING
EARLY SPRING (In February. The March spring process had already occurred)
BAD TRANSPORT: -Sledges slipped badly -Water above the ice appeared too early
BAD FOR HUNTING: -The meat dried out (from air/wind *salgyn*). Much meat lost: a loss in income and less for consumption
SUMMER
MANY PINUS PUMILA CONES: (sables, wood grouses, bears) (period - approx. 3 years: less if there is warm weather)
BAD BERRY HARVEST: (If the snow melts quickly and sudden freezing occurs in spring, berry seeds are destroyed)
COLD SUMMER (usually)
GOOD FOR REINDEER: (also for wild animals), they gain weight and are healthy (during hot periods, they lose weight and become ill: many horseflies drink their blood) (cf. Part II)
GOOD BERRY HARVEST: (If there is no sudden freeze in the spring, berry seeds will be preserved)
AUTUMN
GOOD SABLE HUNTING: enough income for buying food, goods, and additional items
GOOD FOR REINDEER: bears will not attack the reindeer, they can eat their fill of pinus pumila cones.
BAD HUNTING: no berries, sables migrate to places with food supplies (vegetal cover)
GOOD SABLE HUNTING: enough income for buying food, goods, and additional items (sable come to eat berries)
Positive consequences
Negative consequences
© A. Lavrillier, S. Gabyshev, 2015

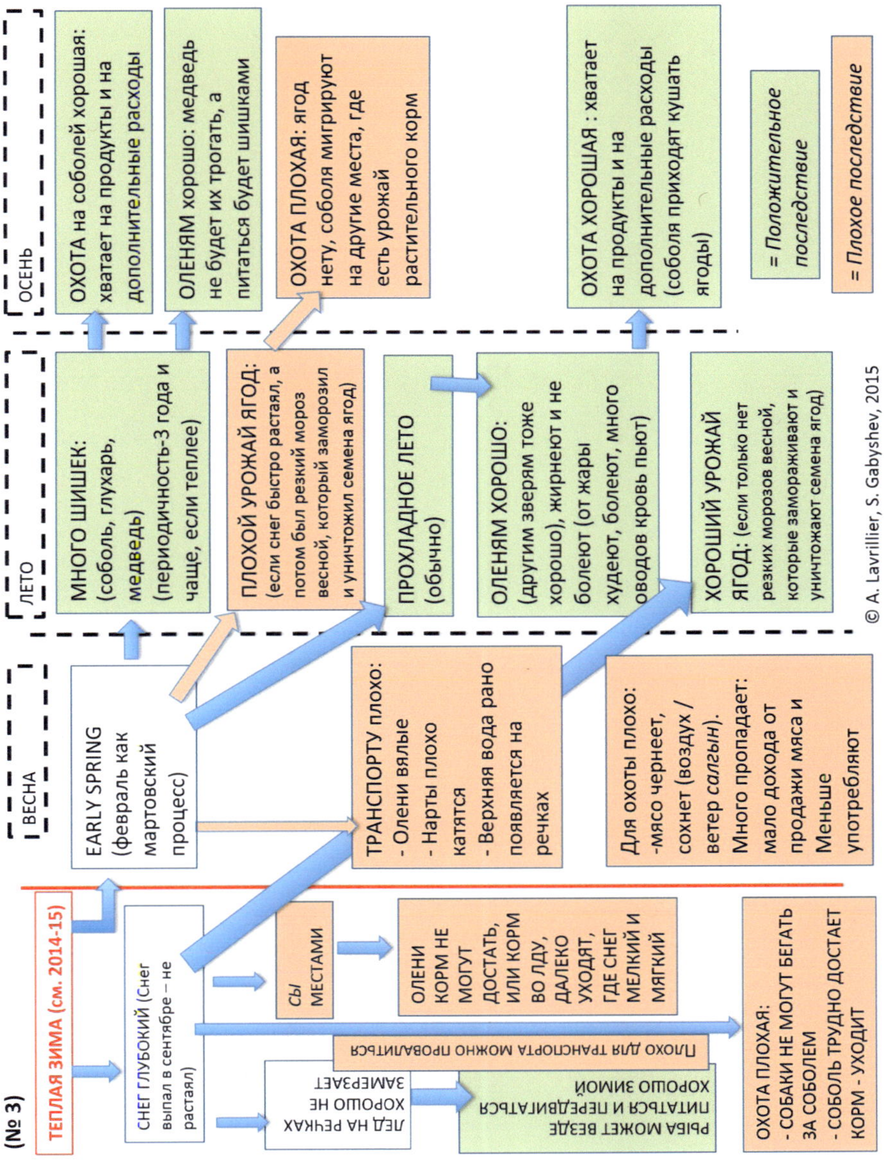
(№ 3)
ТЕПЛАЯ ЗИМА (см. 2014-15)
ВЕСНА
ЛЕТО
ОСЕНЬ
СНЕГ ГЛУБОКИЙ (Снег выпал в сентябре – не растаял)
ЛЕД НА РЕЧКАХ ХОРОШО НЕ ЗАМЕРЗАЕТ
РЫБА МОЖЕТ ВЕЗДЕ ПИТАТЬСЯ И ПЕРЕДВИГАТЬСЯ ХОРОШО ЗИМОЙ
ПЛОХО ДЛЯ ТРАНСПОРТА МОЖНО ПРОВАЛИТЬСЯ
СЫ МЕСТАМИ
ОЛЕНИ КОРМ НЕ МОГУТ ДОСТАТЬ, ИЛИ КОРМ ВО ЛДУ, ДАЛЕКО УХОДЯТ, ГДЕ СНЕГ МЕЛКИЙ И МЯГКИЙ
ОХОТА ПЛОХАЯ: - СОБАКИ НЕ МОГУТ БЕГАТЬ ЗА СОБОЛЕМ - СОБОЛЬ ТРУДНО ДОСТАЕТ КОРМ - УХОДИТ
EARLY SPRING (февраль как мартовский процесс)
ТРАНСПОРТУ плохо: - Олени вялые - Нарты плохо катятся - Верхняя вода рано появляется на речках
Для охоты плохо: -мясо чернеет, сохнет (воздух / ветер *салгын*). Много пропадает: мало дохода от продажи мяса и Меньше употребляют
МНОГО ШИШЕК: (соболь, глухарь, медведь) (периодичность-3 года и чаще, если теплее)
ПЛОХОЙ УРОЖАЙ ЯГОД: (если снег быстро растаял, а потом был резкий мороз весной, который заморозил и уничтожил семена ягод)
ПРОХЛАДНОЕ ЛЕТО (обычно)
ОЛЕНЯМ ХОРОШО: (другим зверям тоже хорошо), жиреют и не болеют (от жары худеют, болеют, много оводов кровь пьют)
ХОРОШИЙ УРОЖАЙ ЯГОД: (если только нет резких морозов весной, которые замораживают и уничтожают семена ягод)
ОХОТА на соболей хорошая: хватает на продукты и на дополнительные расходы
ОЛЕНЯМ хорошо: медведь не будет их трогать, а питаться будет шишками
ОХОТА ПЛОХАЯ: ягод нету, соболя мигрируют на другие места, где есть урожай растительного корм
ОХОТА ХОРОШАЯ : хватает на продукты и на дополнительные расходы (соболя приходят кушать ягоды)
= *Положительное последствие*
= *Плохое последствие*
© A. Lavrillier, S. Gabyshev, 2015

4 CONCLUSION
by A. Lavrillier

Co-Producing and Transferring Knowledge: Trying to Find the Best Way

For this research, one of the greatest difficulties was to find a proper way to transfer Evenki knowledge into academic science: the former is oral, very detailed, and inseparable from the vast natural environment to which it pertains. The fact that it is oral was an additional obstacle. Another huge problem was the fact that the knowledge consists of many complex concepts which do not exist in Western languages. Thus, we had to find approximate translations for each concept rather than for each phenomenon. The reader has seen some illustrative examples with *arbun solgu, арбун солгу / arbu edighu, арбун эдиһу* in the typology of landscapes, *idia, идйа* and the compass in the typology of airs and winds, and in the typology of snow and ice. Since this knowledge is vast, encompassing as it does the entire environment, it is all the more difficult to document and publish. All of these issues made illustrations and multilingual descriptions crucial. Furthermore, this book is not comprehensive: we prepared texts about other aspects of Evenki knowledge that were simply too large to publish here.

Co-production Process, IP Conceptualisation, and Scientific Understanding

As the reader has undoubtedly noticed, we have used a lot of diagrams to document knowledge and analyse environmental changes. There are several reasons for this. One is that English, like Russian, does not contain the words necessary to concisely express the relevant concepts and processes. In addition, the explanations in Evenki themselves contain many other word-concepts that require pages of text and multiple diagrams to explain.

In the process of co-producing knowledge, it was ethically and epistemologically important to identify the author who conceived of the diagrams by placing their name first in the copyright labels. In all cases, one of the authors (Lavrillier or Gabyshev) came up with the idea for a diagram: we then either finalised it together or had Gabyshev check and approve the additions made. Interestingly, if one looks at all the copyright labels together, one can see that the diagrams which contain only drawings and figures were mostly completed on the initiative of Gabyshev or other herder-hunters, while diagrams with pictures were produced on Lavrillier's initiative (an anthropologist). In the beginning, diagrams with figures and drawings were drawn by the herders on paper or in the snow: we then jointly entered these into a computer programme. After around a year of collective work, S. Gabyshev started to create diagrams via the computer.

The idea of using images as the basis for the diagrams initially came from Lavrillier, who noticed that herder-hunters transmit a lot of knowledge when walking in the forest and responding to questions about the surrounding environment. By using lines and pictures in our diagrams, we were able to simulate the Evenki pointing something out with their fingers: thus, we turned an oral narrative given in the Siberian forest into a written narrative created in 'laboratory' conditions.

This also offered us a way to 'translate' indigenous knowledge for non-native and scientific minds: this knowledge is detailed, complex, and full of elements which are left unsaid because they are obvious to the nomads. It took a lot of time for Lavrillier to understand such elements and for Gabyshev and the other herders to identify what the anthropologist did not understand and why. Each diagram represents months of work (in some cases, even years).

The fact that the herder-hunters sought to create abstract diagrams while the anthropologist tried to capture real images means that, epistemologically speaking, the nomads have conceptualised their knowledge (their environment): the anthropologist, however, remains at the stage of observing and coming to terms with the surrounding reality. In other words, nomads have translated the sequences of reality which they observe (or their ancestors observed) into an organised system of knowledge composed of concepts, equations, rules, etc. In their minds, they have a systematised understanding of their environment. In contrast, the anthropologist initially did not know the environment: even after understanding the systematised concept thanks to the diagrams created jointly with the herders, she needed to connect the concept to the real environment by demonstrating it via pictures.

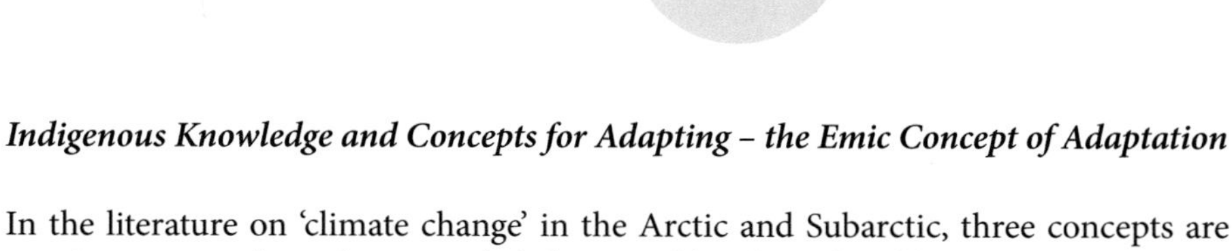

Indigenous Knowledge and Concepts for Adapting – the Emic Concept of Adaptation

In the literature on 'climate change' in the Arctic and Subarctic, three concepts are omnipresent as a basis for research: 'adaptation', 'resilience', and 'vulnerability' (Berkes and Jolly 2001; Ford et al. 2006; Oskal et al. 2009, among many others). With respect to these concepts, Lavrillier came to the following conclusion about the notion of adaptation between 2006 and 2012. Evenki narratives and rituals offer contradictory perspectives of these notions, ranging from optimistic beliefs to millenarian predictions. Meanwhile, their economic practices prove that the Evenki are highly adaptable. Such is not new to them: they have always had to adapt to historical pressures and continue to do so on a daily basis. Nuttall has also stated that Arctic peoples always grow up 'to be prepared for change, to see the world as one of constant surprise' (Nuttall 2009: 298; Nuttall et al. 2005). The Nenets seem to have somewhat similar risk management behaviour and a comparable understanding of 'adaptation' (Stammler-Gossmann 2010). The Evenki like to say: '*For ages we have been adapting and coping, and we will continue to do so for as long as possible*'. I witnessed a nomad telling a journalist about the uncertain future of indigenous peoples in Russia: '*Perhaps we are dying, but we won't die*' (Lavrillier 2013).

Between 2012 and 2016, the authors identified new aspects of the notion of adaptation that nuance the features described above. This was thanks to the Evenki language: the verb used by nomads to talk about adapting in spontaneous discourse and semi-structured interviews is *tatta-da (mamma-ðа)*. This verb has a double meaning: 1) to get used to a new place, situation, constraint, or person (with a connotation of attachment), and 2) to learn and receive both intellectual and practical knowledge and skills (e.g., knowledge about something, mastering a language, a technique, a way of life). So, the first meaning reflects the following: 'to find a way to act in the face of new pressure; a way of doing that will become a habit'. Consequently, there is a connotation of a lasting interaction embedded in this notion of adaptation. It seems here that this emic notion of adaptation does not envisage repeated and frequent new pressures (or anomalies). The second meaning reflects two ideas: that existing indigenous knowledge is indispensable for adapting and that adapting requires the production of new knowledge. For this, Evenki typologies and concepts are crucial tools.

Almost every year between 2012 and 2016, the nomads complained that anomalies were becoming too frequent, rapid, cumulative, and irregular (i.e., they too often belong to our category of 'unprecedented anomalies'). The nomads complained that they were struggling to adapt to each new pressure before it is too late (e.g. in terms of the Evenki concept *tatta-da (mamma-ðа)*, 'they struggle to learn in time, how to adapt to this new situation'). Despite these increasing pressures, however, the Evenki say: '*Since we have indigenous knowledge, reindeer, and landscapes, we can adapt*' (cf. Snow and ice typology, part III; Lavrillier and Gabyshev 2017).

Diverse Landscapes, Indigenous Knowledge, and Mobility in Adaption Processes

Throughout the descriptive texts, diagrams, analyses, and case studies in this book, we argue that the diverse topography where this Evenki group nomadise is a tool they use to adapt. If there are bad snow conditions in one place, nomads know they can move to another landscape type where they will find soft and shallow snow. In contrast, if they need a hard snow layer for hunting, they know that they can find this in a specific landscape type (cf. Topographic typology, Vegetal cover typology, Snow and ice typology). This is distinct from the situation reported in the famous Nenets case study, where bad snow conditions proved very difficult to face because of how flat their territory is (among many others Bartsch 2010; Forbes et al. 2009).

Thus, their tool for adaptation is the diverse topography, which offers many different types and depths of snow cover. However, nomads can only use these tools for enhancing their economic activities and adapting to climate anomalies thanks to their traditional knowledge. Without this, it would be impossible to adapt. Another crucial element for adapting and benefiting from such topographical diversity is the ability to move from one place to another. This they can do not because they have rights to the land (they do not), but because extractive industries are not interested in some of the regions in which they live. However, this is a topic we have developed in other publications (cf. also about interactions of various drivers of change, Lavrillier et al. 2016).

Studying and Understanding Climate and Environmental Changes Requires Indigenous Concepts

The reader should now understand that it is impossible to document, study, and explain the changes in climate and the environment if we do not first document and understand traditional Evenki knowledge. From this, we can reach two conclusions. First, it is almost impossible to understand Evenki observations of change without knowing indigenous typologies, which is one of the reasons for the existence of this book. Second, without knowing and understanding Evenki typologies of the environment and climate, documentation of change may run the risk of being simplistic or superficial. Having undergone four years of intensive courses at the 'nomadic university' (with S. Gabyshev as my supervisor and the other nomads as lecturers), I can now see that my first paper on climate change (2013) written in 2011 discusses environmental and climate changes very simplistically compared to what Gabyshev and I have managed in this book. This is particularly obvious in terms of the changes in the snow. This suggests that academic environmental sciences may be interested in using Evenki and other indigenous concepts and typologies in order to expand their study of climate and environmental changes, fill in the gaps of their work, or begin new projects.

The interdisciplinarity of TEK and the question of classification

While we presented knowledge about the vegetal cover separately from knowledge about the snow in this book, we can see that knowledge about the former relates to the snow cover and fauna and that knowledge about the latter also relates to the sun, vegetal cover, and so on (cf. Topographic typology, Vegetal cover typology, Snow and ice typology; diagram *Arbun solgu arbun adiyu;* and so on). Thus, the typologies we presented separately are jointly engaged by the Evenki when they observe and analyse their surroundings. This would be like an academic scientist combining elements from glaciology, geomorphology, vegetal and animal biology, climatology, and atmospheric physics for a study.

Regarding the analysis of the emic classification system, we can say the following. The idea to display the larger typologies (i.e. vegetal cover, topography, snow and ice, precipitation, etc.) separately came from Lavrillier, who sorted out the term-concepts as soon as they were identified during fieldwork and writing sessions with the nomads. The Evenki also make distinctions between these big categories, but they are not as explicit as the distinctions between term-concepts. In other words, the Evenki do not consider, for example, the horsetail, a slope, and soft snow as belonging to one and the same whole, but neither do they place them into separate categories. For instance, they do not say 'in the vegetal cover category, we have horsetail and A, B, C plants' or 'in the topographic category, we have X, Y, Z'.

Thus, in terms of the cognitive features of traditional knowledge, this might suggest that such large categories ('typologies') are either entirely obvious to the Evenki or less relevant for them than an analysis made by manipulating many concepts (from diverse typologies) to study the interactions between elements of the environment. They are certainly interested in the interplay between the climate, the vegetal cover, and the animal realm. If I (Lavrillier) ask myself why I felt it was necessary to classify term-concepts into large separate typologies, I would say that I did it instinctively (conditioned as I am by the Western scientific tradition) in order to try to see the inner structure of indigenous knowledge, to understand Evenki knowledge, and to present it to Western readers used to big categories.

Sometimes, delineating the categories was not easy or obvious: for instance, should marshes or tussock fields be classified as topography or vegetal cover? Thus, our choices may surprise some readers. Perhaps it can be argued that classification into big typologies is more Western, while the inner classification is more Evenki: the classification system into typologies presented in this book is the result of a real co-production that merges Western and Evenki patterns.

Breaking a myth

As we can see in the different sections on Evenki climatology, the romanticised image of the Siberian nomad entirely unafraid of harsh cold and storms who perhaps even likes such weather has little in common with reality: the Evenki are not big fans of travelling in temperatures that fall to -45°C/-50°C. During storms, considerable snowfalls, and very harsh cold, the nomads prefer to stay at home and prepare wood or make various items. They do so to preserve their health and lives. Indeed, this book shows that the natural environment is extremely dangerous. Indeed, in addition to the predators, the natural environment includes many dangerous places and instants. The knowledge we present here not only relates to hunting and herding, but also to survival. (cf. Snow and ice typolology, Topographic typology)

Hypotheses and 'modelling'

The diagram on the transformation of landscape due to climate change, (cf. Part III: Landscape transf. due to climate change, Gabyshev interview 2014) demonstrates that the nomads are interested in more than analysing the present (of landscape, weather, vegetal cover). The diagram discusses how generally warmer temperatures have led to a decline in the formation of icing blisters (*bukte, буктэ*), which, over the course of several dozen years, will affect the development of the forest and cause the disappearance of large river basins (*amnunna, амнунна*). Thus, nomads produce hypotheses on how the weather, climate, environment, and vegetal and animal realms will change, regardless of how these changes affect their own survival. For instance, their hypothesis on future landscape transformation is not directly related to their own interests (although an *amnunna, амнунна* is an essential landscape type for the Evenki). Their hypotheses on the past state of the landscape also do not have any specific significance for their economy: they are of use only in producing predictions about future transformations. This logical chain seems very close to what scientists do when they produce 'modelling', although nomadic modelling is of course not based on numbers and chemical formulae.

As noted in regards to other cognitive operations, this modelling is both collective and individual: it is both inherited and contemporarily produced knowledge.

Firstly, it (as with other hypotheses presented in this book and in other publications) proves the ability of the Evenki to produce hypotheses for future and past transformations. Secondly, as with other parts of their knowledge, for instance about butterflies, the indigenous knowledge of the Evenki is not only 'practical' or embedded in economic activities, as is often claimed when characterising indigenous knowledge.

The Evenki regularly, if not constantly, 'model' (cf. Part III: inter-annual comparisons and modelling) at least 8–10 months ahead (and sometimes several years ahead). Indeed, several times throughout the year (especially when they observe anomalies), they produce hypotheses for events expected in the climate and the natural environment. They immediately produce several hypotheses based on observed phenomenon or received knowledge, engaging the typologies we presented. For instance, from each assumption they advance several hypotheses for possible developments. Each hypothesis is related not only to the climate, but also to the vegetal and animal realms and to the consequences for the Evenki. Sometimes, they produce hypotheses for three or more years ahead; this can be compared to 'modelling' the inter-influences between climate, landscape, and humans.

In this book, we present only part of our work on Evenki anomaly analysis, inter-annual-comparisons, and forecasting and modelling; the rest will be published elsewhere.

Climate change has been observed by the Evenki reindeer herders of Siberia for several decades. They are front-line witnesses of the profound effects that industrial activity and climate change have had on the environment, and they want their knowledge and observations to be heard. Siberian peoples, as with many other Arctic indigenous peoples, face climate change together with other interplaying global changes. The BRISK project was an experiment in co-production and mutual learning between climatologists, anthropologists, ecologists, geographers, and nomadic reindeer herders. The objective was not to compare the results of each science and knowledge system, but to allow them to complement each other in order to improve our understanding of complex environmental systems in Arctic and Subarctic regions.

The results show the importance of interaction between climatological, biological, social, and economic aspects in IP knowledge. As this book (and the climatologists C. Claud and M. Rojo) proves, the co-production between IP knowledge and science raises new research perspectives (Lavrillier, Rojo Conference 2014). In addition to what we wrote in the introduction regarding the impacts on the herders concerned, the effects of this co-production are various. Some results from the Evenki part of BRISK project were used by A. Lavrillier to support her university courses and for research training through several Masters (1 and 2) internships and PhD programmes. In addition, Lavrillier and Gabyshev have developed a transdisciplinary seminar programme on reindeer herders and climate change for French students.

In this framework, the methodology and protocol for monitoring climate and environmental changes have been adapted to environmental issues, both in the nomadic world and in villages and towns. We have noticed rising interest among Evenki for participating in transdisciplinary research. Through this, the network of Siberian Evenki observatories was extended to several other regions and place types. In addition, according to the nomads concerned, this book represents a quarter of Evenki TEK. Consequently, we have created several research proposals in order to apply for further funding.

In the framework of UVSQ/OVSQ, we should note the influence of our work and the active participation of reindeer herder-co-researchers: at the present time, the involvement of indigenous peoples or citizens in the sciences is considered to be valuable input. For instance, in the EDU-ARTIC (H2020, OVSQ) project, the Evenki team was invited to create video lectures on the impacts of climate change in Siberia.

Some of the results of the Evenki observatory in the BRISK project have been reinvested into the international scene in order to improve knowledge and aid in the preservation of Arctic biodiversity[1]. This Siberian BRISK community-based observatory was selected to publish in the report of the Indigenous and Local Knowledge Task Force of the Intergovernmental Platform on Biodiversity and Ecosystem Services (IPBES) (UNESCO, UNEP, UNDP, FAO).

From the side of the herder-hunters, as was alluded to in the introduction, I (Gabyshev) stress that it is crucial to transmit our language. To me, language bears all the necessary information, the logic and keys for nomadising and living in our environment. It also conveys the rules of behaviour for sustainable hunting, for respecting the spirits, and for taking care of reindeer. Maintaining our language allows us to keep our culture and our understanding of the natural environment alive. This book can contribute to maintaining or revitalising our knowledge and language, since it is akin to a repository that conveys the science of our ancestors. If Evenki have access to this book, they will undoubtedly read it and become interested in the understanding of nature contained therein. We plan to offer to community members some hard copies and upload it in its electronic format, since many nomads and most villagers have smartphones, computers, or tablets. Some can read just for interest, others can discuss and develop it together during our 'nomadic seminars' or put this knowledge into practice, and all can recall it at any point in the future.

To conclude, we can say that while the approach of classical anthropology focuses on the analysis of social institutions, behaviours, and individuals in order to study a society, throughout this book and its illustrations, the reader has read very few descriptions about Evenki society or individuals: it was almost as if the reader and authors were inside the minds of herder-hunters walking in the Siberian taiga, discovering and decrypting it together with them and the anthropologist. As evoked in the introduction and shown by this book, this method also provides information about the society itself. For instance, knowledge about a landscape type informs us about the rules of camp installation, shamanic rituals, etc. Thus, this specific approach allows us not only to understand the TEK itself, but also to comprehend a nomadic society in depth through the transdisciplinary study of its knowledge of the natural environment.

1 As well as some results from the Sami BRISK observatory (Roturier et al. 2016).

REFERENCES

ACIA, 2005. *Arctic Climate Impact Assessment*. Cambridge: Cambridge University Press.

Alekseev, V. P. 2001. *U istokov naledevedeniia*.http://imz-2000.narod.ru/ygn_26.html.

Bartsch, Annett, Timo Kumpula, Bruce C. Forbes, and Florian Stammler 2010. Detection of Snow Surface Thawing and Refreezing in the Eurasian Arctic with QuikSCAT: Implications for Reindeer Herding. *Ecological Applications* 20(8): 2346–2358.

Berkes, Fikret 1999. *Sacred Ecology: Traditional Ecological Knowledge and Resource Management*. Philadelphia: Taylor & Francis.

Berkes, Fikret, and Diana Jolly 2001. Adapting to Climate Change: Social-Ecological Resilience in a Canadian Western Arctic Community. *Conservation Ecology* 5(2): 18.

Bogoslovskaia, Liudmila Sergeevna, Vdovin, Boris Innokentevich, and V. V. Golbtseva 2008. Izmeneniia klimata v regione Beringova proliva. Traditsionnnye i nauchnye znaniia [Climate Change in the Bering Strait Region: Integration of Scientific and Indigenous Knowledge], *Ekologicheskoe planirovanie i upravlenie*, 3–4(8–9): 36–48.

Boldyrev, Boris Vasil'evich 1994. *Russko-evenkiiskii slovar'* (Russian-Evenki dictionary). Novosibirsk: Nauka.

Bouchiat, Vincent, Bernard Saugier, and Claire Waelbroeck 1999. In *Taiga-toundra, au Nord la démesure* [Taiga-Tundra, at the North the Excessiveness]. S. Bobbé (dir.). Paris: Autrement.

BRISK 2012. Bridging Indigenous and Scientific Knowledge about the Global Changes in the Arctic: Adaptation and Vulnerability of Societies and the Environment. Project proposal (by Nakashima, Lavrillier and Roué).

Burgess, Philip 1999. *Traditional Knowledge: A Report Prepared for the Arctic Council Indigenous Peoples Secretariat*. Copenhagen: IPS.

Bulgakova, Tatiana 2010. Climate Change, Vulnerability and Adaptation among Nenets Reindeer Herders. In *Community Adaptation and Vulnerability in Arctic Regions*, G. K. Hovelsrud and B. Smit (eds.), 83–105. Berlin: Springer.

Cincius, Vera Ivanovna 1975 and 1977. Sravnitel'nyi slovar' tunguso-man'shchzhurskikh iazykov, I, II. Leningrad: Nauka.

Costard, François, Emmanuèle Gautier, T. Fedorov, P. Konstantinov, and Laure Dupeyrat 2014. An Assessment of the Erosional Potential of Fluvial Thermal Process During Ice Breakups of the Lena River (Siberia). *Permafrost & Periglacial Processes* 25: 162–171.

Crate, Susan 2008. Gone the Bull of Winter? Grappling with the Cultural Implications of and Anthropology's Role(s) in Global Climate Change. *Current Anthropology* 49(4): 569–595.

Dal Molin, Magali 2013. Modélisation par télédétéction de l'évolution du territoire des nomades Evenks dans la région d'Olyokminsk (Sibérie orientale). Master thesis UVSQ.

D'Andrade, Roy 1995. *The Development of Cognitive Anthropology*, Cambridge: Cambridge University Press.

Davydov, Vladimir Nikolaevich 2015. Khod konem: shakhmaty u evenkov-olenevodov Severnogo Baikala i Zabaikal'ia [A tricky move: chess among Evenki reindeer herders in Northern Baikal and Transbaikal]. In *Radlovskii sbornik: Nauchnye issledovaniia i muzeinye proekty MAE RAN v 2014*. IU.K. Chistov (ed.), 241–246 St. Petersburg: MAE RAN.

Donahoe, Brian 2004. *A Line in the Sayans: History and Divergent Perceptions of Property Among the Tozhu and Tofa of South Siberia*. Indiana University, PhD thesis.

Forbes, Bruce C., Manfred Bolter, Ludger Müller-Wille, Janne Hukkinen, Felix Muller, Nicolas Gunslay, and Yulian Konstantinov (eds.) 2006. *Reindeer Management in Northernmost Europe*. Berlin/Heidelberg: Springer.

Forbes, Bruce C. 2008. Equity, Vulnerability and Resilience in Social-Ecological Systems: a Contemporary Example from the Russian Arctic. *Research in Social Problems and Public Policy* 15: 203–236.

Forbes, Bruce, Florian Stammler, Toma Kumpula, Nina Meschtyb, Anu Pajunen, and Elina Kaarlejärvi 2009. High Resilience in the Yamal-Nenets Social-Ecological System, West Siberian Arctic, Russia. *Proceedings of the National Academy of Sciences of the United States of America* 106(52): 22041–22048.

Ford, James D., Barry Smit, and Johanna Wandel 2006. Vulnerability to Climate Change in the Arctic: A Case Study from Arctic Bay, Canada. *Global Environment Change* 16: 145–160.

Ford, James D., Tristan Pearce, Barry Smit, Wandel Johanna, M. Allurut, Shappa, K. Ittusujurat H., and K. Qrunnut 2007. Reducing Vulnerability to Climate Change in the Arctic: The case of Nunavut, Canada. *Arctic* 60(4): 150–166.

Forsyth, James 2000. *A History of Peoples of Siberia. Russia's North Asian Colony 1581–1990*. Cambridge: Cambridge University Press.

Gabyshev, Semen 2014. Interview in *Bulletin Electronique Russie* numéro 67 (30/10/2014) – Ambassade de France en Russie / ADIT. http://www.cearc.fr/content/semen-nikolaevitch-gabyshev-un-%C3%A9leveur-de-rennes-devenu-cochercheur.

Gearheard, Shari, and Jamal Shirley 2007. Challenges in Community-Research Relationships: Learning from Natural Science in Nunavut. *Arctic* 60(1): 62–74.

Gearheard, Shari, Matthew Pocernich, Ronald Stewart, Joelie Sanguya, and Henry P. Huntington 2009. Linking Inuit Knowledge and Meteorological Station Observations to Understand Changing Wind Patterns at Clyde River, Nunavut. Berlin: Springer. *Climatic Change* 100: 267.

Helander-Renvall, Elina 2005. Composite Report on Status and Trends Regarding Knowledge, Innovations and Practices of Indigenous and Local Communities. *Biological Diversity in the Arctic*. UNEP/CBD?WG8J/4/INF/3.

Huntington, Henry P. 1998. Observations on the Utility of the Semi-directive Interview for Documenting Traditional Ecological Knowledge. *Arctic* 51(3): 237–242.

Huntington, Henry, Terry Callaghan, Shari Fox, and Igor Krupnik 2004. Matching Traditional and Scientific Observations to Detect Environmental Change: A Discussion on Arctic Terrestrial Eco-Systems. *Ambio* 33(7): 20–25.

Huntington, Henry P., and Shari Fox 2005. The Changing Arctic: Indigenous Perspectives. *ACIA. Arctic Climate Impact Assessment*, 61–98. Cambridge: Cambridge University Press.

IPCC, 2014: Climate Change 2014: Synthesis Report. Contribution of Working Groups I, II and III to the Fifth Assessment Report of the Intergovernmental Panel on Climate Change. R. K. Pachauri and L. A. Meyer (eds.). Geneva: IPCC.

Kavry, V., and A. Boltunov 2005–2006. Observations of Climate Change Made by Indigenous Inhabitants of the Coastal Regions of Chukotka Okrug. WWF (Report on the WWF Arctic climate change project).

Krupnik, Igor, and Dyanna Jolly (eds.) 2002. *The Earth is Faster Now: Indigenous Observations of Arctic Environment Change*. Fairbanks, Alaska: Arctic Research Consortium of the United States.

Krupnik, Igor, and Ray G. Carlton 2007. Pacific Walruses, Indigenous Hunters, and Climate Change: Bridging Scientific and Indigenous Knowledge. *Deep Sea Research Part II: Topical Studies in Oceanography* 54(23–26): 2946–2957.

Krupnik, Igor, Claudio Aporta, Shari Gearheard, Gita J. Laidler, and Lene Kielsen Holm (eds.) 2010. *SIKU: Knowing Our Ice. Documenting Inuit Sea Ice Knowledge and Use*. Berlin: Springer.

Krupnik, Igor, and Liudmila S. Bogoslovskaia 2017. 'Our Ice, Snow and Winds': From Knowledge Integration to Co-production in the Russian SIKU Project, 2007–2013. In *Oral History Meets Linguistics*, E. Kasten, K. Roller, and J. Wilbur (eds.), 65–82. Fürstenberg/Havel: Kulturstiftung Sibirien.

Lavrillier, Alexandra 1995. *Les différentes fêtes saisonnières chez les Evenks éleveurs de rennes du sud de la Iakoutie* [The Seasonal Feasts of the Evenki Reindeer Herders (Southern Yakutia)]. Master I thesis in Ethnology, University of Paris-X-Nanterre.

— 2003. De l'oubli à la reconstruction d'un rituel collectif. *L'Ikènipkè* des Évenks [From the Oblivion to the Reconstruction of a Collective Ritual]. *Slovo* 28/29, *Sibérie, Paroles et Mémoires*: 169–191.

— 2005. *Nomadisme et adaptations sédentaires chez les Évenks de Sibérie postsoviétique: «jouer» pour vivre avec et sans chamanes* [Nomadism and Adaptations to a Sedentary Lifestyle Among the Evenki of Post-Soviet Siberia: 'Playing' to Live with and without Shamans], PhD thesis. École Pratique des Hautes Études, Section religious sciences, 2 vols..

— 2005–2006. S'orienter avec les rivières chez les Évenks du Sud-Est sibérien. Un système d'orientation spatial, identitaire et rituel [Rivers as a Spatial, Social and Ritual System of Orientation among the Evenki of Southeastern Siberia]. *Études mongoles, sibériennes, centrasiatiques et tibétaines* 36/37: 95–138.

— 2008. Comment les évenks de Sibérie méridionale ont modifié le rituel sur le gibier tué [How the Evenki of Eastern Siberia Have Modified a Hunting Ritual]. *Annales de la Fondation Fyssen* 22: 112–121.

— 2010 [Lavril'e, Aleksandra] 2010. Orientirovat'sia po rekam u evenkov Yugo-Vostoka Sibiri. Sistema prostranstvennoi, sotsialnoi i ritual'noi orientacii [Rivers as a Spatial, Social, and Ritual System of Orientation Among the Evenki of South-Eastern Siberia]. *Etnograficheskoe obozrenie* 6: 115–132.

Lavrillier, Alexandra 2011a. Creation and Persistence of Cultural Landscape Among the Siberian Evenkis: Two Conceptions of 'Sacred' Space. In *Landscape and Culture in Northern Eurasia*. P. Jordan (ed.), 215–231. Walnut Creek: Left Coast Press.

— 2011b. Renne domestique, renne sauvage face au r.chauffement [Domestic Reindeer and Wild Reindeer Facing the Warming]. In *Mondes polaires. Hommes et biodiversités des défis pour la sciences*. M. Raccurt and R. Chernokian (eds.), 142–145. Paris: Publication INEE.

— 2013. Climate Change Among Nomadic and Settled Tungus of Siberia: Continuity and Changes in Economic and Ritual Relationships with the Natural Environment. *Polar Record* 50: 1–12.

— 2016. Indigenous Evenki Analysis of the Snow Cover in the Siberian Taiga. Film, 12 min. http: cearc.fr

Lavrillier, Alexandra, and Semen Gabyshev 2017. An Emic Science of Climate: A Reindeer Evenki Environmental Knowledge and the Notion of 'Extreme Process of Change'. In *Human-Environment Relationships in Siberia and Northeast China: Skills, Rituals, Mobility and Politics among the Tungus Peoples*. A. Lavrillier, A. Dumont, D. Brandišauskas (eds.). Paris: Études mongoles et sibériennes, centrasiatiques et tibétaines 48 (in print).

Lavrillier, Alexandra, and Semen Gabyshev 2015. Community-Based Observatory, Evenk: Amur-Yakutiia. http://www.arcticbrisk.org/node/67 [accessed 15 March 2017]

Lavrillier, Alexandra, Semen Gabyshev, and Maxence Rojo 2016. The Sable for Evenki Reindeer Herders in Southeastern Siberia: Interplaying Drivers of Changes on Biodiversity and Ecosystem Services: Climate Change, Worldwide Market Econonomy and Extractive Industries. In *Knowing our Land and Resources: Indigenous and Local Knowledge of Biodiversity and Ecosystem Services in Europe & Central Asia*. M. Roué and Z. Molnar (eds.), 111–128. Knowledges of Nature 9. Paris: UNESCO.

Lavrillier, Alexandra, and Henri Lecomte 2002. *Évenks: Chants rituels des nomades de la* taïga, [Evenki: Ritual Songs of the Nomads of the Taiga]. Musique du Monde, Sibérie 8. Paris: Buda Records.

Lavrillier, Alexandra, and Dejan Matic 2013. *Even tales of Dar'iia Mikhailovna Osenina / Dar'ia Mikhailovna Osenina nimkarni / Evenskie nimkany Dar'i Mikhailovna Osenina*. Fürstenberg/Havel: Kulturstiftung Sibirien.
Lenclud, Gérard 1996. Le grand partage ou la tentation ethnologique. In *Vers une ethnologie du present*, G. Althabe, D. Fabre et G. Lenclud (dir.), 9–37. Paris: Édition de la Maison des Sciences de l'Homme (MSH).
Lot-Falck, Éveline 1971. Psychopathes et chamanes yakoutes. In *Echanges et communications. Mélanges offerts à CL. Lévi-Strauss*. J. Pouillon and P. Maranda (eds.), 115–129. Paris/La Haye: Mouton.
— 1972. Les incarnations de l'ilbis yakoute. In *Traditions religieuses et para-religieuses des peuples altaïques*: 65–77. Paris: Presse Universitaire de France.
— 1974. Le principe de l'ilbis et ses rapports avec la fureur meurtrière et le chamanisme. In *Sprache, Geschichte und Kultur der Altaischen Völker*. Schriften zur Geschichte und Kultur des alten Orients 5: 363–374. Berlin: Akademie Verlag.
Macaulay, Ann C., Laura E. Commanda, William L. Freeman, Nancy Gibson, Melvina L. McCabe, Carolyn M. Robbins, and Peter L. Twohig 1998. *Responsible Research with Communities: Participatory Research in Primary Care*. North American Primary Care Research Group (NAPCRG). Policy Statement.
Mauss, Marcel 1969. *Œuvres, t. III : Cohésion sociale et divisions de la sociologie*, Paris, Minuit.
Mauss, Marcel 1978 [1950]. *Sociologie et anthropologie*, Paris: PUF.
Mazin, Anatolii Ivanovich 1984. Traditsionnye verovaniia i obriady evenkov-orochonov [Traditional Beliefs and Rituals of the Evenki–Orochons]. Novosibirsk: Nauka.
Nadasdy, Paul 1999. The Politics of TEK: Power and the 'Integration' of Knowledge *Arctic Anthropology* 36: 1–18.
Nakashima, Douglas J., Kirsty Galloway McLean, Hans D. Thulstrup, Ameyali Ramos Castillo, and Jenifer T. Rubis 2012. Weathering Uncertainty: Traditional Knowledge for Climate Change Assessment and Adaptation. Paris/Darwin: United Nations Educational, Scientific and Cultural Organization and United Nations University.
Nelson, Richard K. 1969. Hunters of the Northern Ice. Chicago: University of Chicago Press.
Nuttall, Mark 2009. Living in a World of Movement: Human Resilience to Environment Instability in Greenland. In *Anthropology and Climate Change: From Encounters to Actions*. S.A. Crate and M. Nuttall (eds.), 292–310. Walnut Creek: Left Coast Press.
Nuttall, Mark, Fikret Berkes, Bruce Forbes, Gary Kofinas, Tatiana Vlassova, and George Wenzel 2005. Hunting, Herding, Fishing and Gathering: Indigenous Peoples and Renewable Resource Use in the Arctic. In *ACIA Arctic Climate Impact Assessment: Scientific Report*, 649–690. Cambridge: Cambridge University Press.

Ob izmenenii v sostave i granitsakh Vitimo-Olekminskogo (Evenkiiskogo) okruga, Vostochno-Sibirskogo kraia, o ego administrativnom centre, o sostave raionov i ikh centrakh togo zhe okruga, 1933. *Sovetskii Sever* 6: 116–117.

Oskal, Anders, Matis Turi, Svein D. Mathiesen, and Philip Burgess (eds.) 2009. *EALAT Reindeer Herder's Voice: Reindeer Herding, Traditional Knowledge and Adaptation to Climate Change and Loss of Grazing Land* (Kautokeino/Guovdageadnu: International Centre for Reindeer Husbandry.

Pikunova, Zinaida Nikolaevna, and Irina Romanovna Pikunova 2004. *Buga diarin entsiklopediia. Flora*. St. Petersburg: Prosveshchenie.

Pikunova, Zinaida Nikolaevna and Irina Romanovna Pikunova 2008. *Buga diarin entsiklopediia. Fauna*. Novosibirsk: Nauka.

Quinn, Naomi 2005. *Finding Culture in Talk: A Collection of Methods*. New York: Palgrave Macmillan.

Retter, Gunn-Britt 2015. *The Position of Indigenous Knowledge on the Roadmap for Future Arctic Research*. ISAR-4 – ICARP III. Plenary Keynote Speech, Arctic Science Summit Week, 29 April 2015.

Robbe, Pierre 1994. *Les Inuit d'Ammassalik, chasseurs de l'Arctique*. Paris: Muséum National d'Histoire Naturelle.

Rojo, Maxence 2016. Impacts des évènements météorologiques extremes et du changement climatiquesur les régions arctiques et subarctiques: Perspectives croisées en climatologie et en sciences humaines et sociales. PhD thesis. University of Paris-Saclay.

Romanova Agniia Vasil'evna, and Anna Nikolaevna Myreeva 1971. Fol'klor evenkov Yakutii [Oral Literature of the Evenki of Yakutia). Leningrad: Nauka.

Roturier, Samuel, and Marie Roué 2009. Of Forest, Snow and Lichen: Sami Reindeer Herders' Knowledge of Winter Pastures in Northern Sweden. *Forest Ecology and Management* 258: 1960–1967.

Roturier, Samuel, Jakob Nygård, Lars-Evert Nutti, Mats-Peter Åstot 2016. Reindeer Husbandry in the Boreal Forest: Sami Ecological Knowledge or the Science of "Working with Nature". In *Knowing our Land and Resources: Indigenous and Local Knowledge of Biodiversity and Ecosystem Services in Europe & Central Asia*. M. Roué and Z. Molnar (eds.), 90–108. Knowledges of Nature 9. Paris: UNESCO.

Roncoli, Carla, Todd Crane, and Ben Orlov 2009. Fielding Climate Change in Cultural Anthropology. In *Anthropology and Climate Change: From Encounters to Actions*, S. A. Crate and M. Nuttall (eds.), 87–115. Walnut Creek: Left Coast Press.

Semenov, Viacheslav 1998. Preskazanie Chokhotui. Film, 40 minutes.

Shadrin, Viachelslav 2009. Russian Report. In *Report of the Indigenous Peoples' Global Summit on Climate Change, 20–24 April*: 26–28. Anchorage, Alaska.

Sharakhmatova, Viktoria Nikolaevna 2011. *Nabliudeniia korennykh narodov Severa Kamchatki za izmeneniiami klimata. Otchet* [The Observation of Northern Kamchatkan Indigenous Peoples on Climate Change. Report]. Petropavlovsk-Kamchatskii: Etno-ekologicheskii informatsionnyi tsentr.

Shirokogoroff, Sergei Mikhailovich [1929] 1966. *Social Organization of the Northern Tungus*. Oosterhout: Anthropological Publications.

— 1935. *The Psychomental Complex of the Tungus*. London: Kegan, Paul, Trench, Trubner.

Sirina, Anna 2006. *Katanga Evenkis in 20th Century and the Ordering of their Life-World*. Edmonton: CCI Press.

Sirina, Anna Anatol'evna 2012. *Evenki i eveny v sovremennom mire*. [Evenki and Even in the Contemporary world] Moskva: Izdatel'skaia firma 'Vostochnaia literatura'.

Stammler-Gossmann, Anna 2010. Translating Vulnerability at the Community Level: Case Study from the Russian North. In *Community Adaptation and Vulnerability in Arctic Regions*, G. K. Hovelsrud and B. Smit (eds.), 131–162. London: Springer.

UNESCO 2009. *Climate Change and Arctic Sustainable Development. Scientific, Social, Cultural and Educational Challenges*. Paris: UNESCO.

Vasilevich, Glafira Makarovna, and M. G. Levin 1951. Tipy olenevodstva i ikh proiskhozhdenie [Types of Reindeer Herding and their Origins]. *Sovetskaia etnografiia* 1: 63–87.

Vasilevich, Glafira Makarovna 1958. *Evenkiisko-russkii slovar'* [Evenki-Russian Dictionary]. Moskva: Gosudarstvennoe izdatel'stvo inostrannykh i natsional'nykh slovarei.

— 1936. *Sbornik materialov po evenkiiskomu (tunguskomu) fol'kloru* [Anthology of Materials on the Oral Literature of the Evenki (Tungus)]. Leningrad: Izd. Inst. Nar. Severa.

— 1966. Istoricheskii fol'klor evenkov, skazanie i predanie [Historical folklore of the Evenki, stories and tales]. Moskva/Leningrad: Nauka.

— 1969. Evenki. Istoriko-etnograficheskie ocherki (XVIII-nachalo XX v) [Evenki. Historical Ethnographic Report]. Leningrad: Nauka.

— 2005. *Russko-evenkiiskii slovar'* [Russian-Evenki Dictionary]. Vols. I, II. St. Petersburg: Prosveshchenie.

Vlassova, Tatiana K. 2006. Arctic Residents' Observations and Human Impact Assessments in Understanding Environmental Changes in Boreal Forests: Russian Experience and Circumpolar Perspectives. *Mitigation and Adaptation Strategies for Global Change* 11: 897–909.

Conferences (cited)

Gabyshev, Semen, and Alexandra Lavrillier 2016. *Siberian Evenk Reindeer Herders' Knowledge and Observation of Biodiversity and Ecosystem Services*, ILK Dialogue Workshop for the Europe & Central Asia Assessment, Task Force on Indigenous and Local Knowledge Systems (ILK), Intergovernmental Platform on Biodiversity and Ecosystem Services (IPBES), Expert Group for the Regional Assessment for Europe and Central Asia (ECA). UNESCO, Paris, 11–13 January 2016.

Lavrillier, A. 2015. *Le changement climatique et les peuples autochtones en Eurasie: lier les connaissances et les observations des éleveurs de rennes et des scientifiques en Sibérie*. Film 22 min., conference on Canal Savoir (http://www.canalsavoir.tv/emission).

— 2015. *Climate Change and the Evenk Reindeer Herders*, Russia and Eastern Europe Day, COP21, Indigenous Peoples' Pavilion, Le Bourget, 4 Dec.

— 2013. *Lier les connaissances scientifiques et celles des peuples autochtones sur les changements arctiques: vulnérabilités et adaptations des sociétés et de l'environnement*. Chantier Arctique – Chantier Arctique – National Research Program on the Scientific issues of the Arctic in Environmental Sciences, Ecology and Geosciences, Human and Social Sciences, and Health sciences. Collège de France, Institut d'Océanographie, Paris, 3–6 June 2015.

— 2012. *Global Change Among Siberian Tungus: Methods, Risks and Advantages in Community-Based Observatories and Co-Production of Knowledge*. International Workshop Global Change in the Arctic and Co-Production of Knowledge. An Assessment of Recent Developments. National Museum of Natural History, Paris, 27–29 September, 2012.

Lavrillier, Alexandra, and Semen Gabyshev 2016a. *How Evenki Reindeer Herders in Siberia Conceive Extreme Events*. BRISK project. Bridging Indigenous and Scientific Knowledge about Global Change in the Arctic: Adaptation and Vulnerability of Society and Environment. Indigenous Knowledge and Climate Change, COP 22. Marrakesh, 2–3 November 2016.

— 2016b. *Traditional Ecological Knowledge Facing Climate Change among Reindeer Herders in Siberia*. BRISK project. Nordic-French Seminar: *The Changing Arctic – Environment and Governance*. Tromsø, French Institute Norway, 14 October 2016.

— 2016c. *Siberian Evenk Reindeer Herders' Knowledge and Observation of Biodiversity and Ecosystem Services*. ILK Dialogue Workshop for the Europe and Central Asia Assessment, Task Force on Indigenous and Local Knowledge Systems (ILK), Intergovernmental Platform on Biodiversity and Ecosystem Services (IPBES), Expert Group for the Regional Assessment for Europe and Central Asia (ECA). UNESCO, Paris, 11–13 January 2016.

— 2015. *Siberian Reindeer Herders Facing Climate Change: Observation and Adaptation*, Resilience in a Time of Uncertainty: Indigenous Peoples and Climate Change International Conference. An International Conference contributing to COP21 Climate Conference. UNESCO, Paris, 26–27 November 2015.

— 2014. *A Community-based Transdisciplinary Observatory among the Southern Siberia Evenk: One Year Experience and Co-Productions*. BRISK project. Global Change, Indigenous Community-Based Observing System, and Co-Production of Knowledge for the Circumpolar North. The Roles of Traditional Knowledge in Governance of Natural Resources in the High North with Case Studies from

Reindeer Husbandry and other Indigenous Societies. Kautokeino, Norway, 25–27 March 2014.

Lavrillier, Alexandra, Rojo Maxence, Claud Chantal, and Semen Gabyshev 2015. *Siberian Reindeer Herders Facing Climate Change: Observation and Adaptation, Resilience in a Time of Uncertainty: Indigenous Peoples and Climate Change.* An International Conference contributing to COP21 climate conference. UNESCO, Paris, 26–27 November 2015.

Lavrillier, Alexandra, Rojo Maxence, Claud Chantal, Chondan Rolanda, and Semen Gabyshev 2014. *Development of Two Community-Based Transdisciplinary Observatories in Siberia Among the Evenk and the Tuva-Tozhu Reindeer Herders – Russian Federation.* Climate Change Assessments and the 'Best Available Knowledge': The Challenge of Bridging Disciplines and Knowledge Systems. UNESCO, Paris, 3 November 2014.

Alexandra Lavrillier and Dejan Matic (eds.)

Эвенские нимканы Дарьи Михайловны Осениной
[Even tales of Dar'iia Mikhailovna Osenina]

2013, Fürstenberg/Havel: Kulturstiftung Sibirien
160 pp.,15,5 x 22 cm
Euro 18; paperback
ISBN: 978-3-942883-15-3

Languages and Cultures of the Russian Far East
http://www.siberian-studies.org/publications/lc_E.html

Bulgakova, Tatiana

Nanai Shamanic Culture in Indigenous Discourse

2013, Fürstenberg/Havel: Kulturstiftung Sibirien
264 pp., 15,5 x 22 cm, 5 color photos
Euro 28; paperback
ISBN: 978-3-942883-14-6

Studies in Social and Cultural Anthropology
http://www.siberian-studies.org/publications/studies_E.html

Erich Kasten, Katja Roller and Joshua Wilbur (eds.)

Oral History Meets Linguistics

2017, Fürstenberg/Havel: Kulturstiftung Sibirien
211 pp., 15,5 x 22 cm, 12 color photos
Euro 26; paperback
ISBN: 978-3-942883-30-6

Exhibitions and Symposia
http://www.siberian-studies.org/publications/exsym_E.html